NEW WORLDS IN ASTROPARTICLE PHYSICS

Proceedings of the Fifth International Workshop

NEW WORLDS IN ASTROPARTICLE PHYSICS

Proceedings of the Fifth International Workshop

Faro, Portugal 8 – 10 January 2005

editors

Ana M. Mourão
CENTRA & Instituto Superior Técnico, Lisbon, Portugal

Mário Pimenta
LIP & Instituto Superior Técnico, Lisbon, Portugal

Robertus Potting
CENTRA & Universidade do Algarve, Faro, Portugal

Paulo M. Sá
CENTRA & Universidade do Algarve, Faro, Portugal

World Scientific

NEW JERSEY • LONDON • SINGAPORE • BEIJING • SHANGHAI • HONG KONG • TAIPEI • CHENNAI

Published by

World Scientific Publishing Co. Pte. Ltd.
5 Toh Tuck Link, Singapore 596224
USA office: 27 Warren Street, Suite 401-402, Hackensack, NJ 07601
UK office: 57 Shelton Street, Covent Garden, London WC2H 9HE

British Library Cataloguing-in-Publication Data
A catalogue record for this book is available from the British Library.

NEW WORLDS IN ASTROPARTICLE PHYSICS
Proceedings of the Fifth International Workshop

ISBN 981-256-625-2

Printed in Singapore by World Scientific Printers (S) Pte Ltd

CONTENTS

Neutrino Physics

Cosmological Parameters Measurements

Black Hole Physics

PREFACE

The World Year of Physics started in Portugal with the Fifth Internacional Workshop on New Worlds in Astroparticle Physics, which took place from the 8th to the 10th of January of 2005 at the Campus of Gambelas of the University of the Algarve, in Faro. For three days, full of talks and discussions, the invisible presence of Albert Einstein was felt in almost all the topics: from the invariance of the laws of physics, to black holes and gravitational waves, including the physics of neutrinos and of cosmic rays. In the end progress was achieved, but we certainly have a long way to go.

The symposium was organised by the University of the Algarve, Instituto Superior Técnico, CENTRA (Multidisciplinary Center for Astrophysics), and CFIF (Center for Physics of Fundamental Interactions). Financial support from FCT (Foundation for Science and Technology) under the Programa Operacional Ciência, Tecnologia, Inovação do Quadro Comunitário de Apoio II, from FLAD (Portuguese-American Foundation, Calouste Gulbenkian Foundation, Italian Institute for Cooperation, and GTAE (High Energy Theory Group) is gratefully acknowledged.

Jorge Dias de Deus

Part 1 Overviews in Astroparticle Physics

AN OVERVIEW OF THE STATUS OF WORK ON ULTRA HIGH ENERGY COSMIC RAYS

A. A. WATSON
School of Physics and Astronomy, University of Leeds, Leeds LS2 9JT, UK
a.a.watson@leeds.ac.uk

The present situation with regard to experimental data on ultra high-energy cosmic rays is briefly reviewed. Whilst detailed knowledge of the shape of the energy spectrum is still lacking, and there remains uncertainty as to whether there is a steepening of the spectrum near 1020 eV, it is likely that events above 1020 eV do exist. Evidence for clustering of the directions of some of the highest energy events now seems less certain than had once been claimed. Our knowledge of the mass composition of cosmic rays is deficient at all energies above 1018 eV and it must be improved if we are to discover the origin of the highest energy cosmic rays. Here it is argued, in particular, that there is no compelling evidence to support the common assumption that cosmic rays of the highest energies are protons. Clearly, more data are needed and these are likely to come from the southern branch of the Pierre Auger Observatory in the next few years. The Observatory is briefly described.

1. The Scientific Motivation for Studying the Highest Energy Cosmic Rays

Efforts to discover the origin of the highest energy cosmic rays have been on going for many years. Since the recognition in 1966, by Greisen and by Zatsepin and Kuzmin, that protons with energies above 4 x 1019 eV would interact with the cosmic microwave radiation, there has been great interest in measuring the spectrum, arrival direction distribution and mass composition of ultra high-energy cosmic rays (UHECRs). UHECRs may be defined as those cosmic rays having energies above 1019 eV. Specifically, it was pointed out that if the highest energy particles are protons and their sources are universally distributed, there should be a sharp steepening of the energy spectrum in the range from 4 to 10 x 1019 eV. This predicted feature has become known as the GZK 'cut-off'. If the UHECR were mainly Fe nuclei then there would also be a steepening of the spectrum. However, it is harder to predict the details of this feature as the relevant diffuse infrared photon field is poorly known: the steepening is expected to set in at higher energy if the cosmic rays leave the sources as heavy nuclei.

Early instruments built to study this energy region (at Volcano Ranch (USA), Haverah Park (UK), Narribri (Australia) and Yakutsk (USSR)), were designed long before the 1966 predictions and when the flux above 10^{19} eV was poorly known. Although of relatively small area (~10 km^2) sufficient exposure was eventually accumulated to measure the rate of cosmic rays above 10^{19} eV accurately and to give the first indications that there might be cosmic rays with energies above 10^{20} eV, well above the GZK cut-off. No convincing evidence of anisotropies above 10^{19} eV was established. Over the same period, it also came to be accepted that the problem of acceleration of protons and nuclei to such energies in known astrophysical sources is a major one.

The present data on UHECR are dominated by measurements from the surface detector array known as AGASA that was operated by groups led by the Institute for Cosmic Ray Research in Japan and by the fluorescence detector known as HiRes operated by several groups from the USA, including the pioneering team from the University of Utah. These projects have also given indications of trans-GZK particles but by the early 1990s it was apparent that even areas of 100 km^2 operated for many years could not measure the properties of UHECR in adequate detail. In 1991 the design of a suitable instrument, and the task of assembling an international collaboration to fund and construct it, were begun. The collaboration and the funding situation were considered sufficiently robust for work on the Pierre Auger Observatory to begin in Malargüe, Mendoza Province, Argentina in March 1999. This southern part of the observatory is seen as the first of the two that are required to provide full sky coverage. In what follows the data on energy spectrum and arrival directions from the AGASA and HiRes instruments will be discussed. In addition the important issue of the mass composition above 10^{17} eV will be examined. Finally the status and prospects of the Auger Observatory will be briefly outlined.

2. The Present Observational Situation

2.1. *Energy Spectrum Measurements*

During the planning and construction of the Pierre Auger Observatory, observations continued with the AGASA surface detector array and the two fluorescence detectors (the HiRes and Fly's Eye) of the University of Utah. The Japanese detector was an array of 111 x 2.2 m^2 plastic scintillators spread over 100 km^2. This instrument ceased operation in January 2004 when an exposure of about 1600 km^2 steradian years had been reached. Eleven events with energies above 10^{20} eV have been reported [1]. By contrast, the fluorescence method uses the scintillation light produced in the atmosphere by the secondary shower cascade and permits a calorimetric estimate of the energy in a manner familiar from accelerator experiments, although there are difficulties associated with the variable transmission properties of the atmosphere and with the accuracy of knowledge of the fluorescence yield. The fluorescence instruments have also seen events with energies above 10^{20} eV but a rate lower than that seen by the Japanese group is found. Nevertheless, the highest energy event ever recorded (3 x 10^{20} eV) was reported by the Fly's Eye group and what is now clear is that there are cosmic rays above 10^{20} eV, seen with both techniques, and that the rate of such events is of order 1 per km^2 per steradian per century. A useful summary of the experimental situation is shown in Fig. 1.

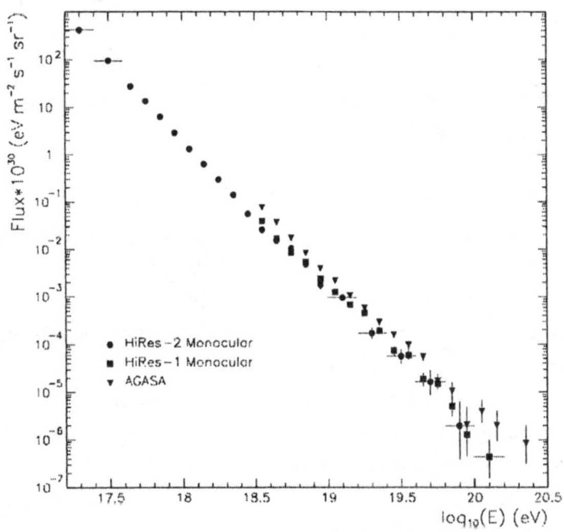

Figure 1. The energy spectra as reported by the AGASA [1] and HiRes [2] groups. This clear presentation of the spectra is due to D R Bergman (Rutgers University).

Many questions remain about the detailed shape of the spectrum. It is clear that the HiRes and AGASA spectra could be reconciled if the energy scale of one or other was adjusted by 30%, or if each was moved by 15%. Spectra derived from arrays of particle detectors suffer from the difficulty that the energy of each primary cosmic ray must be inferred using models of particle physics interactions at energies well beyond those of present, or envisaged, accelerators. Thus, there is a systematic error in these energy assignments that is, inherently, unknowable. In addition, the inputs to models also require an assumption about the mass of the initiating primary particle. The AGASA group assumed the primaries to be protons: if iron nuclei had been assumed then the assigned energies would be reduced. The AGASA group have given some indications [table 1 of ref 1] of the sensitivity of the energies to mass and models. Specifically, the difference between iron (with QGSJET) and protons (with SIBYLL) is about 15%. Assuming iron nuclei reduces the energy.

There is a small sensitivity to model and mass in the energies derived using the fluorescence technique because of so-called 'missing energy', that energy that is carried by muons and neutrinos into the ground. The correction for this missing energy is ~ 10% at 10^{20} eV and is slightly larger for Fe nuclei than for protons, the species assumed in the construction of the spectra reported in [2].

In addition, the possibility that there are uncertainties in the flux measurements should not be overlooked. At the lower end of the AGASA spectrum the aperture is changing quite rapidly with energy [1] and uncertainties in the lateral distribution function that describes the fall-off of signal with distance, may lead to uncertainties in the aperture determination. At the highest energies, the AGASA aperture is limited by requiring that shower cores fall inside the area bounded by the detectors and is known precisely. By contrast, for fluorescence detectors, the aperture continues to grow with energy and there remains considerable uncertainly about the HiRes aperture, even in the case of stereo operation. Further data are expected from the HiRes group and, in particular, from their period of stereo operation.

Despite the above caveats, it appears certain that trans-GZK events do exist. The Utah group reported an event of 300 EeV in 1993 [3] from the Fly's Eye detector, the AGASA group have claimed an event of 210 EeV [4] together with several other events with energies reported above 100 EeV, while a stereo HiRes event has been reported at 220 EeV [5]. We understand that this latter event is not included in the HiRes spectra included in Fig. 1 as it was recorded during a short period of good atmospheric conditions on a night that was otherwise rather unstable.

2.2. *Arrival Direction Results*

The situation concerning the arrival direction distribution of UHECR is not clear-cut either. For some time the AGASA group [6] have reported clustering on an angular scale of 2.5°, from a data set of 59 events above 4×10^{19} eV. The clusters are claimed to occur much more frequently than expected by chance with an estimate of 10^{-4} given for the chance probability. A search of the HiRes data [7] has not revealed clusters with the same frequency as claimed by AGASA.

Recently, Finley and Westerhoff [8] have presented an analysis using the directions of 72 events recently released by the AGASA group. They have taken the 30 events described in [9] as the trial data set and used the additional 42 events to search for pairs, adopting the criteria established by the AGASA group. Two pairs were found: such a result is estimated as having a probability of 19% of occurring by chance.

A further search for clusters has been made by the HiRes group using 27 events from their own data and 57 from AGASA above 4×10^{19} eV [10]. Using a novel likelihood search, the authors state that "no statistically significant clustering of events consistent with a point source is found": the most significant signal found is the AGASA triplet. If the energy scales of the two instruments are then normalised by reducing the HiRes threshold to 3×10^{19} eV, the HiRes sample is increased to 40 events. An event close to the AGASA triplet is found in the resulting sample but, as they state very clearly, it is not possible to evaluate a valid chance probability for this observation. They have identified a direction (close to $\alpha = 169°$ and $\delta = 57°$) and stated that if 2 events from the next 40 observed with the same energy selection fall in a bin of 1° the chance probability will be 10^{-5}.

It is clear that only further data will resolve the controversies over the energy spectrum and over the clusters in arrival direction. The AGASA array has closed having achieved an exposure of ~ 1600 km^2 steradian years. The HiRes instrument is expected to take data for another few years.

3. Interpretation of the Existing Data

Much has been written in attempts to explain the particles that exist beyond the GZK cut-off. If these are protons, the existence of such UHECR is seen as an enigma. They must come from nearby (at 1020 eV about 50% are expected from within 20 Mpc) and, adopting an extragalactic field of a few nanogauss, point sources would be expected to be detectable. However, none are seen and a wide variety of explanations has been offered. Amongst the many mechanisms proposed are the decay of topological defects or other massive relics of the big bang. Even more exotic is the suggestion of a violation of Lorentz invariance at very high energy in such a manner that the energy-loss mechanism against the CMB is not effective (although this still leaves open the question of how the particles are accelerated to very high energies in the first place). If the primaries were iron nuclei then the situation would be slightly easier to understand. The higher charge would mean that acceleration could occur more readily up to the observed energies and that bending, even in a weak magnetic field, would obscure the directions of the sources. It is thus crucial to review the evidence on the mass composition, as without such data it will be hard to draw conclusions about the origin of the particles, even when the spectral and clustering issues are clarified.

4. The Mass of UHECR

Our knowledge about the mass of primary cosmic rays at energies above 10^{17} eV is rudimentary. Different methods of measuring the mass give different answers and the conclusions are usually dependent upon the model calculations that are assumed. Results from some of the techniques that have been used in attempts to assess the mass composition are now described and the conclusions drawn reviewed. Some of these techniques will be applicable with the Pierre Auger Observatory.

4.1. *The Elongation Rate*

The elongation rate is the term used to describe the rate of change of depth of shower maximum with primary energy. The term was introduced by Linsley [11] and, although his original conclusions have been superseded by the results of detailed Monte Carlo studies to some extent, the concept is useful for organising and thinking about data. Figure 2 shows a summary of measurements of the depth of maximum together with predictions from a variety of model calculations [12]. It is clear that if certain models are correct that one might infer that the primaries above 10^{19} eV are dominantly protons but that others suggest a mixed composition. In particular, the QGSJET set of models (basic QGSJET01 and the 5 options discussed in [12]) and the Sibyll 2.1 model force contrary conclusions.

4.2. *Fluctuations in Depth of Maximum*

A way to break this degeneracy has long been seen in the magnitude of fluctuations in the position of depth of maximum. If a group of showers is selected having a narrow range of energies, then fluctuations about the mean of X_{max} would be expected to be larger for protons than for iron nuclei. A recent study of this has been reported by the HiRes group [13]. Their data consist of

728 events in the range $10^{18.0}$ to $10^{19.4}$. It is argued, using the Sibyll or the QGSJET models, that the fluctuations are so large that a large fraction of protons is indicated. The proton fractions deduced with the respective models are 60 and 80% respectively. However, the HiRes data have been analysed assuming a standard US atmosphere for some of the events. This is unlikely to represent reality, as it is probable that the atmosphere deviates from the standard conditions from night to night and even during a night of observation. This view is strengthened by the results of balloon flights made from Malargüe [14], which have shown that the atmosphere changes in a significant way from night to night, and from summer to winter. If a standard atmosphere is used, some of the fluctuations observed in X_{max} may be incorrectly attributed to shower, rather than to atmospheric, variations. Thus, it may be premature to draw conclusions about the presence of protons from this, and similar earlier analyses.

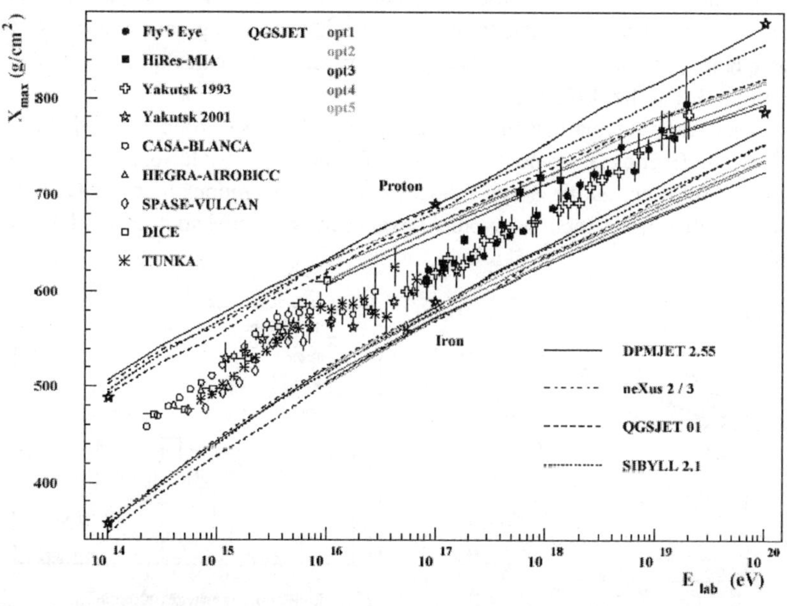

Figure 2. The depth of maximum, as predicted using various models, compared with measurements. The predictions of the five modifications of QGSJET, discussed in [12], from which this diagram is taken, lie below the dashed line that indicates the predictions of QGSJET01.

4.3. *Mass from Muon Density Measurements*

It is well known that a shower produced by an iron nucleus will contain a larger fraction of muons at the observation level than a shower of the same energy created by a proton primary. Many efforts to uncover the mass spectrum of cosmic rays have attempted to make use of this fact. However, although the differences are predicted to be relatively large (~70% more muons in an iron event than a proton event, on average), there are large fluctuations and, again, there are differences between what is predicted by particular models. Thus, the QGSJET set predicts more muons than the Sibyll family (the difference arising from different predictions as to the pion multiplicities produced in nucleon-nucleus and pion-nucleus collisions that in turn arise from differences in the assumptions about the parton distribution within the nucleon) [15]. A recent set of data from the AGASA group [16] is shown in Fig. 3. There are 129 events above 10^{19} eV, of which 19 have energies greater than 3 x 10^{19} eV. Measurements of muon densities at distances between 800 and 1600 m were used to derive the muon density at 1000 m with an average accuracy of 40%. This quantity is compared with the predictions of model calculations. It is clear from Fig. 3 that the difference between the proton and iron predictions is small, especially when fluctuations are considered. The AGASA group conclude that at 10^{19} eV the fraction of Fe nuclei is <40%. Further, the conclusions are sensitive to the model used: as the Sibyll model predicts fewer muons than the QGSJET model, higher iron fractions would have been inferred had that model been adopted.

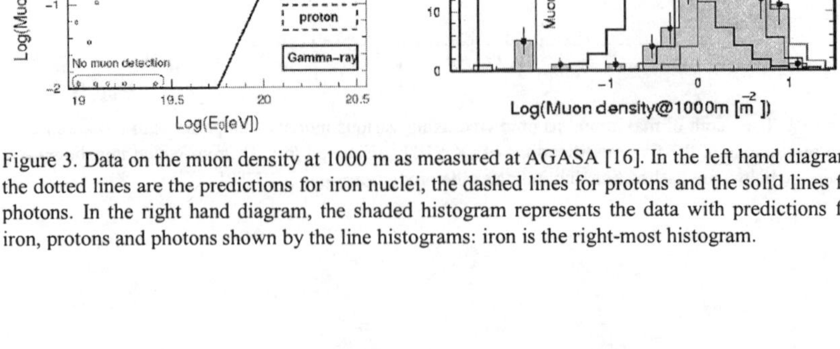

Figure 3. Data on the muon density at 1000 m as measured at AGASA [16]. In the left hand diagram, the dotted lines are the predictions for iron nuclei, the dashed lines for protons and the solid lines for photons. In the right hand diagram, the shaded histogram represents the data with predictions for iron, protons and photons shown by the line histograms: iron is the right-most histogram.

In my view [17], the 5 events above 10^{20} eV for which such measurements are possible, are fitted as well by iron nuclei as by protons. A similar, but quantitative, analysis of this point has been given in [18] where a limit of 50% at the 90% confidence level is set for events above the GZK cut-off.

At lower energies, there are muon data from the Akeno array and from AGASA. Different analyses have been made of these. The AGASA group [19] claim that the measurements are consistent with a mass composition that is unchanging between 10^{18} and 10^{19} eV. Dawson *et al.* [20], in an effort to reconcile the data with earlier fluorescence results (now superseded) have argued that with a different model, the mean mass is lower at the higher energies. In the context of the present discussion, it is worth noting that 50 – 60% of iron is quite consistent with both the AGASA and Akeno data for a range of models and with efforts to account for systematic uncertainties. It might be productive to revisit these data using the latest versions of the QGSJET and Sibyll models. A fuller discussion is given below in Sec. 4.7.

4.4. *Mass Estimates from the Lateral Distribution Function*

The rate of fall of particle density with distance from the shower axis provides another parameter that can be used to extract the mass composition. Showers with lateral distribution functions (LDFs) that are steeper than average will arise from showers that develop later in the atmosphere, and vice versa. A detailed measurement of the LDFs of showers produced by primaries of energy greater than 10^{17} eV was made at Haverah Park using a specially constructed 'infilled array' in which 30 additional water tanks of 1 m^2 were added at the centre of the array on a grid with spacing of 150 m. When the work was completed in 1978, the data could not be fitted with the shower models then available for any reasonable assumption about the primary mass. Recently [21], this data set has been re-examined using the QGSjet98 model.

The appropriateness of this model was established by showing that it adequately described data on the time spread of the Haverah Park detector signal over a range of zenith angles and distances near the core (<500 m). Here the difference predicted between the average proton and iron shower is only a few nanoseconds and the fit achieved is good. Density data were fitted by a function $\rho(r) \sim r^{-(\eta + r/4000)}$, where η is the steepness parameter. The spread of η is compared with predictions for different primary masses in Fig. 4. The proton fraction, assuming a proton-iron mixture, is found to be independent of energy in the range 3 x 10^{17} to 10^{18} eV and is (34± 2) %. If this fraction is evaluated with QGSJET01, in which a different treatment of diffractive processes is adopted from that in QGSJET98, then the fraction increases to 48%. The fraction is larger because the later model predicts shower maxima that are higher in the atmosphere and accordingly, to match the observed fluctuations, the proton fraction must be increased. The difference in the deduced ratio thus has a systematic uncertainty from the models that is larger than the statistical uncertainty. Although the necessary analysis has not been made, it is clear that the Sibyll 2.1 model would require a smaller fraction of protons.

A similar analysis has been carried out using data from the Volcano Ranch array. As with the Haverah Park information, no satisfactory interpretative analysis was possible when the measurements were made. With QGSJET01, the fraction of protons is estimated as (25±5%) between 5 and 10 x 10^{18} eV [22].

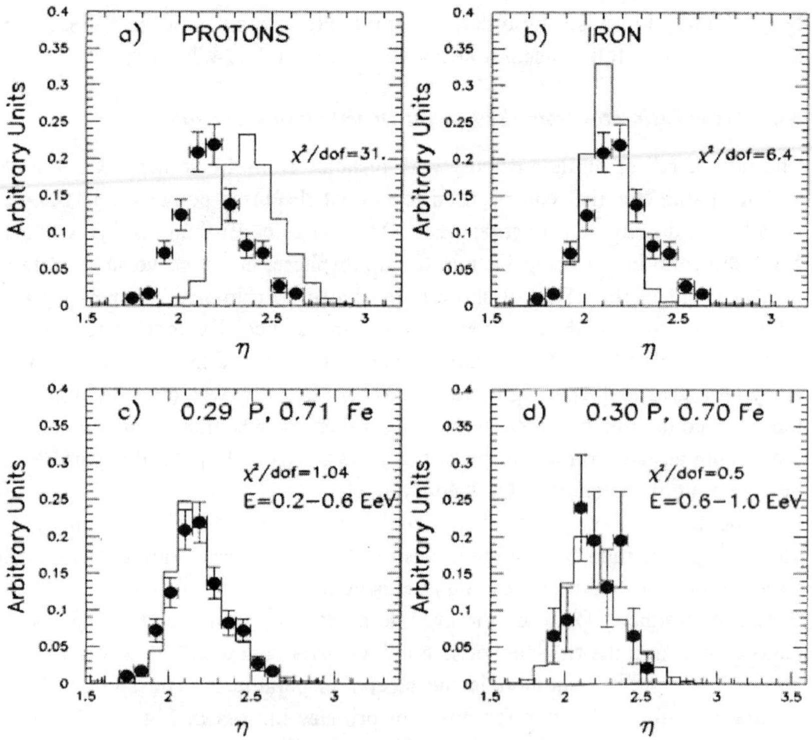

Figure 4. The experimental measurements of the steepness parameter, η, compared with predictions made using the QGSJET98 model assuming different mass mixtures [21]. The lower set of diagrams illustrates the insensitivity of the mass mixture on energy.

4.5. *Mass from the Thickness of the Shower Disk*

The particles in the shower disc do not arrive at a detector simultaneously, even on the shower axis. The arrival times are spread out because of geometrical effects, velocity differences, and because of delays caused by multiple scattering and geomagnetic deflections. The first particles to arrive (except very close to the shower axis) are the muons as they are scattered rather little and geometrical effects dominate. At Haverah Park four detectors, each of 34 m^2, provided a useful tool for studying the thickness of the shower disc, which will depend upon the development of the cascade. Recently, an analysis of 100 events of mean energy $\sim 10^{19}$ eV has shown that the magnitude of the risetime is indicative of a large fraction (~80%) iron nuclei at this energy [23]. It is expected that this type of study will be considerably extended using the Pierre Auger Observatory, in which each water tank is equipped with 25 ns flash ADCs.

4.6. *Limits to Photon Primaries*

It is unlikely that the majority of the events claimed to be near 10^{20} eV have photons as parents as some of the showers seem to have normal numbers of muons (the tracers of primaries that are nuclei) (see Fig. 3). Furthermore, the cascade profile of the most energetic fluorescence event is not entirely consistent with that of a photon primary [24, 25]. This approach can be used when specific shower profiles are available. An alternative method of searching for photons has been developed using showers incident at very large zenith angles. Deep-water tanks have a good response to such events out to beyond 80º. At such angles the bulk of the showers detected are created by nucleonic primaries but they are distinctive in that the electromagnetic cascade stemming from neutral pions has been completely suppressed by the extra thickness of atmosphere penetrated. At 80° the atmospheric thickness to be penetrated is ~ 5.7 atmospheres. At Haverah Park, such large zenith angle showers were observed and found to have the shower disc was found to have a very small time spread. A complication for the study of inclined showers is that the muons, in their long traversal of the atmosphere, are very significantly bent by the geomagnetic field. A detailed study of this has been made and it has been shown that the rate of triggering of the Haverah Park array at large angles can be predicted [26]. In addition, it was shown that the energy of the primaries could be estimated with reasonable precision so that an energy spectrum could be derived. The concept of using the known, and mass independent, spectrum deduced by the fluorescence detectors to predict the triggering rate as a function of the mass of the primary has led to a demonstration that the photon flux at 10^{19}

eV is less than 40% of the nucleonic component [27]. In addition to this novel approach, a more traditional attack on the problem by the AGASA group, searching for showers, which have significantly fewer muons than normal, has given the same answer [28]. These experimental limits are in contrast to the predictions of large photon fluxes from the decay of super-heavy relic particles, one of the exotic candidates that have been invoked to explain the enigma of the highest energy cosmic rays that is evident if the highest energy events really are created by protons [29].

4.7. *Summary of Data on Primary Mass Above 0.3 EeV*

In Fig. 5, taken from [22], the results taken from various reports of the Fe fraction are shown. It is disappointing that the data from Volcano Ranch and from Haverah Park are not in better agreement as a similar quantity, the lateral distribution function of the showers, was measured at each array and the same model - QGSJET98 - was used to interpret the data, although with different propagation codes (AIRES and CORSIKA respectively). The difference is not understood: at 10^{18} eV the estimates of the fraction of Fe are separated by over 2 standard deviations.

In Fig. 5, there are also data from the Akeno/AGASA and the Fly's Eye experiments. The Akeno/AGASA groups measured the muon densities in showers, normalised at 600 m. The energy thresholds for Akeno and AGASA were 1 and 0.5 GeV respectively. The Fly's Eye data are deduced from measurements of the depth of shower maximum. In an effort to reconcile differing claims made by the two groups of the trend of mass composition with energy, Dawson *et al.* [20] reassessed the situation used a single model, SIBYLL 1.5 on both data sets. SIBYLL 1.5 was an early version of the SIBYLL family that evolved to SIBYLL 1.6 and 1.7. It is the estimates of the Fe fractions from [20] that are shown in Fig. 5. There are major discrepancies between these estimates and between those from Volcano Ranch and Haverah Park. However, the predictions of the muon density and of the depth of shower maximum made with the version of SIBYLL used differ significantly from those that would be derived now using QGSJET98 or 01 (or with the later SIBYLL version, 2.1). We now discuss this point in some detail.

An extremely useful set of comparisons of the predictions from SIBYLL 1.7 and 2.1 with those from QGSJET98 has been given in [15]. We understand that SIBYLL 1.6 and SIBYLL 1.7 differ only in that the neutral pions were allowed to interact in the latter model and it is not believed that this will make a serious difference to the predictions at energies below 10^{19} eV [30]. Therefore, in what follows, we regard the SIBYLL 1.7 and the QGSJET98 differences as being identical to those that exist between SIBYLL 1.6 (or 1.5) and QGSJET98, for which no similar comparisons are available. It is convenient to make comparisons at 10^{18} eV. More detailed cross-checks, over a range of energies, would require more extensive knowledge of features of the Fly's Eye and Akeno/AGASA systems than we possess.

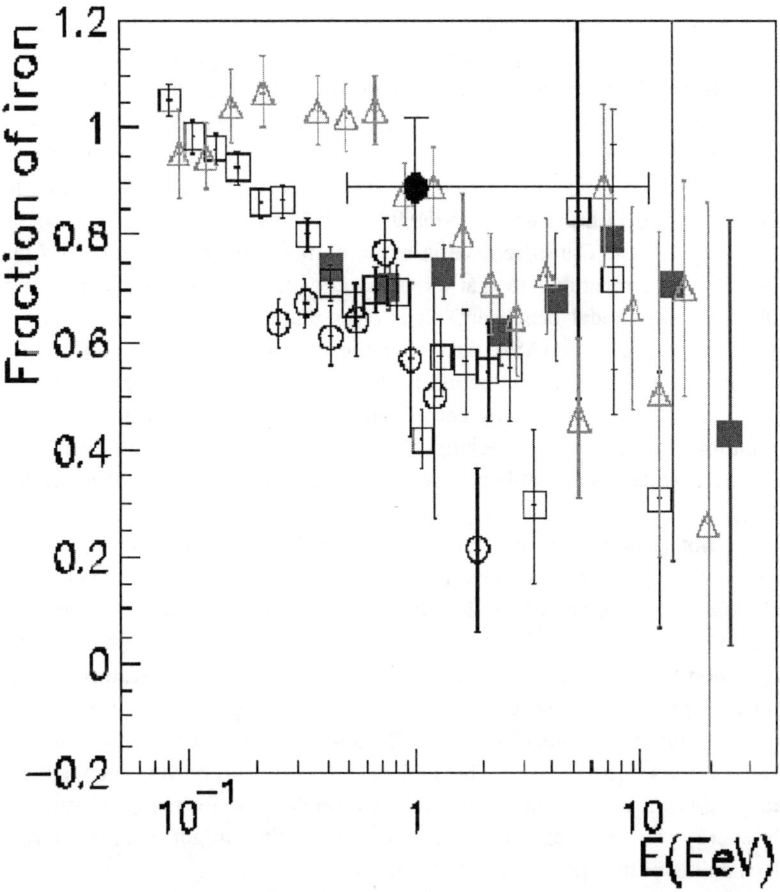

Figure 5. The fraction of Fe nuclei as a function of energy as reported from various experiments. Fly's Eye (△), AGASA A100 (■), AGASA A1 (□), Haverah Park (○) and Volcano Ranch (•) (Figure taken from [22]).

Turning first to the data from the depth of maximum, we note that at 10^{18} eV the measured value of X_{max} is ~675 g cm^{-2}, with an error that is less than the size of the data point (< 10 g cm^{-2}). The predictions for proton primaries made with SIBYLL 1.7 and QGSJET98 are 760 and 730 g cm^{-2} respectively[18]. Thus, a mass composition less dominated by Fe is favoured compared with the ~90% estimated in [20]. The choice of SIBYLL 2.1 would alter this argument rather little as the predicted depth [15] at 10^{18} eV is 740 g cm^{-2}.

A qualitative statement about the shift expected in the Fe fraction, as estimated from the measurement of muon densities at 600 m, from changes in the model can be made using information in [15]. Although the calculations do not exactly match the energies of the Akeno/AGASA measurements (> 0.3 GeV is computed and > 0.5 GeV measured), ratios between the predictions of different models are not strongly dependent upon energy threshold. What is of importance is the ratio of the number of muons predicted, at 10^{18} eV, for SIBYLL 1.7, SIBYLL 2.1 and QGSJET98. At 10^{18} eV, these numbers are in the ratios 1: 1.17: 1.44. The difference in muon number between SIBYLL 1.7 and QGSJET98 is comparable to that expected between proton and Fe primaries (~50%, but also model dependent). It is clear that the more recent models, if applied to the Akeno/AGASA data after the manner of the analysis of [20], would lead to a significant reduction in the predicted fraction of Fe nuclei. To pursue this further would require knowledge of the predicted densities at 600 m, information that is presently lacking. We note that the shift in the Fe fraction from the muon data is probably substantially larger than it is when using the data on X_{max}.

It is not clear how to use the information reported from the HiRes-MIA experiment [31] in which muons and X_{max} were observed simultaneously. As with Akeno/AGASA, the muon density at 600 m was determined. The problem is that while the papers describe the data as being consistent with a mass composition that becomes lighter with energy, this appears, on scrutiny of Figs. 1 and 2 of [31], to be true only for the X_{max} data. The muon data, which are compared with predictions of QGSJET98, look to be consistent with a constant and heavy mass from 5 x 10^{16} to beyond 10^{18} eV. It would be very interesting to establish that the same model gives different predictions for the mass variation with energy for different measured quantities: this might lead to further understanding of the appropriate model to use.

The difficulties with which one is faced with when trying to compare data are demonstrated by the above discussions. Measurements from different experiments are rarely analysed contemporaneously and the shifts in the inferences from the use of different models can be substantial.

5. The Pierre Auger Observatory

It must be clear from what has been said above that a vast increase in available data is needed to clarify the issues discussed (along with a better understanding of what hadronic models are applicable at the highest energies). To this end the Pierre Auger Observatory has been designed to measure the properties of the highest energy cosmic rays with unprecedented statistics and precision. It has been planned as an instrument with sites in the Northern and Southern Hemisphere. Each site will contain 1600 water-Cherenkov detectors spread out over 3000 km^2. Each tank will contain 12 tonnes of water. The water tanks will be overlooked by a set of 4 fluorescence detectors capable of detecting the faint light produced from N_2 molecules excited by the shower particles as they traverse the atmosphere. A single fluorescence detector comprises 6 telescopes, each having a camera of 440 photomultipliers that views an 11 m^2 mirror. Light falls on the mirrors after passing through filters that also act as windows to shield the instrumentation from dust and rain. The filters, which are 1.7 m in diameter, transmit in the 350 to 450 nm range. Each telescope views a sky area of 30° x 30°. The southern part of the Observatory, sited on Pampa Amarilla near Malargüe, Argentina, is nearing completion. As at December 2004, 2 of the 4 fluorescence detectors are fully operational and are overlooking 557 water tanks. All of these devices are taking data.

The Pierre Auger Observatory is now the largest shower detector ever constructed and by March 2005 will have achieved an exposure that is comparable to that of the AGASA array. A description of a prototype instrument comprising 32 water-Cherenkov detectors and two fluorescence telescopes that was deployed and operated to demonstrate the techniques and the ability of the Collaboration has been given in [32]. Distinctive and novel features of the water detectors of the Auger Observatory are the FADC records that are obtained from each of the three 9” photomultipliers that view the individual water volumes. In Fig. 6 the FADC records in a near vertical shower (13°) are compared with those in one at 76° from the zenith where the shower penetrated ~ 4 times as much matter. These events were registered with the prototype array. The broad FADC traces, the curved shower front (4 km as compared with 27 km in these examples) and the steep fall-off of signal size with distance from the shower core evident in the near vertical event are the signatures that will be sought in inclined events and, if any are found, used to assess the events as neutrino candidates [33].

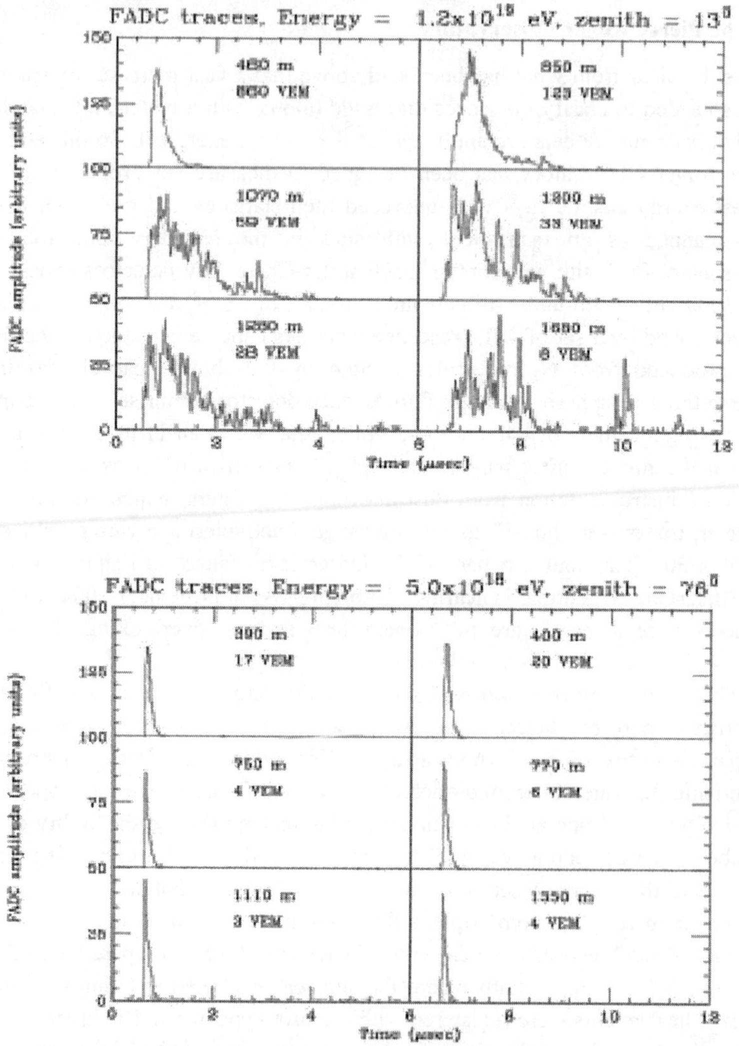

Figure 6. The upper plot shows FADC traces from a 13° shower: the radius of curvature was 4 km, and the density ratio was 134 across a distance ratio of 3.7. By contrast the lower figure shows the FADC traces in a shower of 76° which has a radius of curvature of 27 km and a density ratio of 7.5 across a distance ratio of 3.5. Picture from Auger Collaboration data: from [33].

Some appreciation of the present performance of the surface detector of the Auger Observatory can be had by considering a few events. In Fig. 7 data for an event that arrived at 34° from the vertical are shown. 14 detectors were above trigger level. The fall-off of the signal size with distance is what was anticipated for a shower produced by a high-energy baryonic cosmic ray: the energy is currently estimated as being ~10^{19} eV. In Fig. 8 data for a 32-fold event that arrived at 72° are displayed. Here the scale of the footprint on the ground is dramatic. Detectors separated by over 15 km have been triggered. Inspection of the radius of curvature and the FADC traces of this event suggest that it has been created by a baryonic rather than a neutrino primary. It is harder to estimate the energy of the primary but it is certainly well above 10^{19} eV.

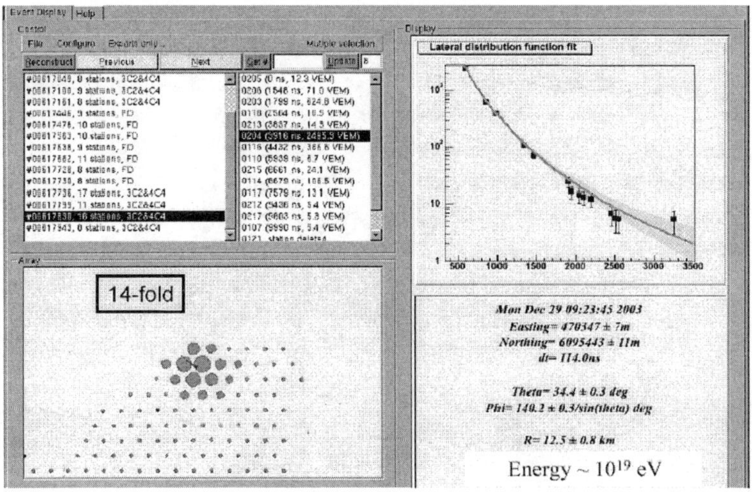

Figure 7. A 14-fold event at 34° from the zenith. The figure shows the on-line event display. The sequence of events is shown in the top left corner with the density pattern of the triggered detectors below. The lateral distribution and some parameters of the event are in the right hand panels.

While the phenomenon of Cherenkov light is essential and totally beneficial in the case of the surface detectors, this is not the case for the fluorescence detection. The fluorescence light is emitted isotropically (it is this feature that allows showers to be seen at distances as great as 30 km). However, Cherenkov light is also produced by every electron and muon in the shower that is above the Cherenkov threshold (21 MeV and 4 GeV respectively at STP). The light output per metre of track depends on the air density but at an altitude of 5 km is about 15 photons per metre and correspondingly larger at greater depths. The total Cherenkov light at any depth builds up as the shower propagates through the atmosphere because of the transparency of the air. In some events, the shower trajectory may be moving towards a fluorescence telescope so that air-Cherenkov light falls directly onto the mirrors. In other cases some fraction of the Cherenkov light will be scattered into the aperture by Rayleigh scattering or because of the aerosol content of the air. The aerosol density is a function of altitude and can also be a function of distance from the telescope: the density depends on the atmospheric conditions.

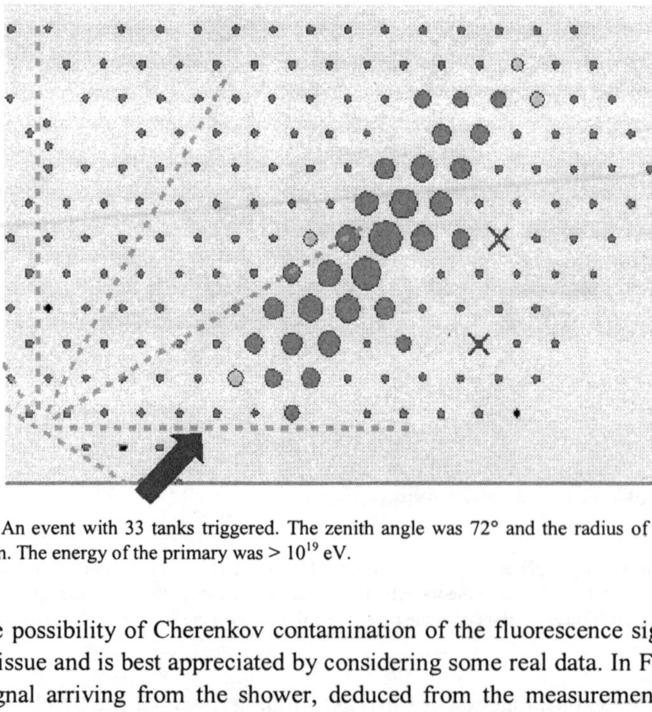

Figure 8. An event with 33 tanks triggered. The zenith angle was 72° and the radius of curvature was 37 km. The energy of the primary was > 10^{19} eV.

The possibility of Cherenkov contamination of the fluorescence signal is a serious issue and is best appreciated by considering some real data. In Fig. 9 the light signal arriving from the shower, deduced from the measurements made with the photomultipliers in one of the fluorescence detector telescopes, is shown as a function of time of receipt at the detector. In addition, estimates of the direct Cherenkov contribution and of the Cherenkov light that arrives because of Rayleigh and aerosol scattering are displayed. The estimates of these contributions are made by an iterative method once an initial estimate of the geometry of the event has been made. Note how the contribution of the Rayleigh scattered light increases as the air density rises. Similarly there is more scattered light from aerosol contamination close to the ground. In Fig. 10 the longitudinal profile of this event is displayed with the inferred number of particles shown as a function of atmospheric depth. The position of the shower maximum is clearly seen. By integrating the number of particles under the shower development curve (the track length integral) an estimate of the electromagnetic energy in the shower is obtained. In addition an additive correction of ~10% must be made to allow for the energy carried by high energy muons and neutrinos. The event shown is typical for one of this energy. In practice it is probable that events where the Cherenkov light contamination is above 30% will not be used in any science analysis. It is clear that careful monitoring of the state of the atmosphere in needed and the Auger procedures to do this are described in [32].

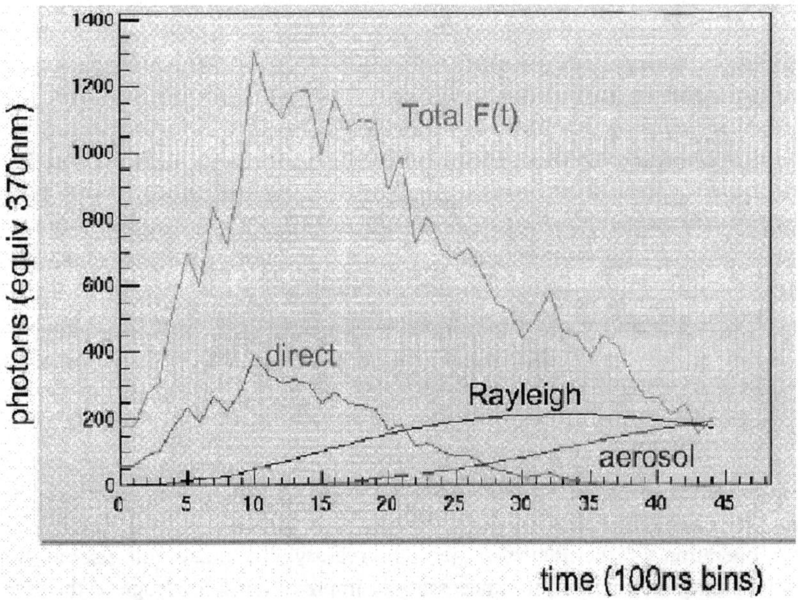

Figure 9. The light profile of a typical shower, showing the contamination arising from direct and scattered Cherenkov light.

While this observatory will mainly be used to survey the Southern sky, it is expected to give a guide as to which of the spectra so far reported from the Northern Hemisphere is correct and to the reality of clustering of high energy events: the exposure achieved by the early 2005 will be comparable to that made at AGASA. The immediate prospect, therefore, is for science data to be reported in mid-2005.

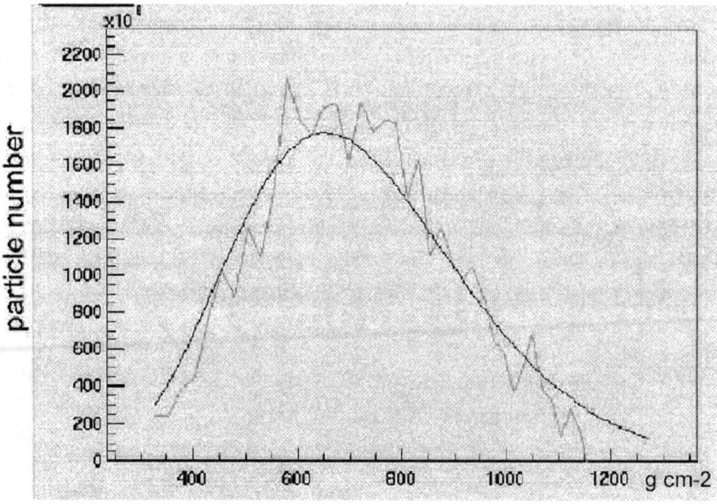

Figure 10. The particle number as a function of atmospheric depth for the event of Fig. 7. The primary energy is ~ 3 x 10^{18} eV and the depth of maximum is at ~650 g cm^{-2} where the particle number is ~1.8 x 10^{9}.

6. Conclusions

The question of spectral shape of the UHECRs remains uncertain and, along with the issue of the clustering of the arrival directions, may only be resolved by the operation of the Pierre Auger Observatory. To make full use of this forthcoming information, it is necessary to improve our knowledge of the mass of the cosmic rays above 10^{19} eV. Such evidence as there is does not support the common assumption that all of these cosmic rays are protons: there may be a substantial fraction of iron nuclei present. However, photons do not appear to dominate at the highest energies.

Acknowledgements

I would like to thank the organisers, and in particular Mário Pimenta, for inviting me to the excellent 2005 Faro meeting. Work on UHECRs at the University of Leeds is supported by PPARC, UK.

References

1. M. Takeda *et al.*, *Astropart. Phys.* **19**, 447 (2003) and astro-ph/0209422.
2. R. U. Abbasi *et al.*, *Phys. Rev. Lett.* **92,** 151101 (2004) and astro-ph/0501317.
3. D. Bird *et al.*, *Astrophys. J.* **441,** 144 (1995).
4. N. Hayashida *et al.*, *Phys. Rev. Lett.* **73**, 3491 (1994).
5. G. Loh, NEEDS Workshop, Karlsruhe 2002, http://www-ik.fzk.de/~needs/needs/loh.pdf.
6. M. Teshima *et al.*, Proceedings of the 28th Int. Cosm. Ray Conf. (Tsukuba) **1,** 437 (2003).
7. R. U. Abbasi *et al.*, *Astrophys. J.* **610,** L73 (2004) and astro-ph/0404137.
8. C. B. Finley and S. Westerhoff, *Astropart. Phys.* **21**, 359 (2004) and astro-ph/0309159.
9. N. Hayashida *et al.*, *Astrophys. J.* **522**, 225 (1999).
10. R. U. Abbasi *et al.*, *Astrophys. J.* **623,** 164 (2005) and astro-ph/0412617.
11. J. Linsley, Proceedings of the 15th Int. Cosm. Ray Conf. (Plovdiv) **12,** 89 (1977).
12. M. Zha, J. Knapp and S. Ostapchenko, Proceedings of the 28th Int. Cosm. Ray Conf. (Tsukuba) **2**, 515 (2003).
13. R. U. Abassi *et al.*, *Astrophys. J.* **622**, 910 (2005) and astro-ph/0407622v3.
14. B. Keilhauer *et al.*, Proceedings of the 28th Int. Cosm. Ray Conf. (Tsukuba) **2**, 879 (2003).
15. J. Alvarez-Muñiz *et al.*, *Phys. Rev. D* **66,** 033011 (2002) and astro-ph/0205302.
16. K. Shinosaki *et al.*, Proceedings of the 28th Int. Cosm. Ray Conf. (Tsukuba) **1,** 401 (2003).
17. A. A. Watson, *Nucl. Phys. B (Proc. Suppl.)* **136**, 290 (2004) and astro-ph/0408110.
18. M. Risse *et al.*, astro-ph/0502418.
19. N. Hayashida *et al.*, *J. Phys. G* **21,** 1101 (1995).
20. B. R. Dawson *et al.*, *Astropart. Phys.* **9,** 331 (1998).
21. M. Ave *et al.*, *Astropart. Phys.* **19**, 61 (2003) and astro-ph/0203150.
22. M. T. Dova *et al.*, *Astropart. Phys.* **21,** 597 (2004) and astro-ph/0312463.
23. M. Ave *et al.*, Proceedings of the 28th Int. Cosm. Ray Conf. (Tsukuba) **1,** 349 (2003).
24. F. Halzen *et al.*, *Astropart. Phys.* **3,** 151 (1995).
25. M. Risse *et al.*, *Astropart. Phys.* **21,** 479 (2004) and astro-ph/0401629.

26. M. Ave *et al.*, *Astropart. Phys.* **14,** 109 (2000) and astro-ph/0003011.
27. M. Ave *et al.*, *Phys. Rev. Lett.* **85,** 2244 (2000) and astro-ph/0007386.
28. K. Shinosaki *et al.*, *Astrophys. J.* **571**, L117 (2002).
29. V. Berezinsky, M. Kachelreiss and A. Vilenkin, *Phys. Rev. Lett.* **79**, 4302 (1997); K. Benakli, J. Ellis and D. V. Nanopolous, *Phys. Rev. D* **59**, 047301 (1999); M. Birkel and S. Sarkar, *Astropart. Phys.* **9**, 297 (1998); J. R. Chisholm and E. W. Kolb, *Phys. Rev. D* **69**, 085001 (2004) and hep-ph/0306288; D. J. H. Chung, E. W. Kolb and A. Riotto, *Phys. Rev. Lett.* **81,** 4048 (1998); N. A. Rubin, M. Phil. Thesis, University of Cambridge, 1999; S. Sarkar and R. Toldra, *Nucl. Phys. B* **621,** 495 (2002).
30. T. Stanev, private communication, February 2004.
31. T. Abu-Zayyad *et al.*, *Phys. Rev. Lett.* **84**, 4276 (2000); T. Abu-Zayyad *et al.*, *Astrophys. J.* **557**, 686 (2001).
32. Auger Collaboration: J. Abraham *et al.*, *Nucl. Instrum. Meth. A* **523**, 50 (2004).
33. J. W. Cronin, Proceedings of the TAUP meeting (Seattle, 2003), *Nucl. Phys. B (Proc. Suppl.)* **138**, 465 (2005).

GRAVITATIONAL WAVES FROM COMPACT SOURCES

K. D. KOKKOTAS AND N. STERGIOULAS
Department of Physics,
Aristotle University of Thessaloniki,
54124 Thessaloniki, Greece
E-mail: kokkotas@auth.gr, niksterg@auth.gr

We review sources of high-frequency gravitational waves, summarizing our current understanding of emission mechanisms, expected amplitudes and event rates. The most promising sources are gravitational collapse (formation of black holes or neutron stars) and subsequent ringing of the compact star, secular or dynamical rotational instabilities and high-mass compact objects formed through the merger of binary neutron stars. Significant and unique information for the various stages of the collapse, the structure of protoneutron stars and the high density equation of state of compact objects can be drawn from careful study of gravitational wave signals.

1. Introduction

The new generation of gravitational wave (GW) detectors is already collecting data, improving the achieved sensitivity by at least one order of magnitude compared to operating resonant bar detectors. Broadband GW detectors are sensitive to frequencies between 50 and a few hundred Hz. In their advanced stage, the current GW detectors will have a broader bandwidth but will still not be sensitive enough to frequencies above 500 to 600Hz. Nevertheless, improved sensitivity can be achieved at high frequencies through narrow-band operation [1,2]. In addition, there are proposals for constructing wide band resonant detectors in the kHz band [3].

In this short review we will discuss some of the sources that are in the high frequency band ($\gtrsim 500 - 600$Hz), where the currently operating interferometers are sensitive enough only if they are narrowbanded. Since there exist a variety of GW sources with very interesting physics to be explored, this high-frequency window deserves special attention. If either resonant or narrow-band interferometers achieve the required sensitivity, a plethora of unique information can be deduced from detected signals.

2. Gravitational collapse

Core collapse. One of the most spectacular astrophysical events is the core collapse of massive stars, leading to the formation of a neutron star (NS) or a black hole (BH). The outcome of core collapse depends sensitively on several factors: mass, angular momentum and metallicity of progenitor, existence of a binary companion, high-density equation of state, neutrino emission, magnetic fields, etc. Partial understanding of each of the above factors is emerging, but a complete and consistent theory for core collapse is still years away.

Roughly speaking, isolated stars more massive than $\sim 8 - 10 M_\odot$ end in core collapse and $\sim 90\%$ of them are stars with masses $\sim 8 - 20 M_\odot$. After core bounce, most of the material is ejected and if the progenitor star has a mass $M \lesssim 20 M_\odot$ a neutron star is left behind. On the other hand, if $M \gtrsim 20 M_\odot$ fall-back accretion increases the mass of the formed proton-neutron star (PNS), pushing it above the maximum mass limit, which results in the formation of a black hole. Furthermore, if the progenitor star has a mass of roughly $M \gtrsim 45 M_\odot$, no supernova explosion is launched and the star collapses directly to a BH[4].

The above picture is, of course, greatly simplified. In reality, the metallicity of the progenitor, the angular momentum of the pre-collapse core and the presence of a binary companion will decisively influence the outcome of core collapse[5]. Very massive stars lose mass through strong stellar winds. The mass loss rate is sensitive to the metallicity of the star and can be very high, allowing a 60 $M_\odot$ star to leave behind a typical 1.4 $M_\odot$ neutron star, instead of a much more massive black hole. Since mass-loss is a complex phenomenon, current observations are still insufficient to constrain the different possible outcomes of core collapse for stars with $M \gtrsim 45 M_\odot$. Through mass transfer or common envelope episodes, a binary companion can cause a massive star to enter the rapid mass-losing (Wolf-Rayet) phase earlier, making wind mass loss more effective. Roughly half of all stars are in suffiently close binaries that binary interactions must be taken into account when trying to predict to outcome of core collapse. Rotation influences the collapse by changing dramatically the properties of the convective region above the proto-neutron star core. Centrifugal forces slow down infalling material in the equatorial region compared to materiall falling in along the polar axis, yielding a weaker bounce. This asymmetry between equator and poles also strongly influences the neutrino emmission and the revival of the stalled shock by neutrinos [6,7].

The supernova event rate is 1-2 per century per galaxy [8] and about 5-40% of them produce BHs in delayed collapse (through fall-back accretion), or direct collapse[9].

Initial rotation rates. Of considerable importance is the *initial rotation rate* of proto-neutron stars, since (as will be detailed in the next sections) most mechanisms for emission of detectable gravitational waves from compact objects require very rapid rotation at birth (rotational periods of the order of a few milliseconds or less). Since most massive stars have non-negligible rotation rates (some even rotate near their break-up limit), simple conservation of angular momentum would suggest a proto-neutron star to be strongly differentially rotating with very high rotation rates and this picture is supported by numerical simulations of rotating core collapse[10,11].

On the other hand, observationally, we know of pulsars (that have been associated with a supernova remnant) rotating only as fast as with a 16ms period, which suggest a period of at least several milliseconds at birth[12]. A possible explanation for this discrepancy is that the pre-supernova core has been slowed down by e.g. magnetic torques[13]. The most recent evolutionary models of rotating core-collapse progenitors (including the effect of magnetic torques) suggest that progenitors of neutron stars with typical masses of $\sim 1.4M_\odot$ are indeed slowly rotating, producing remnants with periods at birth of the order of 10 ms. Nevertheless, the same study finds that very massive progenitors evolve so rapidly that the angular momentum transfer out of the core, by magnetic torques, is diminished, yielding heavy proto-neutron stars with rotational periods of the order of a few milliseconds (provided mass-loss takes place to prevent black hole formation)[14]. Binary interactions can also accelerate some phases of progenitor evolution, allowing for fast initial spins[15,16]. If the magnetic torques operate efficiently, it is clear that rapid rotation at birth will have an event rate much smaller than the usual galactic supernova rate. Still, strong emission mechanisms (e.g. bar-mode instabilities) could yield detectable signals at acceptable event rates.

Other ways to form a rapidly rotating proto-neutron star would be through *fall-back accretion*[17], through the *accretion-induced collapse of a white dwarf*[18,19,20,21] or through the merger of binary white dwarfs in globular clusters[22]. It is also relevant to take into account current gamma-ray-burst models. The *collapsar*[23] model requires high rotation rates of a proto-black hole[16]. In addition, a possible formation scenario for magnetars involves a rapidly rotating protoneutron star formed through the collapse of a very massive progenitor and some observational evidence is already emerging[24].

Gravitational wave emission. Gravitational waves from core collapse have a rich spectrum, reflecting the various stages of this event. The initial signal is emitted due to the *changing axisymmetric quadrupole moment* during collapse. In the case of neutron star formation, the quadrupole moment typically becomes larger, as the core spins up during contraction. In contrast, when a rapidly rotating neutron star collapses to form a Kerr black hole, the axisymmetric quadrupole moment first increases but is finally reduced by a large factor when the black hole is formed.

A second part of the gravitational wave signal is produced when gravitational collapse is halted by the stiffening of the equation of state above nuclear densities and the core bounces, driving an outwards moving shock. The dense fluid undergoes motions with relativistic speeds ($v/c \sim 0.2-0.4$) and a rapidly rotating proto-neutron star thus oscillates in several of its axisymmetric *normal modes of oscillation.* This quasi-periodic part of the signal could last for hundreds of oscillation periods, before being effectively damped. If, instead, a black hole is directly formed, then black hole quasi-normal modes are excited, lasting for only a few oscillation periods. A combination of neutron star and black hole oscillations will appear if the proton-neutron star is not stable but collapses to a black hole.

In a rotating proto-neutron star, nonaxisymmetric processes can yield additional types of gravitational wave signals. Such processes are *dynamical instabilities*, *secular gravitational-wave driven instabilities* or *convection* inside the proto-neutron star and in its surrounding hot envelope. *Anisotropic neutrino emission* is accompanied by a gravitational wave signal. *Nonaxisymmetries* could already be present in the pre-collapse core and become amplified during collapse[25]. Furthermore, if there is persistent fall-back accretion onto a proto-neutron star or black hole, these can be brought into *ringing.*

Below, we discuss in more detail those processes which result in high frequency gravitational radiation.

2.1. *Neutron star formation*

Core collapse as a potential source of GWs has been studied for more than three decades (some of the most recent simulations can be found in [26,27,20,11,28,29,30,31,32,33]). The main differences between the various studies are the progenitor models (slowly or rapidly rotating), equation of state (polytropic or realistic), gravity (Newtonian or relativistic) and neutrino emission (simple, sophisticated or no treatment). In general, the gravitational wave signal from neutron star formation is divided into a core bounce signal, a signal due to convective motions and a signal due to anisotropic neutrino emission.

Core bounce signal. The core bounce signal is produced due to rotational flattening and excitation of normal modes of oscillations, the main contributions coming from the axisymmetric quadrupole ($l = 2$) and quasi-radial ($l = 0$) modes (the latter radiating through its rotationally acquired $l = 2$ piece). If detected, such signals will be a unique probe for the high-density EOS of neutron stars[34,35]. The strength of this signal is sensitive to the available angular momentum in the progenitor core. If the progenitor core is rapidly rotating, then core bounce signals from Galactic supernovae ($d \sim 10$kpc) are detectable even with the initial LIGO/Virgo sensitivity at frequencies $\lesssim$1kHz. In the best-case scenario, advanced LIGO could detect signals from distances of 1Mpc, but not from the Virgo cluster ($\sim$15Mpc), where the event rate would be high. The typical GW amplitude from 2D numerical simulations [11,28] for an observer located in the equatorial plane of the source is[31]

$$h \approx 9 \times 10^{-21} \varepsilon \left(\frac{10\text{kpc}}{d} \right) \tag{1}$$

where $\varepsilon \sim 1$ is the normalized GW amplitude. For such rapidly rotating initial models, the total energy radiated in GWs during the collapse is $\lesssim 10^{-6}$–$10^{-8} M_\odot c^2$. If, on the other hand, progenitor cores are slowly rotating (due to e.g. magnetic torques[13]), then the signal strength is significantly reduced, but, in the best case, is still within reach of advanced LIGO for galactic sources.

Normal mode oscillations, if excited in an equilibrium star at a small to moderate amplitude, would last for hundreds to thousands of oscillation periods, being damped only slowly by gravitational wave emission or viscosity. However, the proto-neutron star immediately after core bounce has a very different structure than a cold equilibrium star. It has a high internal temperature and is surrounded but an extended, hot envelope. Nonlinear oscillations excited in the core after bounce can penetrate into the hot envelope. Through this damping mechanism, the normal mode oscillations are damped on a much shorter timescale (on the order of ten oscillation periods), which is typically seen in the core collapse simulations mentioned above.

Convection signal. The post-shock region surrounding a proto-neutron star is convectively unstable to both low-mode and high-mode convection. Neutrino emission also drives convection in this region. The most realistic 2D simulations of core collapse to date[30] have shown that the gravitational wave signal from convection significantly exceeds the core bounce signal for slowly rotating progenitors, being detectable with advanced LIGO for galactic sources, and is detectable even for nonrotating collapse. For slowly rotating collapse, there is a detectable part of the signal in the high-frequency range of 700Hz-1kHz, originating from convective motions that dominate around 200ms after core bounce. Thus, if both core a bounce signal and a convection signal would be detected in the same frequency range, these would be well separated in time.

Neutrino signal. In many simulations the gravitational wave signature of anisotropic neutrino emission has also been considered[36,37,38]. This type of signal can be detectable by advanced LIGO for galactic sources, but the main contribution is at low frequencies for a slowly rotating progenitor[30]. For rapidly rotating progenitors, stronger contributions at high frequencies could be present, but would probably be burried within the high-frequency convection signal.

Numerical simulations of neutron star formation have gone a long way, but a fully consistent 3D simulation including relativistic gravity, neutrino emission and magnetic fields is still missing. The combined treatment of these effects might not change the above estimations by orders of magnitude but it will provide more conclusive answers. There are also issues that need to be understood such as pulsar kicks (velocities exceeding 1000 km/s) which suggest that in a fraction of newly-born NSs (and probably BHs) the formation process may be strongly asymmetric[39]. Better treatment of the microphysics and construction of accurate progenitor models for the angular momentum distributions are needed. All these issues are under investigation by many groups.

2.2. *Neutron star ringing through fall-back accretion*

A possible mechanism for the excitation of oscillations in a proto-neutron star after core bounce is the fall-back accretion of material that has not been expelled by the revived supernova shock. The isotropy of this material is expected to be broken due to e.g. rotation or nonaxisymmetric convective motions, thus a large number of oscillation modes will be excited as this material falls back onto the neutron star. This process is, of course, complex and the detectability of gravitational waves from these oscillations will depend on several factors, such as the fall-back accretion rate, the degree of asymmetry of the fall-back material the structure of the proto-neutron star envelope, the presence of magnetic fields etc.

Recently, the ringing of a neutron star through fall-back accretion has been modelled through relativistic 2D nonlinear hydrodynamical simulations[40]. Quadrupolar shells of matter were accreted on a static neutron star (in the approximation that the background spacetime remains unchanged). Gravitational waves were then extracted through the Zerilli-Moncrief formalism. The gravitational wave signal from such a process comprises a *narrow peak* at the $l = 2$ normal mode frequency of the neutron star and a very *broad peak*, featuring *interference fringes*, centered at a much higher frequency. Since the frequency of the broad peak is still too low to be identified with a w-mode of the star, the interpretation for this part of the signal is that it is related to the motion of the fluid shell and the reflection of the gravitational-wave pulse from this motion in the external Zerilli potential, which also creates the interference fringes. The accretion of a quadrupolar shell containing 1% of the mass of the star releases gravitational waves with a total energy similar to the energy emitted immediately after core bounce. It is thus interesting to consider this mechanism in more detail, since the excitation of the normal modes in the neutron star happens when the star has already cooled somewhat (compared to the proto-neutron star immediately after core bounce) which simplifies the identification of observed oscillations with normal modes of cold neutron star models. The formation of a dense torus as as a result of stellar gravitational collapse, binary neutron star merger or disruption. Such a system either becomes unstable to the runaway instability or exhibit a regular oscillatory behavior, resulting in a quasi-periodic variation of the accretion rate as well as of the mass quadrupole suggesting a new sources of potentially detectable gravitational waves[41].

2.3. *Black hole formation*

The gravitational-wave emission from the formation of a Kerr BH is a sum of two signals: the *collapse signal* and the *BH ringing*. The collapse signal is produced due to the changing multipole moments of the spacetime during the transition from a rotating iron core or proto-neutron star to a Kerr BH. A uniformly rotating neutron star has an axisymmetric quadrupole moment given by[42]

$$Q = -a\frac{J^2}{M} \tag{2}$$

where a depends on the equation of state and is in the range of $2-8$ for $1.4M_\odot$ models. This is several times larger in magnitude than the corresponding qudrupole moment of a Kerr black hole ($a = 1$). Thus, the *reduction* of the axisymmetric quadrupole moment is the main source of the collapse signal. Once the BH is formed, it continues to oscillate in its axisymmetric $l = 2$ QNM, until all oscillation energy is radiated away and the stationary Kerr limit is approached.

The numerical study of rotating collapse to BHs was pioneered by Nakamura[43] but first waveforms and gravitational-wave estimates were obtained by Stark and Piran[44] . These simulations we performed in 2D, using approximate initial data (essentially a spherical star to which angular momentum was artificially added). A new 3D computation of the gravitational wave emission from the collapse of unstable uniformly rotating relativistic polytropes to Kerr BHs[45] finds that the energy emitted is

$$\Delta E \sim 1.5 \times 10^{-6}(M/M_\odot), \tag{3}$$

significantly less than the result of Stark and Piran. Still, the collapse of an unstable $2M_\odot$ rapidly rotating neutron star leads to a characteristic gravitational-wave amplitude $h_c \sim 3 \times 10^{-21}$, at a frequency of ~ 5.5kHz, for an event at 10kpc. Emission is mainly through the "+" polarization, with the "×" polarization being an order of magnitude weaker.

Whether a BH forms promptly after collapse or a delayed collapse takes place depends sensitively on a number of factors, such as the progenitor mass and angular momentum and the high-density EOS. The most detailed investigation of the influence of these factors on the outcome of collapse has been presented recently in[33], where it was found that shock formation increases the threshold for black hole formation by $\sim 20-40\%$, while rotation results in an increase of at most 25%.

2.4. *Black hole ringing through fall-back or hyper-accretion*

Single events. A black hole can form after core collapse, if fall-back accretion increases the mass of the proto-neutron star above the maximum mass allowed by axisymmetric stability. Material falling back after the black hole is formed excites the black hole quasi-normal modes of oscillation. If, on the other hand, the black hole is formed directly through core collapse (without a core bounce taking place) then most of the material of the progenitor star is accreted at very high rates ($\sim 1 - 2M_{\odot}$/s) into the hole. In such *hyper-accretion* the black hole's quasi-normal modes (QNM) can be excited for as long as the process lasts and until the black hole becomes stationary. Typical frequencies of the emitted GWs are in the range 1-3kHz for $\sim 3 - 10M_{\odot}$ BHs.

The frequency and the damping time of the oscillations for the $l = m = 2$ mode can be estimated via the relations [46]

$$\sigma \approx 3.2\text{kHz}\ M_{10}^{-1} \left[1 - 0.63(1 - a/M)^{3/10}\right] \tag{4}$$

$$Q = \pi\sigma\tau \approx 2\,(1 - a)^{-9/20} \tag{5}$$

These relations together with similar ones either for the 2nd QNM or the $l = 2$, $m = 0$ mode can uniquely determine the mass M and angular momentum parameter a of the BH if the frequency and the damping time of the signal have been accurately extracted [47,48,49]. The amplitude of the ring-down waves depends on the BH's initial distortion, i.e. on the nonaxisymmetry of the blobs or shells of matter falling into the BH. If matter of mass μ falls into a BH of mass M, then the gravitational wave energy is roughly

$$\Delta E \gtrsim \epsilon\mu c^2(\mu/M) \tag{6}$$

where ϵ is related to the degree of asymmetry and could be $\epsilon \gtrsim 0.01$ [50]. This leads to an effective GW amplitude

$$h_{\text{eff}} \approx 2 \times 10^{-21} \left(\frac{\epsilon}{0.01}\right) \left(\frac{10\text{Mpc}}{d}\right) \left(\frac{\mu}{M_{\odot}}\right) \tag{7}$$

Resonant driving. If hyper-accretion proceeds through an accretion disk around a rapidly spinning Kerr BH, then the matter near the marginally bound orbit radius can become unstable to the magnetorotational (MRI) instability, leading to the formation of large-scale asymmetries[51]. Under certain conditions, *resonant driving* of the BH QNMs could take place. Such a continuous signal could be integrated, yielding a much larger signal to noise ratio than a single event. For a $15M_{\odot}$ nearly maximal Kerr BH created at 27Mpc the integrated signal becomes detectable by LIGO II at a frequency of ~ 1600Hz, especially if narrow-banding is used[51].

3. Rotational instabilities

If proto-neutron stars rotate rapidly, nonaxisymmetric *dynamical instabilities* can develop. These arise from non-axisymmetric perturbations having angular dependence $e^{im\phi}$ and are of two different types: the *classical bar-mode* instability and the more recently discovered *low-$T/|W|$ bar-mode* and *one-armed spiral* instabilities, which appear to be associated to the presence of corotation points. Another class of nonaxisymmetric instabilities are *secular instabilities*, driven by dissipative effects, such as fluid viscosity or gravitational radiation.

3.1. *Dynamical instabilities*

Classical bar-mode instability. The classical $m = 2$ bar-mode instability is excited in Newtonian stars when the ratio $\beta = T/|W|$ of the rotational kinetic energy T to the gravitational binding energy $|W|$ is larger than $\beta_{\rm dyn} = 0.27$. The instability grows on a dynamical time scale (the time that a sound wave needs to travel across the star) which is about one rotational period and may last from 1 to 100 rotations depending on the degree of differential rotation in the PNS.

The bar-mode instability can be excited in a hot PNS, a few milliseconds after core bounce, or, alternatively, it could also be excited a few tenths of seconds later, when the PNS cools due to neutrino emission and contracts further, with β becoming larger than the threshold β_{dyn} (β increases roughly as $\sim 1/R$ during contraction). The amplitude of the emitted gravitational waves can be estimated as $h \sim MR^2\Omega^2/d$, where M is the mass of the body, R its size, Ω the rotation rate and d the distance of the source. This leads to an estimation of the GW amplitude

$$h \approx 9 \times 10^{-23} \left(\frac{\epsilon}{0.2}\right) \left(\frac{f}{3\text{kHz}}\right)^2 \left(\frac{15\text{Mpc}}{d}\right) M_{1.4} R_{10}^2. \qquad (8)$$

where ϵ measures the ellipticity of the bar, M is measured in units of $1.4M_\odot$ and R is measured in units of 10km. Notice that, in uniformly rotation Maclaurin spheroids, the GW frequency f is twice the rotational frequency Ω. Such a signal is detectable only from sources in our galaxy or the nearby ones (our Local Group). If the sensitivity of the detectors is improved in the kHz region, signals from the Virgo cluster could be detectable. If the bar persists for many ($\sim$ 10-100) rotation periods, then even signals from distances considerably larger than the Virgo cluster will be detectable. Due to the requirement of rapid rotation, the event rate of the classical dynamical instability is considerably lower than the SN event rate.

The above estimates rely on Newtonian calculations; GR enhances the onset of the instability, $\beta_{\rm dyn} \sim 0.24$ [52,53] and somewhat lower than that for large compactness (large M/R). Fully relativistic dynamical simulations of this instability have been obtained, including detailed waveforms of the associated gravitational wave emission. A detailed investigation of the required initial conditions of the progenitor core, which can lead to the onset of the dynamical bar-mode instability in the formed PNS, was presented in[31]. The amplitude of gravitational waves was due to the bar-mode instability was found to be larger by an order of magnitude, compared to the axisymmetric core collapse signal.

Low-$T/|W|$ instabilities. The *bar-mode* instability may be excited for significantly smaller β, if centrifugal forces produce a peak in the density off the source's rotational center[54]. Rotating stars with a high degree of differential rotation are also dynamically unstable for significantly lower $\beta_{\rm dyn} \gtrsim 0.01$ [55,56]. According to this scenario the unstable neutron star settles down to a non-axisymmetric quasi-stationary state which is a strong emitter of quasi-periodic gravitational waves

$$h_{\rm eff} \approx 3 \times 10^{-22} \left(\frac{R_{\rm eq}}{30{\rm km}}\right) \left(\frac{f}{800{\rm Hz}}\right)^{1/2} \left(\frac{100{\rm Mpc}}{d}\right) M_{1.4}^{1/2}. \tag{9}$$

The bar-mode instability of differentially rotating neutron stars is an excellent source of gravitational waves, provided the high degree of differential rotation that is required can be realized. One should also consider the effects of viscosity and magnetic fields. If magnetic fields enforce uniform rotation on a short timescale, this could have strong consequences regarding the appearance and duration of the dynamical nonaxisymmetric instabilities.

An $m = 1$ *one-armed spiral* instability has also been shown to become unstable in proto-neutron stars, provided that the differential rotation is sufficiently strong [54,57]. Although it is dominated by a "dipole" mode, the instability has a spiral character, conserving the center of mass. The onset of the instability appears to be linked to the presence of corotation points[58] (a similar link to corotation points has been proposed for the low-$T/|W|$ bar mode instability[59]) and requires a very high degree of differential rotation (with matter on the axis rotating at least 10 times faster than matter on the equator). The $m = 1$ spiral instability was recently observed in simulations of rotating core collapse, which started with the core of an evolved $20M_\odot$ progenitor star to which differential rotation was added[60]. Growing from noise level ($\sim 10^{-6}$) on a timescale of 5ms, the $m = 1$ mode reached its maximum amplitude after $\sim$ 100ms. Gravitational waves were emitted through the excitation of an $m = 2$ nonlinear harmonic at a frequency of $\sim$ 800Hz with an amplitude comparable to the core-bounce axisymmetric signal.

3.2. *Secular gravitational-wave-driven instabilities*

In a nonrotating star, the forward and backward moving modes of same $(l, |m|)$ (corresponding to $(l, +m)$ and $(l, -m)$) have eigenfrequencies $\pm|\sigma|$. Rotation splits this degeneracy by an amount $\delta\sigma \sim m\Omega$ and both the prograde and retrograde modes are dragged forward by the stellar rotation. If the star spins sufficiently rapidly, a mode which is retrograde (in the frame rotating with the star) will appear as prograde in the inertial frame (a nonrotating observer at infinity). Thus, an inertial observer sees GWs with positive angular momentum emitted by the retrograde mode, but since the perturbed fluid rotates slower than it would in the absence of the perturbation, the angular momentum of the mode in the rotating frame is negative. The emission of GWs consequently makes the angular momentum of the mode increasingly negative, leading to the instability. A mode is unstable when $\sigma(\sigma - m\Omega) < 0$. This class of *frame-dragging instabilities* is usually referred to as Chandrasekhar-Friedman-Schutz[61,62] (CFS) instabilities.

f-mode instability. In the Newtonian limit, the $l = m = 2$ f-mode (which has the shortest growth time of all polar fluid modes) becomes unstable when $T/|W| > 0.14$, which is near or even above the mass-shedding limit for typical polytropic EOSs used to model uniformly rotating neutron stars. Dissipative effects (e.g. shear and bulk viscosity or mutual friction in superfluids)[63,64,65,66] leave only a small instability window near mass-shedding, at temperatures of $\sim 10^9$K. However, relativistic effects strengthen the instability considerably, lowering the required β to $\approx 0.06 - 0.08$ [67,68] for most realistic EOSs and masses of $\sim 1.4M_\odot$ (for higher masses, such as hypermassive stars created in a binary NS merger, the required rotation rates are even lower).

Since PNSs rotate differentially, the above limits derived under the assumption of uniform rotation are too strict. Unless uniform rotation is enforced on a short timescale, due to e.g. magnetic braking[69], the f-mode instability will develop in a differentially rotating background, in which the required $T/|W|$ is only somewhat larger than the corresponding value for uniform rotation[70], but the mass-shedding limit is dramatically relaxed. Thus, in a differentially rotating PNS, the f-mode instability window is huge, compared to the case of uniform rotation and the instability can develop provided there is sufficient $T/|W|$ to begin with.

The f-mode instability is an excellent source of GWs. Simulations of its nonlinear development in the ellipsoidal approximation[71] have shown that the mode can grow to a large nonlinear amplitude, modifying the background star from an axisymmetric shape to a differentially rotating ellipsoid. In this modified background the f-mode amplitude saturates and the ellipsoid becomes a strong emitter of gravitational waves, radiating away angular momentum until the star is slowed-down towards a stationary state. In the case of uniform density ellipsoids, this stationary state is the Dedekind ellipsoid, i.e. a nonaxisymmetric ellipsoid with internal flows but with a stationary (nonradiating) shape in the inertial frame. In the ellipsoidal approximation, the nonaxisymmetric pattern radiates gravitational waves sweeping through the LIGO II sensitivity window (from 1kHz down to about 100Hz) which could become detectable out to a distance of more than 100Mpc.

Two recent hydrodynamical simulations[72,73] (in the Newtonian limit and using a post-Newtonian radiation-reaction potential) essentially confirm this picture. In [72] a differentially rotating, $N = 1$ polytropic model with a large $T/|W| \sim 0.2 - 0.26$ is chosen as the initial equilibrium state. The main difference of this simulation compared to the ellipsoidal approximation comes from the choice of EOS. For $N = 1$ Newtonian polytropes it is argued that the secular evolution cannot lead to a a stationary Dedekind-like state does not exist. Instead, the f-mode instability will continue to be active until all nonaxisymmetries are radiated away and an axisymmetric shape is reached. This conclusion should be checked when relativistic effects are taken into account, since, contrary to the Newtonian case, relativistic $N = 1$ uniformly rotating polytropes *are* unstable to the $l = m = 2$ f-mode[67] – however it has not become possible, to date, to construct relativistic analogs of Dedekind ellipsoids.

In the other recent simulation [73] , the initial state was chosen to be a uniformly rotating, $N = 0.5$ polytropic model with $T/|W| \sim 0.18$. Again, the main conclusions reached in [71] are confirmed, however, the assumption of uniform initial rotation limits the available angular momentum that can be radiated away, leading to a detectable signal only out to about $\sim$ 40Mpc. The star appears to be driven towards a Dedekind-like state, but after about

10 dynamical periods, the shape is disrupted by growing short-wavelength motions, which are suggested to arise because of a shearing type instability, such as the elliptic flow instability[74].

r-mode instability. Rotation does not only shift the spectra of polar modes; it also lifts the degeneracy of axial modes, give rise to a new family of *inertial* modes, of which the $l = m = 2$ r-mode is a special member. The restoring force, for these oscillations is the Coriolis force. Inertial modes are primarily velocity perturbations. The frequency of the r-mode in the rotating frame of reference is $\sigma = 2\Omega/3$. According to the criterion for the onset of the CFS instability, the r-mode is unstable for any rotation rate of the star[75,76]. For temperatures between $10^7 - 10^9$K and rotation rates larger than 5-10% of the Kepler limit, the growth time of the unstable mode is smaller than the damping times of the bulk and shear viscosity[77,78]. The existence of a solid crust or of hyperons in the core [79] and magnetic fields [80,81], can also significantly affect the onset of he instability (for extended reviews see [82,83]). The suppression of the r-mode instability by the presence of hyperons in the core is not expected to operate efficiently in rapidly rotating stars, since the central density is probably too low to allow for hyperon formation. Moreover, a recent calculation[84] finds the contribution of hyperons to the bulk viscosity to be two orders of magnitude smaller than previously estimated. If accreting neutron stars in Low Mass X-Ray Binaries (LMXB, considered to be the progenitors of millisecond pulsars) are shown to reach high masses of $\sim 1.8M_\odot$, then the EOS could be too stiff to allow for hyperons in the core (for recent observations that support a high mass for some millisecond pulsars see [85]).

The unstable r-mode grows exponentially until it saturates due to nonlinear effects at some maximum amplitude α_{max}. The first computation of nonlinear mode couplings using second-order perturbation theory suggested that the r-mode is limited to very small amplitudes (of order $10^{-3} - 10^{-4}$) due to transfer of energy to a large number of other inertial modes, in the form of a cascade, leading to an equilibrium distribution of mode amplitudes[86]. The small saturation values for the amplitude are supported by recent nonlinear estimations [87,88] based on the drift, induced by the r-modes, causing differential rotation. On the other hand, hydrodynamical simulations of limited resolution showed that an initially large-amplitude r-mode does not decay appreciably over several dynamical timescales[89], but on a somewhat longer timescale a catastrophic decay was observed[90] indicating a transfer of energy to other modes, due to nonlinear mode couplings and suggesting that a hydrodynamical instability may be operating. A specific resonant 3-mode coupling was identified[91] as the cause of the instability and a perturbative analysis of the decay rate suggests a maximum saturation amplitude $\alpha_{max} < 10^{-2}$. A new computation using second-order perturbation theory finds that the catastrophic decay seen in the hydrodynamical simulations[90,91] can indeed be explained by a parametric instability operating in 3-mode couplings between the r-mode and two other inertial modes[92,93,94]. Whether the maximum saturation amplitude is set by a network of 3-mode couplings or a cascade is reached, is, however, still unclear.

A neutron star spinning down due to the r-mode instability will emit gravitational waves of amplitude

$$h(t) \approx 10^{-21}\alpha \left(\frac{\Omega}{1\text{kHz}}\right)\left(\frac{100\text{kpc}}{d}\right) \tag{10}$$

Since α is small, even with LIGO II the signal is undetectable at large distances (VIRGO cluster) where the SN event rate is appreciable, but could be detectable after long-time integration from a galactic event. However, if the compact object is a strange star, then the instability may not reach high amplitudes ($\alpha \sim 10^{-3} - 10^{-4}$) but it will persist for a few hundred years (due to the different temperature dependence of viscosity in strange quark matter) and in this case there might be up to ten unstable stars in our galaxy at any time [95]. Integrating data for a few weeks could lead to an effective amplitude $h_{\text{eff}} \sim 10^{-21}$ for galactic signals at frequencies $\sim 700 - 1000$Hz. The frequency of the signal changes only slightly on a timescale of a few months, so that the radiation is practically monochromatic.

Other unstable modes. The CFS instability can also operate for core g-mode oscillations[96] but also for w-mode oscillations, which are basically spacetime modes[97]. In addition, the CFS instability can operate through other dissipative effects. Instead of the gravitational radiation, any radiative mechanism (such as electromagnetic radiation) can in principle lead to an instability.

3.3. *Secular viscosity-driven instability*

A different type of nonaxisymmetric instability in rotating stars is the instability driven by viscosity, which breaks the circulation of the fluid [98,99]. The instability is suppressed by gravitational radiation, so it cannot act in the temperature window in which the CFS-instability is active. The instability sets in when the frequency of a *prograde* $l = -m$ mode goes through zero in the rotating frame. In contrast to the CFS-instability, the viscosity-driven instability is not generic in rotating stars. The $m = 2$ mode becomes unstable at a high rotation rate for very stiff stars and higher m-modes become unstable at *larger* rotation rates.

In Newtonian polytropes, the instability occurs only for stiff polytropes of index $N < 0.808$ [99,100]. For relativistic models, the situation for the instability becomes worse, since relativistic effects tend to suppress the viscosity-driven instability (while the CFS-instability becomes stronger). For the most relativistic stars, the viscosity-driven bar mode can become unstable only if $N < 0.55$[101]. For $1.4 M_\odot$ stars, the instability is present for $N < 0.67$.

An investigation of the viscosity-driven bar mode instability, using incompressible, uniformly rotating triaxial ellipsoids in the post-Newtonian approximation[102] finds that the relativistic effects increase the critical $T/|W|$ ratio for the onset of the instability significantly. More recently, new post-Newtonian [103] and fully relativistic calculations for uniform-density stars [104] show that the viscosity-driven instability is not as strongly suppressed by relativistic effects as suggested in [102]. The most promising case for the onset of the viscosity-driven instability (in terms of the critical rotation rate) would be rapidly rotating strange stars [105], but the instability can only appear if its growth rate is larger than the damping rate due to the emission of gravitational radiation - a corresponding detailed comparison is still missing.

4. Accreting neutron stars in LMXBs

Spinning neutron stars with even tiny deformations are interesting sources of gravitational waves. The deformations might results from various factors but it seems that the most interesting cases are the ones in which the deformations are caused by accreting material. A class of objects called Low-Mass X-Ray Binaries (LMXB) consist of a fast rotating neutron star (spin $\approx 270 - 650$Hz) torqued by accreting material from a companion star which has filled up its Roche lobe. The material adds both mass and angular momentum to the star, which, on timescales of the order of tenths of Megayears could, in principle, spin up the neutron star to its break up limit. One viable scenario [106] suggests that the accreted material (mainly hydrogen and helium) after an initial phase of thermonuclear burning undergoes a non-uniform crystallization, forming a crust at densities $\sim 10^8 - 10^9 \mathrm{g/cm}^3$. The quadrupole moment of the deformed crust is the source of the emitted gravitational radiation which slows-down the star, or halts the spin-up by accretion.

An alternative scenario has been proposed by Wagoner[107] as a follow up of an earlier idea by Papaloizou-Pringle[108]. The suggestion was that the spin-up due to accretion might excite the f-mode instability, before the rotation reaches the breakup spin. The emission of gravitational waves will torque down the star's spin at the same rate as the accretion will torque it up, however, it is questionable whether the f-mode instability will ever be excited for old, accreting neutron stars. Following the discovery that the r-modes are unstable at any rotation rate, this scenario has been revived independently by Bildsten [106] and Andersson, Kokkotas and Stergioulas [109]. The amplitude of the emitted gravitational waves from such a process is quite small, even for high accretion rates, but the sources are persistent and in our galactic neighborhood the expected amplitude is

$$h \approx 10^{-27} \left(\frac{1.6\text{ms}}{P}\right)^5 \frac{1.5\text{kpc}}{D}. \tag{11}$$

This signal is within reach of advanced LIGO with signal recycling tuned at the appropriate frequency and integrating for a few months[1]. This picture is in practice more complicated, since the growth rate of the r-modes (and consequently the rate of gravitational wave emission) is a function of the core temperature of the star. This leads to a thermal runaway due to the heat released as viscous damping mechanisms counteract the r-mode growth [110]. Thus, the system executes a limit cycle, spinning up for several million years and spinning down in a much shorter period. The duration of the unstable part of the cycle depends critically on the saturation amplitude α_{max} of the r-modes [111,112]. Since current computations [86,88] suggest an $\alpha_{max} \sim 10^{-3} - 10^{-4}$, this leads to a quite long duration for the unstable part of the cycle of the order of $\sim 1Myear$.

The instability window depends critically on the effect of the shear and bulk viscosity and various alternative scenarios might be considered. The existence of hyperons in the core of neutron stars induces much stronger bulk viscosity which suggests a much narrower instability window for the r-modes and the bulk viscosity prevails over the instability even in temperatures as low as 10^8K [79]. A similar picture can be drawn if the star is composed of "deconfined" u, d and s quarks - a strange star [113]. In this case, there is a possibility that the strange stars in LMXBs evolve into a quasi-steady state with nearly constant rotation rate, temperature and mode amplitude [95] emitting gravitational waves for as long as the accretion lasts. This result has also been found later for stars with hyperon cores [114,115]. It is interesting that the stalling of the spin up in millisecond pulsars (MSPs) due to r-modes is in good agreement with the minimum observed period and the clustering of the frequencies of MSPs [111].

5. Binary mergers

Depending on the high-density EOS and their initial masses, the outcome of the merger of two neutron stars may not always be a black hole, but a hypermassive, differentially rotating compact star (even if it is only temporarily supported against collapse by differential rotation). A recent detailed simulation[116] in full GR has shown that the hypermassive object created in a binary NS merger is nonaxisymmetric. The nonaxisymmetry lasts for a large number of rotational periods, leading to the emission of gravitational waves with a frequency of 3kHz and an effective amplitude of $\sim 6-7\times 10^{-21}$ at a large distance of 50Mpc. Such large effective amplitude may be detectable even by LIGO II at this high frequency.

The tidal disruption of a NS by a BH [117] or the merging of two NSs [118] may give valuable information for the radius and the EoS if we can recover the signal at frequencies higher than 1 kHz.

6. Gravitational-wave asteroseismology

If various types of oscillation modes are excited during the formation of a compact star and become detectable by gravitational wave emission, one could try to identify observed frequencies with frequencies obtained by mode-calculations for a wide parameter range of masses, angular momenta and EOSs. [35,119,120,121,122,123]. Thus, *gravitational wave asteroseismology* could enable us to estimate the mass, radius and rotation rate of compact stars, leading to the determination of the "best-candidate" high-density EoS, which is still very uncertain. For this to happen, accurate frequencies for different mode-sequences of rapidly rotating compact objects have to be computed.

For slowly rotating stars, the frequencies of $f-$, $p-$ and $w-$ modes are still unaffected by rotation, and one can construct approximate formulae in order to relate observed frequencies and damping times of the various stellar modes to stellar parameters. For example, for the fundamental oscillation ($l=2$) mode (f-mode) of non-rotating stars one obtains [35]

$$\sigma(\text{kHz}) \approx 0.8 + 1.6 M_{1.4}^{1/2} R_{10}^{-3/2} + \delta_1 m\bar{\Omega} \quad (12)$$

$$\tau^{-1}(\text{secs}^{-1}) \approx M_{1.4}^{3} R_{10}^{-4}\left(22.9 - 14.7 M_{1.4} R_{10}^{-1}\right) + \delta_2 m\bar{\Omega} \quad (13)$$

where $\bar{\Omega}$ is the normalized rotation frequency of the star, and δ_1 and δ_2 are constants estimated by sampling data from various EOSs. The typical frequencies of NS oscillation modes are larger than 1kHz. Since each type of mode is sensitive to the physical conditions where the amplitude of the mode is largest, the more oscillations modes can be identified through gravitation waves, the better we will understand the detailed internal structure of compact objects, such as the existence of a possible superfluid state of matter[124].

If, on the other hand, some compact stars are born rapidly rotating with moderate differential rotation, then their central densities will be much smaller than the central density of a nonrotating star or same baryonic mass. Correspondingly, the typical axisymmetric oscillation frequencies will be smaller than 1kHz, which is more favorable for the sensitivity window of current interferometric detectors[125]. Indeed, axisymmetric simulations of rotating core-collapse have shown that if a rapidly rotating NS is created, then the dominant frequency of the core-bounce signal (originating from the fundamental $l = 2$ mode or the $l = 2$ piece of the fundamental quasi-radial mode) is in the range 600Hz-1kHz[11].

If different type of signals are observed after core collapse, such as both an axisymmetric core-bounce signal and a nonaxisymmetric one-armed instability signal, with a time separation of the order of 100ms, this would yield invaluable information about the angular momentum distribution in the proto-neutron stars.

Acknowledgments

This work has been supported by the EU program ILIAS(ENTApP) and the GSRT program Heracleitus.

References

1. Cutler C and Thorne K S 2002 gr-qc/0204090
2. Schnabel R, Harms J, Strain K A and Danzmann K 2003 Class. Quantum Grav. 21, S1045
3. Bonaldi M, Cerdonio M, Conti L, Prodi G A, Taffarello L, Zendri J P 2003 Phys. Rev. D 68, 102004
4. Fryer C L 1999 *Astrophys. J.* **522** 413
5. Fryer C L 2003 *Class. Quantum Gravity* **20** S73
6. Fryer C L and Warren M S 2004 *Astrophys. J.* **601** 391
7. Burrows A, Walder R, Ott C D and Livne E 2005 *Nucl. Phys. A* **752** 570
8. Cappellaro E, Turatto M, Tsvetov D Yu, Bartunov O S, Pollas C, Evans R and Hamuy M 1999 *A&A* **351** 459

9. Fryer C L and Kalogera V 2001 *Astrophys. J.* **554**, 548
10. Fryer C L and Heger A 2000 *Astrophys. J.* **541** 1033
11. Dimmelmeier H, Font J A and Müller E 2002 *A&A* **393** 523
12. Marshall F E, Gotthelf E V, Zhang W, Middleditch J and Wang Q D 1998 *Astrophys. J.* **499** L179
13. Spruit H C 2002 *Astron. Astrophys.* **381** 923
14. Heger A, Woosley S E and Spruit H C 2004 astro-ph/0409422
15. Podsiadlowski Ph, Langer N, Poelarends A J T, Rappaport S, Heger A and Pfalh E 2004 *Astrophys. J.* **612** 1044
16. Petrovic J, Langer N, Yoon S C and Heger A 2005 astro-ph/0504175
17. Watts A L and Andersson N 2002 *MNRAS* **333** 943
18. Hillebrandt W, Wolff R G and Nomoto K 1984 *Astrophys. J.* **133** 175
19. Liu Y T and Lindblom L 2001 *MNRAS* astro-ph/0012198
20. Fryer C L, Holz D E, Hughes S A 2002 *Astrophys. J.* **565** 430
21. Yoon S C and Langer N 2005 astro-ph/0502133
22. Middleditch J 2004 *Astrophys. J.* **601** L167
23. Woosley S E 1993 *Bulletin of the American Astronomical Society* **25** 894
24. Gaensler B M et al. 2005 *Nature* **434** 1104
25. Fryer C L, Holz D E,Hughes S A 2004 *Astrophys. J.* **609** 288
26. Zwerger T, Müller E 1997 *A&A* **320** 209
27. Rampp M, Müller E and Ruffert M 1998 *A&A* **332** 969
28. Ott C D, Burrows A, Livne E, Walder R, 2003 astro-ph/0307472
29. Kotake K, Yamada S and Sato K, 2003 astro-ph/0306430
30. Müller E, Rampp M, Buras R, Janka H-Th and Shoemaker D H 2004 *Astrophys. J.* **603** 221
31. Shibata M and Sekiguchi Y 2004 *Phys. Rev. D* **69** 084024
32. Shibata M and Sekiguchi Y 2005 *Phys. Rev. D* **71** 044017
33. Sekiguchi Y and Shibata M 2005 *Phys. Rev. D* **71** 084013
34. Andersson N, Kokkotas K D 1996 *Phys. Rev. Lett.* **77** 4134
35. Andersson N, Kokkotas K D 1998 *MNRAS* **299** 1059
36. Epstein R 1978 *Astrophys. J.* **223** 1037
37. Burrows A and Hayes J 1996 *Phys. Rev. Lett.* **76** 352
38. Müller E and Janka H T 1997 *Astron. Astrophys.* **317** 140
39. Hoeflich P, Khokhlov A, Wang L, Wheeler J C and Baade D 2002, IAU Symposium 212 on Massive Stars, D. Reidel Conf. Series, ed. E. van den Hucht, astro-ph/0207272
40. Nagar A, Diaz G, Pons J A and Font J A 2004 *Phys. Rev. D* **69** 124028
41. Zanotti O, Rezzolla L and Font J A 2003 *MNRAS* **341** 832
42. Laarakkers W G and Poisson E 1999 *Astrophys. J.* **512** 282
43. Nakamura T 1981 *Prog. Theor. Phys.* **65** 1876
44. Stark R F and Piran T 1985 *Phys. Rev. Lett.* **55** 891
45. Baiotti L, Hawke I, Rezzolla L and Schnetter E 2005 *Phys. Rev. Lett.* **94** 131101
46. Echeverria E 1980 *Phys. Rev. D* **40** 3194
47. Finn L S 1992 *Phys. Rev. D* **8** 3308
48. Nakano H, Takahashi H, Tagoshi H and Sasaki M 2003 *Phys. Rev. D* **68**

102003
49. Dryer O, Kelly B, Krishnan B, Finn L S, Garrison D, Lopez-Aleman R 2004 *Class. Quantum Grav.* **21** 787
50. Davis M, Ruffini R, Press W H and Price R H 1971 **27** 1466
51. Araya-Góchez R A 2004 *MNRAS* **355** 336
52. Shibata M, Baumgarte T W and Shapiro S L 2000 *Astrophys. J.* **542** 453
53. Saijo M, Shibata M, Baumgarte T W and Shapiro S L 2001 *Astrophys. J.* **548** 919
54. Centrella J M , New K C B., Lowe L L , Brown J D 2001 *Astrophys. J. Lett.* **550** 193
55. Shibata M, Karino S, Eriguchi Y 2002 *MNRAS* **334** L27
56. Shibata M, Karino S, Eriguchi Y 2003 *MNRAS* **343** 619
57. Saijo M, Baumgarte T W and Shapiro S L, astro-ph/0302436
58. Saijo M and Yoshida Y 2005 astro-ph/0504002
59. Watts A L, Andersson N and Jones D I 2005, ApJ, 618, L37
60. Ott C D, Ou S, Tohline J E, Burrows A 2005 astro-ph/0503187
61. Chandrasekhar S 1970 *Phys. Rev. Lett.* **24** 611
62. Friedman J L and Schutz B F 1978 *Astrophys. J.* **222** 281
63. Cutler C and Lindblom L 1987 *Astrophys. J.* **314** 234
64. Lindblom L and Detweiler S 1979 *Astrophys. J.* **232** L101
65. Ipser J R and Lindblom L 1991 *Astrophys. J.* **373** 213
66. Lindblom L and Mendell G 1995 *Astrophys. J.* **444** 804
67. Stergioulas N, Friedman J L, 1998 *Astrophys. J.* **492** 301
68. Morsink S, Stergioulas N and Blattning S 1999 *Astrophys. J.* **510** 854
69. Liu Y T and Shapiro S L 2004 *Phys. Rev. D* **69** 044009
70. Yoshida S, Rezzolla L, Karino S and Eriguchi Y, 2002 *Astrophys. J. Lett.* **568** 41
71. Lai D and Shapiro S L 1995 *Astrophys. J.* **442** 259
72. Shibata M and Karino S 2004 *Phys. Rev. D* **70** 084022
73. Ou S, Tohline J E and Lindblom L 2004 *Astrophys. J.* **617** 490
74. Lifschitz A and Lebovitz N 1993 *Astrophys. J.* **408** 603
75. Andersson N 1998 *Astrophys. J.* **502** 708
76. Friedman J L and Morsink S 1998 *Astrophys. J.* **502** 714
77. Linblom L, Owen B J and Morsink S M 1998 *Phys. Rev. Lett.* **80** 4843
78. Andersson N, Kokkotas K D and Schutz B F 1999 *Astrophys. J.* **510** 846
79. Lindblom L, Owen B, 2002 *Phys. Rev. D* **65** 063006
80. Rezzolla L, Lamb, F K, Markovic D and Shapiro S L 2001 *Phys. Rev. D* **64** 104013
81. Rezzolla L, Lamb, F K, Markovic D and Shapiro S L 2001 *Phys. Rev. D* **64** 104014
82. Andersson N and Kokkotas K D 2001 *Int. J. Modern Phys.* **D10** 381
83. Andersson N 2003 *Class. Quantum Grav.* **20** R105
84. van Dalen E N E and Dieperink A E L 2003 *Phys. Rev. C* **69** 025802
85. Nice, D. J., Splaver, E. M., Stairs, I. H., 2003, astro-ph/0311296
86. Schenk A K, Arras P, Flanagan E E, Teukolsky S A, Wasserman I 2002 *Phys. Rev. D* **65** 024001

87. Sa P M 2004 *Phys. Rev. D* **69** 084001
88. Sa P M and Tome B 2005 *Phys. Rev. D* **71** 044007
89. Stergioulas N and Font J A 2001 *Phys. Rev. Lett.* **86** 1148
90. Gressman P, Lin L M, Suen W M, Stergioulas N and Friedman J L 2002 *Phys. Rev. D.* **66** 041303(R)
91. Lin L M and Suen W M 2004 gr-qc/0409037
92. Brink J, Teukolsky S A and Wasserman I 2004 *Phys. Rev. D.* **70** 121501
93. Brink J, Teukolsky S A and Wasserman I 2004 *Phys. Rev. D.* **70** 124017
94. Brink J, Teukolsky S A and Wasserman I 2005 *Phys. Rev. D.* **71** 064029
95. Andersson N, Jones D I and Kokkotas K D 2002 *MNRAS* **337** 1224
96. Lai D 1999 *MNRAS* **307** 1001
97. Kokkotas K D, Ruoff J and Andersson N 2004 *Phys. Rev. D* **70** 043003
98. Roberts P H and Stewartson K 1963 *Astrophys. J.* **137** 777
99. James R A 1964 *Astrophys. J.* **140** 552
100. Skinner D and Lindblom L 1996 *Astrophys. J.* **461** 920
101. Bonazzola S, Frieben J and Gourgoulhon E 1998 *Astron. Astrophys. J* **331** 280
102. Shapiro S L and Zane S 1996 *Astrophys. J.* **460** 379
103. Di Girolamo T and Vietri M 2002 *Astrophys. J.* **581** 519
104. Gondek-Rosińska D and Gourgoulhon E 2002 *Phys. Rev. D* **66** 044021
105. Gondek-Rosińska D, Gourgoulhon E and Haensel P 2003 *Astron. Astrophys.* **412** 777
106. Bildsten L. 1998 *Astrophys. J.* **501** L89
107. Wagoner R.V., 1984 *Astrophys. J.* **278** 345
108. Papaloizou J., Pringle J.E., 1978 *MNRAS* **184**, 501
109. Andersson N, Kokkotas K D and Stergioulas N 1999 *Astrophys. J.* **307** 314
110. Levin Y., 1999 *Astrophys. J.* **517** 328
111. Andersson N, Jones D I, Kokkotas K D and Stergioulas N 2000 *Astrophys. J. Lett.* **534** 75
112. Heyl J 2002 *Astrophys. J. Lett.* **574** L57
113. Madsen J., 1998 *Phys. Rev. Lett.* **81**, 3311
114. Wagoner R V 2002 *Astrophys. J. Lett. 578* L63
115. Reisenegger A., Bonacic A., 2003 *Phys. Rev. Lett.* **91** 201103
116. Shibata M 2005 gr-qc/0504082
117. Vallisneri M 2000 *Phys. Rev. Lett.* 84 3519
118. Faber J A, Glandcement P, Rasio F A and Taniguchi K 2002 *Phys. Rev. Lett.* **89** 231102
119. Benhar O, Berti E and Ferrari V 1999 *MNRAS* **310** 797
120. Kokkotas K D, Apostolatos T, Andersson N 2001 *MNRAS* **320** 307
121. Sotani H, Kohri K and Harada T 2004 *Phys. Rev. D* **69** 084008
122. Sotani H and Kokkotas K D 2004 *Phys. Rev. D* **70** 084026
123. Benhar O, Ferrari V and Gualtieri L 2004 *Phys. Rev. D* **70** 124015
124. Andersson N and Comer G L 2001 *Phys. Rev. Lett.* **87**, 241101
125. Stergioulas N, Apostolatos T A and Font J A 2004 *MNRAS* **352**, 1089

NEUTRINO PHYSICS AND ASTROPHYSICS

ENRIQUE FERNANDEZ*

IFAE/Univ. A. de Barcelona,
Campus UAB, Edif. Cn
08193 Bellaterra, Barcelona, Spain
e-mail: enrique.fernandez@ifae.es

These notes are an introduction to neutrino physics intended for doctoral students entering the field of particle physics and astrophysics. After a brief recall of the neutrino properties the concept of neutrino oscillations is introduced. This is followed by an account of the main neutrino oscillation results. The prospects for observing neutrinos for astrophysical sources are also described. The notes summarize the talk given at the Fifth International Workshop "New Worlds in Astroparticle Physics" and are not intended as a review of the existing experiments and their latest results.

1. Introduction

Neutrino physics is going through a very interesting period. Neutrino oscillations in atmospheric neutrinos were clearly observed by the SuperKamiokande collaboration in 1998, a result which implies that neutrinos have mass. The long standing solar neutrino puzzle has been finally solved by the SNO collaboration in 2003, also in terms of neutrino oscillations. These results, supported by several other experiments, have renewed the interest in neutrinos in general, and have stimulated many plans for future experiments. At the same time, the very large and difficult experiments needed for observing neutrinos from astrophysical sources have proved technically feasible, giving hope to the observation of positive signals in the near term future.

*work partially supported by ministerio de educación y ciencia, programa nacional de física de partículas, project fpa-2003-06921-c02-01

2. Neutrino properties, some history

The history and properties of neutrinos can be found in many sources [1]. Neutrinos were introduced by Pauli in 1930 to account for the continuum spectrum of electrons in beta-decay as well as for inconsistencies in the spin-statistics properties of certain nuclei. In 1934 Fermi incorporated the neutrino (and gave it its name) into an extremely successful theory not only of beta-decay but of all weak interactions, and which has formed the basis of the present Electroweak Theory.

It was soon noticed that Fermi's theory predicted the existence of inverse beta decay, by which a neutrino would interact with matter, emitting an electron or a positron. But, owing to the very small cross section for neutrino interactions, the detection of the neutrino had to wait until 1956, when Reines and Cowan and collaborators first observed the inverse beta decay induced by antineutrinos produced in the fission of neutron-rich nuclei in a nuclear reactor [2]. The reaction was

$$\bar{\nu}_e + p \rightarrow n + e^+ \tag{1}$$

The detector of Reines and Cowan consisted in a tank filled with liquid scintillator and equipped with photomultiplier tubes. The detection of the neutrino is quite indirect. First, the produced positron quickly annihilates with an electron in the liquid giving two back to back photons. These photons Compton-scatter in the liquid and the resulting electrons radiate UV photons that are absorbed by the dopant molecules, causing them to scintillate in the visible. This is the light detected by the photomultiplier tubes of the detector. At the same time the primary neutron produced in the reaction wanders in the liquid until it is captured by a Cadmium nucleus (also added to the scintillator mixture) producing an excited state. The de-excitation releases around 9 MeV in the form of gamma-rays, that also induce scintillation which is recorded by the photomultiplier tubes. The two flashes of light, from the positron annihilation and from the neutron capture, are separated by a characteristic time of a few microseconds. The "delayed coincidence" between those two flashes provided the signature for the detection of the reaction, above the large background of spurious light from many other sources.

After the discovery rapid progress was made in the understanding of the neutrino and of weak interactions. A very important first step was the discovery of parity non-conservation in beta-decay of Cobalt-60 nuclei by C. S. Wu and collaborators in 1957[3] in an experiment that had been suggested by Lee and Yang[4]. Based on this fact Lee and Yang formulated the two-component massless neutrino hypothesis: neutrinos are left-handed particles (their spin is always oriented in the opposite direction than their momentum, that is, their helicity is always negative) while antineutrinos are right-handed[5]. The actual direct measurement of the neutrino helicity was made by Goldhaber[6] and collaborators in a very clever experiment shortly afterwards, and found to be indeed left-handed. At about the same time Feynman and Gell-Mann and Marshak and Sudarsan formulated the V-A theory of the weak interaction[7]. According to this theory, the interaction occurs with the left-handed chirality component of particles and the right-handed chirality component of antiparticles. Chirality and helicity are identical for massless particles, so if neutrinos had mass the right-handed component of neutrinos and the left-handed component of antineutrinos would never be seen, as they would not interact weakly (their only interaction would be gravitational, through their energy). Because of these reasons, and the lack of evidence to the contrary, when the Standard Model was constructed the neutrino was incorporated as a massless particle. But this is something that has to be revised, given the evidence for oscillations.

The next important step in neutrino physics took place in the early sixties. Recall that the neutrinos produced in beta decay, as detected by Reines and Cowan, are antineutrinos produced together with electrons. But neutrinos are also produced in other decays, most notably of muons and charged pions. The latter decay into two particles, namely into a muon and a neutrino. Is this neutrino the same state as that produced in beta decay? This question was asked in the fifties. In particular Feinberg [8] noticed that there should be two kinds of neutrinos, otherwise one could expect a diagram like that of Fig. 1, implying that the muon will decay into an electron and a gamma, a process not observed.

But the question was actually put to test in an experiment done at the Brookhaven National Laboratory in 1962. This experiment was the first in which a beam of neutrinos was produced in an accelerator following a method proposed independently by Pontecorvo and Schwartz [9]. In the Brookhaven experiment, protons produced by the AGS accelerator were made to collide with an internal target, producing hadrons moving mostly in the direction of the proton due to the Lorentz boost. These hadrons were allowed to enter a decay region, 17m long. The neutrinos come mainly from

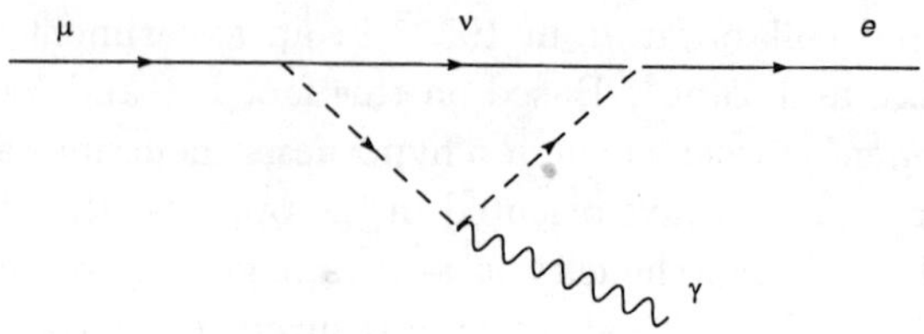

Figure 1. If the neutrinos associated to muon and electron were the same one could expect transitions like that represented in the above diagram.

the decay of pions into a muon and a neutrino, and, to a lesser extent, from the decay of kaons to the same final state. Again because of the Lorentz boost the neutrinos are collimated along the direction of the primary proton. After the decay area, a very thick shield consisting of 13m of iron, filtered most particles. This method of producing the beam has remained essentially unchanged until today. One innovation consists of surrounding the target area with a magnetic field, produced by a device called a "horn". The magnetic field is pulsed in synchrony with the arrival of the protons into the target. This collimates the pions and kaons of a given charge depending on the horn polarity, giving an enhanced flux of neutrinos (or antineutrinos) in the forward direction.

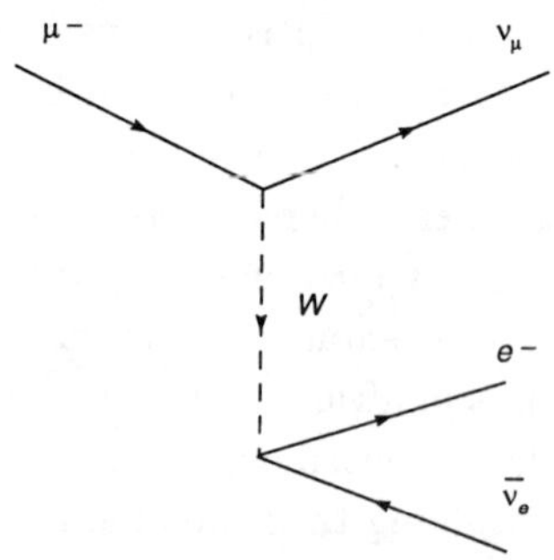

Figure 2. Diagram for mu decay.

The detector for the Brookhaven experiment was located after the shield and consisted of 10 tons of aluminum in the form of 2.5 cm plates separated by spark chambers. After several months of running a total of 34 events were positively identified as neutrino interactions having a muon in the final state, 5 could be attributed to cosmic ray interactions and 6 more had several particles in the final state, distinct from electrons [10]. This result proved that the neutrinos produced with muons in pion and kaon decay have a different nature than those produced in association with electrons in beta-decay. The latter produce electrons when they interact while the former produce muons. The two lepton families were thus established. Muon decay proceeds as in Fig. 2.

The tau lepton was discovered in the seventies by its leptonic decays into an electron and two neutrinos or a muon and two neutrinos [11], as in Fig. 3.

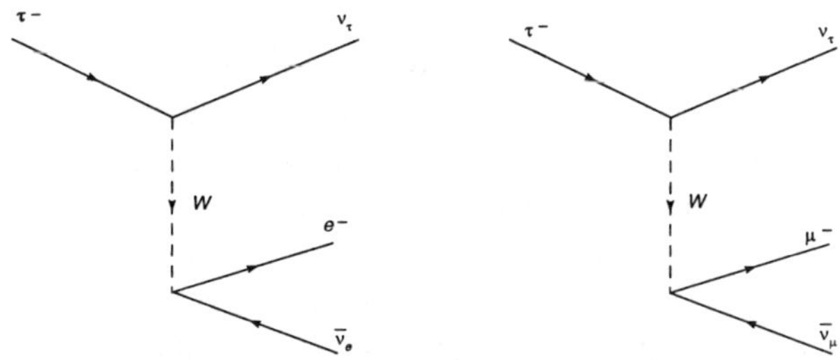

Figure 3. Diagrams for tau decay.

But the observation of tau neutrino interactions, producing a tau in the final state, had to wait until 1998, when they were first detected at Fermilab by the DONUT experiment[12].

Are there more than three neutrino families? The answer to this question comes from the LEP experiments. At LEP, in the first phase called LEP-1, electrons and positrons were made to collide at a center of mass energy close to the Z mass, such that the Z boson was produced resonantly. The Z immediately decays into a fermion-antifermion pair, where fermion stands for any of the Standard Model leptons or quarks (all but the top quark which cannot be produced due to its large mass). By measuring the cross-section (into any final state) as a function of the energy near the peak of the resonance one can determine the total width of the Z. This information, together with the measurement of the peak cross-sections into lepton-antilepton and quark-antiquarks final states, also allows the determination of the partial widths into those states (which add up to the total visible width). The difference between the total and visible widths of the Z, if attributed to neutrinos interacting according to the Standard Model and with a mass sufficiently low as to be produced from Z decays, gives us the number of "standard light neutrino families". This number is[13]

$$N_\nu = 3.00 \pm 0.06 \tag{2}$$

a truly remarkable result.

3. Solar and atmospheric neutrinos

As we have seen in the previous section neutrinos were assumed to be massless in the Standard Model. Experimentally, though, only lower limits on the masses can be given[14]. But, already in the fifties, if was suggested by Pontecorvo[15] that, if neutrinos had mass, they could undergo oscillations, similar to those of neutral kaons. As we explain later this implies that neutrinos change nature as they propagate from the production to the detection point. The first hint that neutrinos could oscillate came from an experiment that started in the mid sixties by R. Davis Jr. and collaborators with the aim of measuring the neutrinos originating in the Sun. This experiment reported a deficit[16] in the solar neutrino flux with respect to the expectations from a solar model calculation of J. Bahcall and collaborators, published in the same issue of the Physical Review Letters[17] in 1968. Other solar neutrino experiments followed and they also saw a lower than expected flux, but both the experiments and the solar model calculations were notoriously difficult so that the discrepancy was met with skepticism. The situation took an unexpected turn in 1998 when the SuperKamiokande collaboration presented results clearly indicative of oscillations, an indication that has now proven correct. Let us briefly describe these developments.

The production of energy and neutrinos in the Sun is due to the fusion reactions taking place deep in the solar nucleus, as first written down by Hans Bethe in the thirties [18]. Over the years a Standard Solar Model has emerged, constantly being refined with new experimental and theoretical insights by its main architect John Bahcall[19] and collaborators. It is against this model that the solar neutrino experiments have been confronted. The model predicts the spectrum of neutrinos due to the different fusion reactions. The latest calculations can be seen in Fig. 4[21].

The Davis pioneer solar neutrino experiment started to take data in 1967. The experiment was located in the Homestake Mine in South Dakota

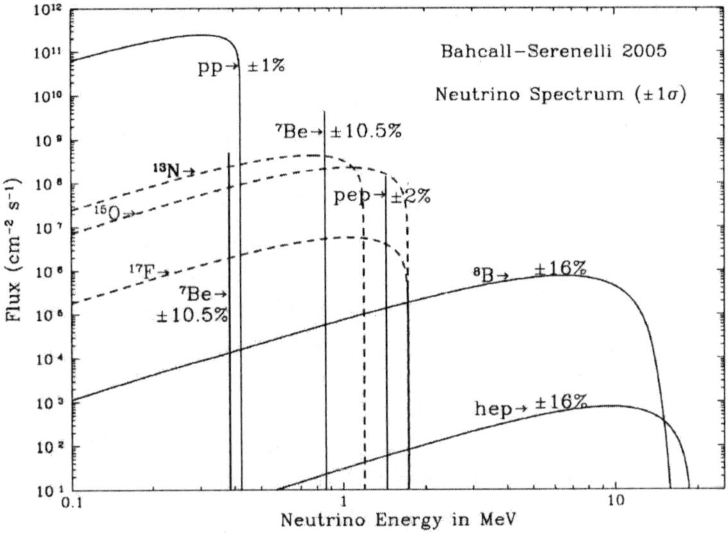

Figure 4. The most recent calculation of the solar neutrino flux.

at about 1400m below the surface. The detector consisted on a large tank of 600 tons of a liquid, perchloroethilene C_2Cl_4, commonly used in dry cleaning. The ν_e's arriving from the Sun interact with the Cl nuclei through inverse beta decay

$$\nu_e +^{37} Cl \rightarrow^{37} Ar + e^- \tag{3}$$

The Argon-37 nucleus is radioactive with a half-life of approximately 35 days, decaying again through beta decay into Chlorine-37. The neutrino threshold energy for this reaction is 0.81 MeV and therefore only the Berilium-7 or higher energy neutrinos are detected by this method, especially those of Boron-8. The detection was "radio-chemical": the radioactive Argon nuclei were extracted from the liquid through a careful chemical process and their decay detected in an external detector. As mentioned above this experiment reported a deficit in 1968[16].

A quite different solar experiment was Kamiokande, consisting of a 3,000 ton tank of very pure water equipped with photomultiplier tubes, located about one kilometer underground in the Kamioka mine, in Japan. (The "nde" at the end of the word stand for Nucleon Decay Experiment, the observation of which was the original goal of the detector). Here the neutrinos are detected when they interact through the reaction

$$\nu_e + e^- \rightarrow \nu_e + e^- \tag{4}$$

The final state electron moves in the water above Cherenkov threshold. The reconstruction of the Chrerenkov light allows the measurement of the incoming neutrino direction and the estimation of its energy. However the threshold for observing this process above backgrounds was 7 MeV, making the experiment sensitive to the Boron-8 neutrinos only. The first results also showed a smaller than expected number of neutrinos[20], but the rate of this reaction is very sensitive to the parameters of the solar model.

Other radio-chemical experiments were started, based on Gallium instead of Chlorine, namely SAGE (from Soviet-American Gallium Experiment) in Baksan (Russia, then Soviet Union) and GALLEX in the Gran Sasso Laboratory (Italy). In these experiment the reaction is

$$\nu_e +^{71} Ga \rightarrow^{71} Ge + e^- \tag{5}$$

The Germanium-71 is also radioactive. The neutrino energy threshold for the above reaction is 0.2332 MeV which is below the pp fusion line. Both of these detectors also reported a less than expected flux[22].

In the nineties Kamiokande was succeeded by SuperKamiokande, a 50 kiloton tank of water, employing the same technique for detecting neutrinos. The results of SuperKamiokande confirmed the previous results with much higher precision: there was a deficit on the number of solar neutrinos[23]. The present situation is illustrated in Fig.5[24]. The three leftmost columns are the results of the experiments just described while the two to the right will be described later. Clearly either the experiments are wrong, or the solar model is wrong, or both, or another explanation has to be found, which turns out to be the case.

At the 1998 international neutrino conference, the SuperKamiokande detector also reported the observation of atmospheric neutrinos[30]. These are neutrinos produced by the interaction of the primary cosmic rays, high in the Earth atmosphere. The charged pions produced in the primary cosmic rays collisions with the nuclei in the atmosphere decay into muons and muon-neutrinos (or antineutrinos). Most of the secondary negative (positive) muons also decay into an electron (positron), an electron antineutrino (neutrino) and a muon neutrino (antineutrino). The ratio of muon neutrinos to electron neutrinos arriving at the surface should then be about 2.

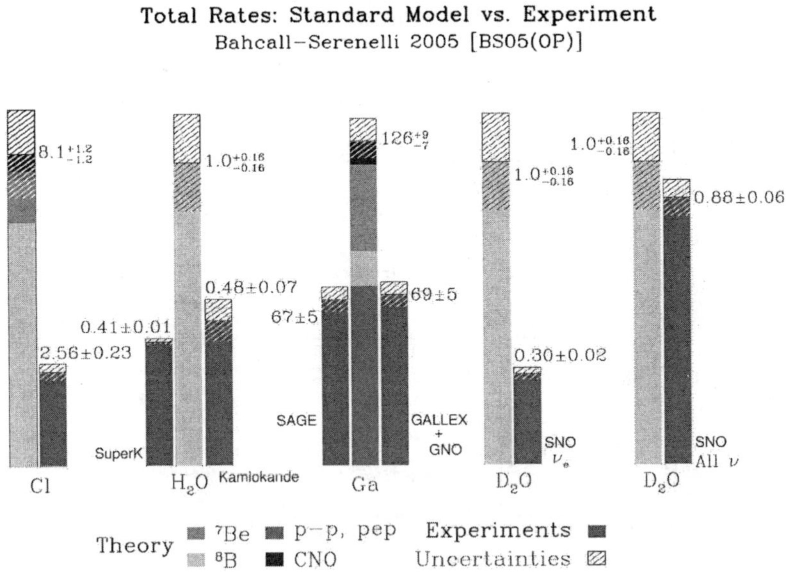

Figure 5. Comparison of measured solar fluxes with the SSM predictions. See main text for an explanation.

Some of the muons do not decay before they reach the surface so the ratio should be slightly larger than 2, an effect that grows with the energy of the primary muon (and thus of the resulting neutrinos).

The ratio of electron to muon neutrinos reported by SuperKamiokande was closer to 1, instead of 2, but the most striking result was the zenith angle dependence of the high energy muon neutrinos (Fig. 6). The zenith angle is related to the distance traveled by the neutrinos before reaching the detector. Zenith angle equal 0 (cosine equal 1) means neutrinos entering the detector from above, thus produced between 10 and 20 kilometers above the detector, while zenith angle equal π (cosine equal -1) means neutrinos produced in the antipodes of the detector, at 12,000 kilometers. Clearly the expected number of neutrinos decreases with distance. Notice that the effect is present for muon neutrinos but not for electron neutrinos.

All the above results can be explained assuming that neutrinos oscil-

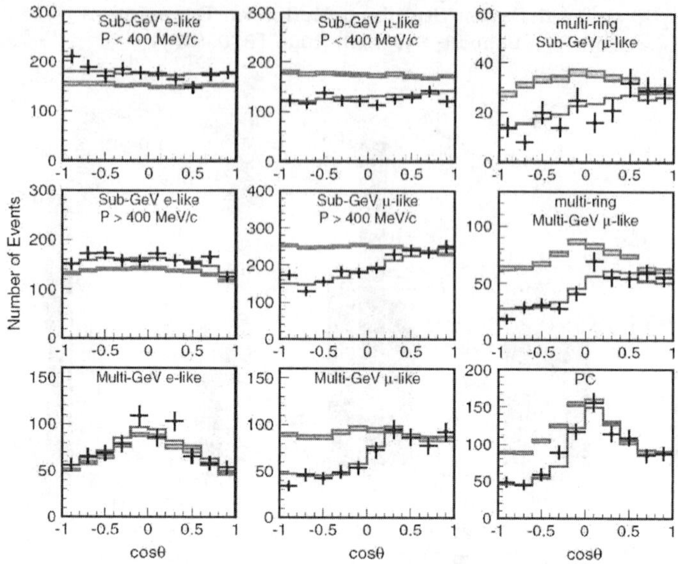

Figure 6. Number of electron and muon atmospheric neutrinos detected in SuperKamiokande as a function of the zenith angle. $\mathrm{Cos}\theta = 1$ corresponds to neutrinos coming from above (near) the detector while $\mathrm{Cos}\theta = -1$ corresponds to neutrinos coming from below (far).

late, a quantum effect through which a neutrino changes its nature as it travels from the production to the interaction points. When the solar neutrinos change nature into muon or tau neutrinos, the latter cannot interact through their charged current reactions, since they are below threshold for the production of muons or taus. The effect is thus a decrease in the number of electron neutrinos, as observed. The atmospheric neutrinos oscillate into tau neutrinos, which produce a tau in their interactions and cannot be identified in SuperKamiokande. The oscillation depends on the distance and thus only those coming from far away show the effect. That the oscillation is into tau neutrinos and not electron neutrinos (over the energies and distances involved) is derived in part from the fact that there is no anomaly on the electron neutrinos seen also in SuperKamiokande.

In the last two years new results have been presented with confirm the above picture. The atmospheric neutrino oscillation has been confirmed by the K2K experiment[25]. In this experiment a beam of muon neutrinos is sent from KEK to SuperKamiokande. The energy and distance of the experiment are such that the relevant parameters describing the oscillation (see next section) are the same as in atmospheric neutrinos.

The most impressive progress has been in Solar neutrinos. The Sudbury Neutrino Observatory (SNO) experiment has dramatically confirmed the solar neutrino oscillation. The SNO detector[26] consists of 1 kiloton of heavy water D_2O. Electron neutrinos can interact in SNO by the charged current (CC) reaction that transforms the neutron of the deuterium nucleus into a proton

$$\nu_e + d \rightarrow p + p + e^- \tag{6}$$

The electron is observed by the Cherenkov light and its direction is strongly correlated with that of the neutrino. But what is unique to SNO is that the three types of neutrinos can interact with the deuterium via the neutral current reaction (NC)

$$\nu_x + d \rightarrow \nu_x + n + p \tag{7}$$

The threshold energy for this reaction is 2.22 MeV, the binding energy of the deuterium nucleus. This reaction is detected through the observation of the neutron. To be able to do so the purity of the detector elements has to be extremely high. A third reaction is the elastic scattering (ES) of neutrinos with electrons (as in SuperKamiokande)

$$\nu_x + e^- \rightarrow \nu_x + e^- \tag{8}$$

This reaction occurs in principle for the three neutrino types but the cross section is higher for electron neutrinos. SNO has been able to measure the three reactions and from them infer a flux of neutrinos from the Sun. The solar flux inferred from these reactions separately is[27]

$$\begin{aligned} \Phi_{CC} &= 1.76^{+0.06+0.09}_{-0.05-0.09} \times 10^6\ cm^{-2}s^{-1} \\ \Phi_{ES} &= 2.39^{+0.24+0.12}_{-0.23-0.12} \times 10^6\ cm^{-2}s^{-1} \\ \Phi_{NC} &= 5.09^{+0.44+0.46}_{-0.43-0.43} \times 10^6\ cm^{-2}s^{-1} \end{aligned} \tag{9}$$

The excess of the NC flux over the other fluxes is a clear indication that neutrinos from the Sun change flavor. Furthermore the total flux from NC is in very good agreement with the total ^8B flux of $5.05^{+1.01}_{-0.81} \times 10^6\ cm^{-2}s^{-1}$ predicted by the SSM[19]. The results are illustrated in Fig. 7, which shows the flux of muon or tau neutrinos versus the flux of electron neutrinos which are deduced from the SNO measurements. As it can be seen the CC reaction gives only the electron-neutrino flux, while the other two reactions

(NC) and (ES) give two different linear combinations of Φ_e and $\Phi_{\mu\tau}$. The three bands intersect at one point, giving a consistent solution. The dotted band is the prediction of the SSM. A long standing problem in physics, the deficit of solar neutrinos with respect to the SSM, was finally solved.

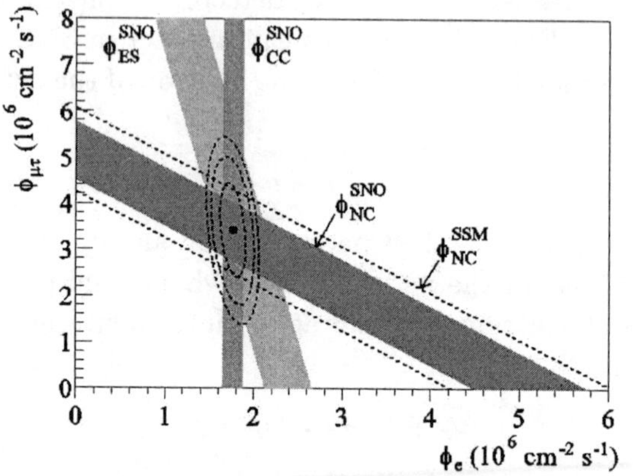

Figure 7. The solar neutrino fluxes inferred from the SNO measurements. See the text for an explanation.

Another experiment has also confirmed the solar neutrino oscillations. This is the KAMLAND experiment[28], located also in the Kamioka mine in Japan. This experiment detects reactor antineutrinos produced in nuclear reactors in Japan, in one kiloton of liquid scintillator. It turns out that the distance and energy of these neutrinos prove the same parameters as the solar neutrinos. The experiment detected 258 antineutrino interactions while 365 ± 24 were expected, with a background of 18 ± 8[29]. The difference is inconsistent with a square distance decrease of the flux and is attributed to oscillations. Furthermore the energy spectrum of these neutrinos, also measured, is also inconsistent with a scaled down spectrum of the sum of the reactors[29].

4. Massive neutrinos

All this impressive results have been thoroughly analyzed by many people in the context of neutrino oscillations which are explained in the next section. Oscillations require that the neutrinos have mass, and thus that the SM be revised. It can be that the neutrino acquires mass like any other fermion through the interaction with the Higgs field, and thus be described by a four component Dirac spinor with two, right-handed and left-handed, particle states and two, right-handed and left-handed, antiparticle states. As already mentioned the right-handed neutrino and the left handed antineutrino will not interact weakly and would be "sterile". Global lepton number will be conserved but there will be mixing between the lepton families similar to that of the quarks. This will have an effect on the oscillations, since the normal neutrinos could oscillate into the sterile states.

But for a neutral particle there is another possibility to acquire mass. The interaction with the Higgs can flip the state from left to right, or vice-versa, changing at the same time the particle into its antiparticle. Such a mass term is called a Majorana mass term and the two component particle is called a Majorana particle. Left handed neutrino and right handed antineutrino would be their own antiparticles and lepton number would not be conserved. It can also be that both Majorana and Dirac mass terms are present. Experimentally the only hope at present to see if the neutrino is Majorana would be the observation of neutrinoless double beta decay[31]. The following discussion on oscillations is not affected by whether the neutrino is a Dirac or a Majorana particle.

5. Neutrino oscillations

If neutrinos have mass the mass eigenstates do not need to be the same as the weak eigenstates. The latter are the states that couple to the W in the weak interaction, by definition ν_e, ν_μ and ν_τ. The mass eigenstates are usually denoted by ν_1, ν_2 and ν_3. This is similar to what happens with the quarks. The differencc is that in the case of the quarks the mass eigenstates are those that constitute the hadrons, and those that we usually speak of when referring to quarks, while in the neutrino case we do not have a direct access to the mass eigenstates, since when a neutrino is produced or interacts it is in a pure weak eigenstate. If mass and weak states are not the same they are related by a transformation

$$|\nu_l>=\sum_{i=1}^{n} U_{li}|\nu_i> \tag{10}$$

If there are only three mass states, then U is a 3×3 unitary matrix. Here we consider only this case for the sake of simplicity in the explanation of the oscillations. All the experimental evidence can be explained with three mass states, except for the results of the LSND experiment, in which a transition from muon to electron neutrinos in a low energy accelerator beam was reported[32]. If this result is correct then one would require at least a fourth mass state. For an excellent review of neutrino oscillations see article of B. Kayser in the PDG[33]. The 3×3 U matrix can be written as follows

$$U = \begin{pmatrix} U_{e1} & U_{e2} & U_{e3} \\ U_{\mu 1} & U_{\mu 2} & U_{\mu 3} \\ U_{\tau 1} & U_{\tau 2} & U_{\tau 3} \end{pmatrix} \tag{11}$$

This mixing matrix is usually called the Maki-Nakagawa-Sakata-Pontecorvo (MNSP) matrix [36], analogous to the CKM matrix for the quarks. Let us assume that a neutrino is produced at $t = 0$ in a pure weak eigenstate, $|\nu_\alpha >$. This state is a mixture of mass eigenstates determined by the mixing matrix. As space and time changes each of the mass eigenstates, ν_j, all produced with the same energy, evolves acquiring a phase $-i(E_j t - p_j x)$, different for the 3 mass eigenstates due to their different masses (e.g. different momenta). When the neutrino is detected through a weak interaction, this quantum-mechanical mixture is projected again into a weak eigenstate, $|\nu_\beta >$, which can be different from $|\nu_\alpha >$. We say that the neutrino has oscillated from one flavor to another. With some algebra one can compute the probability for the transition $|\nu_\alpha >$ to $|\nu_\beta >$:

$$\begin{aligned} P(\nu_\alpha \to \nu_\beta) &= \delta_{\alpha\beta} - 4 \sum_{i>j} \Re(U^*_{\alpha i} U_{\beta i} U_{\alpha j} U^*_{\beta j}) \sin^2[1.27 \Delta m^2_{ij} (L/E)] \\ &\quad + 2 \sum_{i>j} \Im(U^*_{\alpha i} U_{\beta i} U_{\alpha j} U^*_{\beta j}) \sin[2.54 \Delta m^2_{ij} (L/E)] \end{aligned} \tag{12}$$

In the above expression Δm^2_{ij} is in eV^2, L, the distance from the source to the detector, is in km, and E, the neutrino energy, is in GeV.

These equations simplify considerably if certain conditions are met. It could be for example that one of the mass splittings, Δm, is very different from the others. If E and L in a given experiment are such that $1.27 \Delta m^2 (L/E) \approx \pi/2$ then only the corresponding term is relevant in the above expressions. The resulting formulae are like those that one would obtain assuming that only two generations participate in the oscillation. The corresponding situation is called a "quasi-two-neutrino oscillation". It can also be that only two mass states couple significantly to the flavor partner of the neutrino being studied. In that case the equations also become quasi-two-neutrino oscillations. Nature seems to be kind enough to have chosen these situations, the first in the case of atmospheric neutrinos the second in solar neutrinos.

The analysis of solar neutrino data[34] indicates that the electron neutrino couples significantly only to two mass states, chosen as ν_1 and ν_2. The solar neutrino oscillations occur not only because of the mixing but also because electron neutrinos propagate differently through matter than the other two species. When neutrinos propagate through matter they can forward scatter coherently with the medium, via Z exchange. But for electron neutrinos (and only for electron neutrinos) the elastic forward scattering can also proceed via W exchange. As a consequence the flavor transitions, assuming that there is mixing, are modified with respect to those in vacuum. The effect is called the MSW effect, from Mikheyev, Smirnov and Wolfstein[35]. What the analysis of the solar data indicate is that the neutrino born as an electron neutrino in the core of the Sun, is also in a state which is almost the heavier of the two mass eigenstates and it remains in that state until it leaves the Sun. This heavier states is usually chosen as $|\nu_2>$. But since that state is also an eigenstate of the vacuum Hamiltonian it does not change when it propagates freely from the Sun to the Earth. When the solar neutrino interacts on Earth the probability of finding it as an electron neutrino is just the sine-square of the mixing angle appearing in the quasi-two-neutrino oscillations [33]. In the case of the atmospheric neutrinos the analysis indicates that over the relevant energies and distances only one mass splitting is relevant, traditionally chosen as Δm^2_{23}.

A convenient parametrization of the mixing matrix is the following

$$U = \begin{pmatrix} c_{12} & s_{12} & 0 \\ -s_{12} & c_{12} & 0 \\ 0 & 0 & 1 \end{pmatrix} \begin{pmatrix} 1 & 0 & 0 \\ 0 & c_{23} & s_{23} \\ 0 & -s_{23} & c_{23} \end{pmatrix} \begin{pmatrix} c_{13}e^{i\alpha} & 0 & s_{13}e^{i\delta} \\ 0 & e^{i\beta} & 0 \\ s_{13}e^{-i\delta}e^{i\alpha} & 0 & c_{13} \end{pmatrix} \qquad (13)$$

where $c_{ij} = cos\theta_{ij}$, $s_{ij} = sin\theta_{ij}$. The two phases α and β are only present if the neutrino is a Majorana particle. They do not affect oscillations nor the interpretation of present neutrino results. An analysis taking into account not only the oscillation experiments but also results from cosmology can be seen in[37]. The summary is in Fig. 8. Notice that from current experiments we do not know the sign of Δm^2_{23}, that is, we do not know if m_3 is heavier or lighter than m_1 and m_2, as depicted in the figure. Another as yet unknown parameter is θ_{13} for which only the upper limit shown in Fig. 8 is known[38].

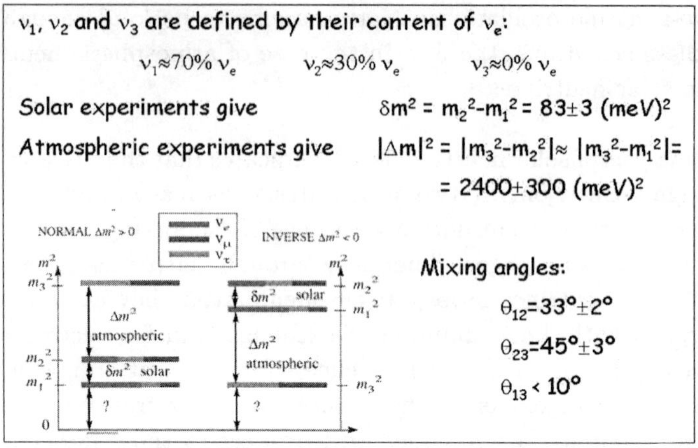

Figure 8. Summary of the neutrino oscillation fits from reference Fogli *et al.*, ref. above

Several experiments are now being prepared or contemplated to further study neutrino oscillations. Among the solar experiments KAMLAND and SNO phase-3 will improve the determination of the solar mass difference and mixing angle. The Borexino experiment, under construction at the Gran Sasso, will measure the Be-7 line. Among the long-baseline experiments MINOS[39], from FNAL to Soudan aims at improving the measurment of the "atmospheric" oscillation parameters and may also improve the limit (or measure) θ_{13}. The CNGS program[40] (CERN to Gran Sasso), with the ICARUS and OPERA experiments, aims at observing the explicit ν_μ to ν_τ transition. To measure (or substantially improve the limit on) θ_{13} requires a new generation of experiments. T2K[41] (Tokai to Kamioka) in Japan has been approved and has sensitivity to values of $sin^2 2\theta_{13}$ above 0.006. In Europe the Double Chooz[42] experiment, proposed to run at the Chooz reator has sensitivity to $sin^2 2\theta_{13}$ for values greater than 0.03. For the latest information on neutrino results see presentations at the XXI International Neutrino Conference that took place in Paris in June 2004[43]

6. Neutrino astrophysics

As we have seen in the previous section neutrinos are copiously produced in the Sun and presumably in other Hydrogen burning stars. Are there any other astrophysical sources of neutrinos? It has been known for quite some time that neutrinos play a crucial role in the explosion of core-collapse Supernovae (Type II, Type Ib and Type Ic)[46]. In these, about 10 % of the mass of the approximately $1.4M_{\odot}$ remnant neutron star is radiated in the form of neutrinos over a period of seconds. This phenomenal neutrino explosion was experimentally confirmed in 1987 with the observation of neutrinos coming from Supernova 1987A. Over a period of 10 seconds the Kamiokande[45] detector in Japan and the IMB[44] detector in the US were able to detect a "burst" of 11 and 8 neutrino interactions respectively. Supernova 1987A exploded in the Large Magellanic Cloud, about 50 kpc away. Supernovae happen in our galaxy once every 40 years on the average. Existing neutrino detectors should be able to detect a galactic supernova explosion would it occur today. Solar and Supernova 1987A neutrinos remain to date the only detected sources of neutrinos beyond Earth.

The existence of very high energy cosmic rays implies with almost complete certainty that neutrinos are produced in other cosmic sources. But since the origin of cosmic rays and their production mechanism are not known one cannot directly relate their observation here on Earth to the likely neutrino flux produced together with them, or in their interactions with matter at the source. A more direct correlation can be established between high energy gamma rays and neutrinos, since both are undeflected as they propagate. If one assumes that the predominant source of high energy gamma rays is hadronic, that is, that the high energy gamma rays come from neutral pion decay, then the expected neutrino flux would be of the same order, from charged pion decay. Fig. 9 illustrates this case. As for the accelerator-produced neutrinos, a cosmic source (for example the accretion jet of a black hole) accelerates protons to high energy. The protons interact with the "target" (for example the radiation enveloping the back hole) photoproducing pions, the neutral component of which produces gammas and the charged component produces neutrinos.

However, it may be that most of the gammas are produced by electromagnetic processes not involving pions, in which case the neutrino flux

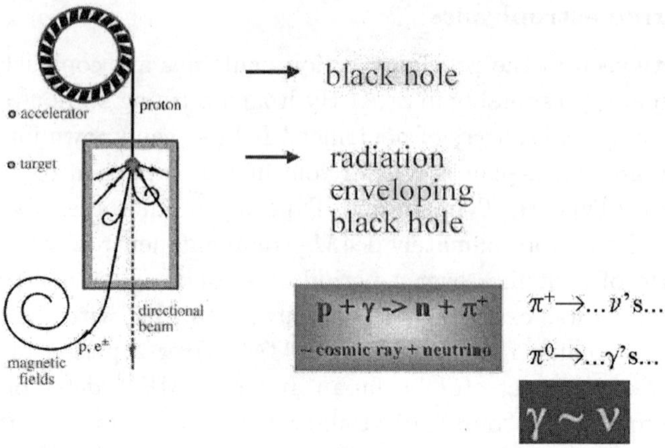

Figure 9. A possible realization of a cosmic neutrino beam. Protons accelerated in the accretion jets of a black hole interact with ambient photons producing secondary pions, which decay producing neutrinos and gammas in similar amounts.

would be much reduced. In any case, the fact that there are very high energy cosmic rays means that some of them will interact as they leave their sources, producing pions and therefore neutrinos. In particular if there are extra-galactic ultra-high energy cosmic rays, above 10^{20} eV, they will interact with the cosmic microwave background photons over relatively short distances (100 Mpc). This has two observational consequences: first that there should not be charged particles above about 5×10^{19} GeV (the so called GZK cut-off[47]). Second, there should be neutrinos produced in these interactions, named "GZK neutrinos".

Some of the physics topics that can be studied with these detectors[50] are the following:

a) Gamma ray bursts (GRB)
Widely discussed current models of GRB sources involve dissipation of the kinetic energy of a relativistically expanding fireball caused by either a cataclysmic collapse of a massive star, or by the coalescence of two compact objects. The resulting shocks may accelerate particles via the Fermi process to ultra-relativistic energies. Accelerated electrons emit non-thermal radiation that can explain the observed MeV gamma-ray spectra. This model has received support from the verification of the predictions of an afterglow at lower (X-ray, optical, radio) energies. However, the model remains

largely phenomenological. The underlying progenitor remains unknown, and the physics of electron coupling and magnetic field amplification by the shocks is not well understood. The observation of neutrinos could help in clarifying what GRB models remain valid.

b) Active galactic nuclei and blazars (AGN)

The presence of jet outflows determines whether AGN's are radio-loud or radio-quiet. Radio-loud AGN with luminous and rapidly variable (hours to days) non-thermal electromagnetic radiation and strong optical polarization are referred to as blazars. These sources are often strong sources of GeV and, in some cases, TeV gamma-radiation. Blazars are thought to be AGN where the jet is nearly aligned with the direction to the observer. One of the major questions that neutrino astronomy can address is whether AGN jets are powerful accelerators of protons and ions to ultra-high energies. If relativistic leptons were mainly accelerated in the jets, then few or no neutrinos would be emitted. If relativistic hadrons are accelerated with comparable power to non-thermal electrons, then observable fluxes of neutrinos would be produced in the jet through pion production by nuclear and photo-hadronic interactions.

c) Microquasars

The jets associated with Galactic microquasars are believed to be ejected by accreting stellar-mass black holes or neutron stars. Much like for AGN, the content of the jets is an open issue. The dominant energy carrier in the jet is at present unknown. The observation of neutrinos would again clarify these issues.

d) Supernovae and supernova remnants

There are suggestions[48] that simultaneous detection of MeV and TeV neutrinos from a supernova may be possible, with the MeV neutrinos detected by Super- Kamiokande, SNO, or future detectors such as UNO or IceCube(see next section). When a type II supernova shock emerges from the progenitor star, it becomes collisionless and may accelerate protons to TeV energy and higher. Inelastic nuclear collisions of these protons produce a burst of TeV neutrinos over durations 1 hour. This occurs about 10 hours after the thermal neutrino burst from the cooling neutron star. A galactic supernova explosion of a red supergiant star would produce 100 muon events in a km^3-scale neutrino detector, thus providing important constraints on both supernova models and neutrino physics. Given the most probable direction for the next galactic supernova, this objective is best achieved by a Northern hemisphere detector.

e) Diffuse neutrino flux
Waxman and Bahcall[49] have shown that cosmic-ray observations set an upper bound to the neutrino flux from sources which, like candidate sources of protons with more than 10^{19} eV, are optically thin for high-energy nucleons to gamma - proton and proton - proton (neutron) interactions. For sources of this type, the energy generation rate of neutrinos can not exceed the energy generation rate of high-energy protons implied by the observed cosmic-ray flux above 10^{19} eV.

f) Hidden sources
Sources may exist where pion production losses prevent the escape of high-energy nucleons and allow only neutrinos (and possibly low-energy gamma rays) to escape. While cosmic-ray data do not constrain the possible neutrino flux from such sources, which may therefore exceed the Waxman-Bahcall bound, the data also do not provide evidence for the existence of such "hidden" sources.

7. Neutrino telescopes

From an observational point of view the hope of detecting neutrinos is at very high energies. The reason is that the cross-section for neutrino interactions grows linearly with energy. Furthermore the detection above background is considerably easier, such that enormous detectors can be envisioned. The possibility of using natural water was already suggested in the 60's[51], and was attempted with the DUMAND[52] project, which did not succeed. It was then suggested to use ice[53], instead of water as the detecting medium.

The idea, illustrated in Fig. 10, is to instrument a large volume of water or ice with an array of photo-detectors. A high energy neutrino interacting through a charged current interaction inside the instrumented volume or in its surroundings will produce a charged lepton which will travel through the water or ice at ultra-relativistic speed, thus producing Cherenkov radiation. The Cherenkov light will propagate in the medium over relatively large distances such that a sparse array of photo-detectors will be sufficient for its detection. The time of arrival of the light to the different photo-detectors

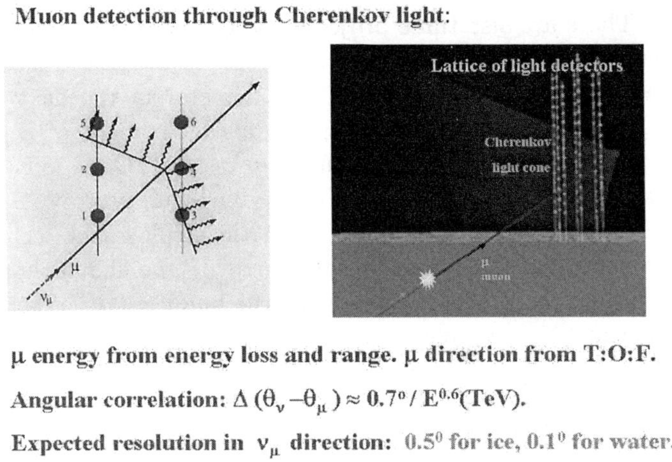

Figure 10. Illustration of the neutrino telescope concept. Neutrinos interacting near the detector volume produce a charged lepton, which in turn produces Cherenkov light detected by the photo-detectors.

permits the reconstruction of the light front and thus of the direction of the lepton, which is highly correlated with that of the original neutrino. The energy of the lepton can be estimated from the light yield.

Two detectors have established the technique in water and ice: the NT-200[54] detector operating in Lake Baikal and the AMANDA[55] detector operating in the South pole. The Lake Baikal experiment takes advantage of the fact that the surface of the lake is frozen for part of the year and serves as a natural platform to deploy strings of photomultiplier tubes. The detector has been in operation for several years and has been able to observe atmospheric neutrinos[54]. AMANDA, in the South Pole, consists of several strings of photomultipliers at high depth. The photo-multipliers are deployed by making a very deep hole, melting the ice. The hole stays liquid for some time during which the string can be deployed. This experiment has also observed atmospheric neutrinos[56].

Both of these techniques plan to be pursued further, towards $1km^3$-scale detectors[57]. In ice, the ICECUBE[58] detector in the South Pole is an approved project, presently being constructed. It is expected to be finishedby the end of this decade. In water there are several ongoing projects. The Lake Baikal experiment will continue to be expanded, adding longer strings[54]. There are also three projects trying to set the foundations for a km^3-scale detector in the Mediterranean sea. ANTARES[59], off the cost of Toulon, France, plans to install a $0.1k^2$ detector in the next few years. NEMO[60], near Catania, in Sicily, plans at deployment of a prototype, exploring the technologies for a future km^3. NESTOR[61] off the cost of Greece, is pursuing the same goal of deploying a full prototype. The three approaches differ, aside from the site, in the deployment technique and in the geometry on the photo-detector array. One should hope that in the near future these efforts joint forces to build a km^3 detector in the Mediterranean.

Acknowledgements

It is a pleasure to thank Mário Pimenta, Gaspar Barreira and their colleagues for the kind invitation to give this talk in such a pleasant environment.

References

1. See for example "Current Aspects of Neutrino Physics", David O. Caldwell (Editor), *Springer-Verlag (2001), ISBN 3-540-41002-3.*
2. F. Reines and C.L. Cowan, *Nature* ***170**, 446 (1956).*
3. C.S. Wu et al. *Phys. Rev.* ***105**, 1413 (1957).*
4. T.D. Lee and C.N. Yang, *Phys. Rev.* ***104**, 254 (1956).*
5. T.D. Lee and C.N. Yang, *Phys. Rev.* ***105**, 1671 (1957).*
6. M. Goldhaber, L. Grodzins and A.W. Sunyar, *Phys. Rev.* ***109**, 10151017 (1958).*
7. R.P. Feynman and M. Gell-Mann, *Phys. Rev.* ***109**, 193 (1958)*; E.C.G. Sudarsan and R. Marshak, *Phys. Rev.* ***109**, 1860 (1958).*
8. G. Feinberg, *Phys. Rev.* ***110**, 1482 (1958).*
9. B. Pontecorvo, *JETP* ***37**, 171 (1959)*; M. Schwartz, *Phys. Rev. Letters* ***4**, 306 (1960).*
10. G. Danby et al., *Phys. Rev. Letters* ***9**, 36 (1962).*
11. M. Perl et al., *Phys. Rev. Letters* ***35**, 1489 (1975).*
12. K. Kodama, et.al., *Physics Letters* ***B 504** (2001) 218-224.*
13. Particle Data Group, S. Eidelman *et al., Physics Letters* ***(592)**, 1 (2004).*
14. S.M. Bilenky, C. Giunti, J.A. Grifols, E. Masso, *Phys. Rep.* ***379**, 69 (2003).*
15. B. Pontecorvo, *Soviet Phys. J. Exp. Theo. Phys.,* ***33**, 429 (1957)*; ibd. ***34**, 172 (1958).*
16. R. Davis Jr., D.S. Harmer and K.C. Hoffman *Phys. Rev. Lett.* ***20**, 1205 (1968).*
17. J.N.Bahcall, N. Bahcall and G. Shaviv *Phys. Rev. Lett.* ***20**, 1209 (1968).*

18. H. A. Bethe, *Phys. Rev.*, ***55***, *(1939)*.
19. See for example J. N. Bahcall, Solar Models and Solar Neutrinos: Current Status, *Proceedings of the Nobel Symposium 2004, Enkoping, Sweden, August 19-24, 2004, eds. L. Bergstrom, O. Botner, P. Carlson, P. O Hulth, and T. Ohlsson, hep-ph/0412068*, and references therein. Also see http://www.sns.ias.edu/ jnb.
20. K.S. Hirata et al *Phys. Rev. Lett.* ***63***, *16 (1989)*.
21. J. N. Bahcall, A. Serenelli, and S. Basu, *ApJ Letters* ***621***, *L85 (2005)*.
22. J.N. Abdurashitov et al. (SAGE Collaboration) *Phys. Rev. Lett.* ***77***, *4708 (1996)*; W. Hampel *et al. (GALLEX Collaboration), Phys. Letters* ***B388***, *384 (1996)*.
23. Y. Fukuda *et al. Phys. Rev. Lett.* ***81***, *1158 (1998)*.
24. See reference[19] above.
25. S. Andringa, Report on K2K, These Proceedings.
26. The SNO Collaboration, *Nucl. Instr. and Methods,* ***A449***, *172 (2000)*.
27. A. Bellerive, Lepton-Photon Symposium 2003; Q.R. Ahmad *et al., Phys. Rev. Letters* ***89***, *011301 (2002)*; S.N. Ahmed *et al., nucl-ex/0309004*.
28. J. Shirai at *Proc. Neutrino-2002 Conference, Nucl. Phys. B (Proc. Suppl.)* ***118***, *15 (2003)*.
29. T. Araki *et al. (KamLAND Collaboration), Phys. Rev. Lett.* ***94*** *081801 (2005)*.
30. Y. Fukuda *et al. Phys. Rev. Lett.* ***81***, *1562 (1998)*.
31. See for example S.M. Bilenki, S. Pascoli, S.T. Peskov *Phys. Rev.* ***D64***, *053010 (2001)*.
32. C. Athanassopoulos et al., Phys. Rev. Lett., **81**, 1774 (1998).
33. B. Kayser, Neutrino Mass, Mixing, and Flavor Changing, in PDG-2004[13].
34. See for example J.N. Bahcall, M.C. Gonzalez-Garcia and C. Peña-Garay hep-ph/0212147, published in *JHEP**02**,09(2003)*, and references therein.
35. S. P.Mikheev and A. Yu. Smirnov, *Sov. J. Nucl. Phys.,* ***42***, *913 (1985)*; L. Wolfstein, *Phys. Rev.* ***D17***, *2369 (1978)*.
36. Z. Maki, M. Nakagawa, S. Sakata, *Prog. Theor. Phys.* ***28***, *870 (1962)*; B. Pontecorvo, *Soviet Phys. J. Exp. Theo. Phys.,* ***26***, *984 (1968)*.
37. G.L. Fogli, E. Lisi, A. Marrone, A. Melchirri, A. Palazzo, P. Serra, J. Silk, *Phys. Rev.* ***D70*** *113003 (2004)*.
38. M. Appolonio *et al.* (Chooz collaboration), *Eur. Phys. J* ***C27***, *331 (2003)*; F. Boehm *et al.* (Palo Verde collaboration), *Nucl. Phys. (Proc. Supp.)* ***77***, *166 (1999)*.
39. The MINOS technical design report can be found in http://www.hep.anl.gov/ndk/hypertext/minos_tdr.html.
40. Information on the CNGS experiments can be found in http://operaweb.web.cern.ch/operaweb/index.shtml (OPERA) and http://pcnometh4.cern.ch/bodyintro.html (ICARUS).
41. Information on T2K can be found in http://neutrino.kek.jp/jhfnu/.
42. Information on the "Double-Chooz" experiment can be found in http://doublechooz.in2p3.fr/dc_overview.htm.
43. Recent neutrino results can be obtained from the talks presented at the XXI

International Neutrino Conference in Paris, in June 2004 *Nuclear Physics B (Proceedings Supplements)* ***143**, 367 (2005).*
44. M. Bionta *et al., Phys. Rev. Lett.* ***58**, 1494 (1987).*
45. K. Hirata *et al., Phys. Rev. Lett.* ***58**, 1490 (1987).*
46. See for example J.F. Beacom *Proc. Neutrino-2002 Conference, Nucl. Phys. B (Proc. Suppl.)* ***118**, 307 (2003).*
47. K. Greisen Phys. Rev. Lett. **16** 748 (1966); G.T. Zatsepin and V.A. Kuz'min *JETP Lett., 4, 78 (1966).*
48. A. Loeb and and E. Waxman, *Phys. Rev. Lett.* ***87**, 071101 (2001).*
49. E. Waxman and J. N. Bahcall, *Phys. Rev.* ***D59**, 023002 (1999).* See also K. Mannheim, R. J. Protheroe and J. P. Rachen *Phys. Rev.**D63**, 023003 (2001)*; J. N. Bahcall and E. Waxman, *Phys. Rev. D64, 023002 (2001).*
50. J. G. Learned and K. Mannheim, *Annu. Rev. Nucl. Part. Sci.* ***50**, 679 (2000).*
51. M.A. Markov *Proceedings of the International Conference in High Energy Physics at Rochester, E.C.G. Sudarshan, J.H.Tinlot and A.C. Melissinos, Eds. (1960).*
52. Information on DUMAND project can be obtained from http://www.phys.hawaii.edu/dmnd/dumand.html.
53. F. Halzen and J. Learned, *Proceedings of the International Symposium on Very High Energy Cosmic Ray Interactions, Univ. of Lodz Publ., M. Giler Ed., (1989)*; F. Halzen, J. Learned and T. Stanev, *AIP Conf. Proceedings* ***198**, 39 (1989).*
54. See for example G. Domogatsky, in *Proceedings of the XXth International Conference on Neutrino Physics and Astrophysics Neutrino 2002, Nucl. Phys.* ***B118C** (Proc. Suppl.) 496 (2003).*
55. See for example R. Wischnewski contribution to TAUP(2001) *Nucl. Phys.* ***B110** 510 (2002).*
56. J. Ahrens *et al., Physical Review* ***D66**, 012005 (2002).*
57. E. Fernandez *et. al., HENAP Report (2002).* The report can be obtained from the PaNAGIC website: http://www.lngs.infn.it/site/exppro/panagic /section_indexes/frame_particles.htm
58. A. Achterberg *et al., in Proceedings of Neutrino 2004 Conference, Nuclear Physics B (Proceedings Supplements)* ***143**, 367 (2005)* and references therein.
59. The ANTARES Technical Design Report can be obtained from http://antares.in2p3.fr/Publications/TDR/v1r0/.
60. Information on NEMO project can be obtained from http://nemoweb.lns.infn.it/project.htm.
61. Information on NESTOR project can be obtained from http://www.cc.uoa.gr/ nestor/nestor.htm.

BLACK HOLES AND FUNDAMENTAL PHYSICS

JOSÉ P. S. LEMOS
Centro Multidisciplinar de Astrofísica - CENTRA
Departamento de Física, Instituto Superior Técnico,
Universidade Técnica de Lisboa,
Av. Rovisco Pais 1, 1049-001 Lisboa, Portugal.
E-mail: lemos@fisica.ist.utl.pt

We give a review of classical, thermodynamic and quantum properties of black holes relevant to fundamental physics.

1. Introduction

The term black hole is usually associated with a large astrophysical object that has formed due to huge gravitational fields that can arise in the center of massive concentrations (see, e.g., Ref. [1]). However, the black hole is an object in itself which should be studied within the domain of physics, irrespective of the interactions with exterior astrophysical plasmas which excite, and are excited by, the strong gravitational fields of the black hole. Here we want to understand a black hole as a physical object. This program was consciously initiated by Wheeler [2] back in the 1950s. We have not yet understood it entirely, but we have come very far, if we think that, back in 1960, Wheeler, Kruskal and others [3] managed to understand, for the first time, the global causal structure of the complete manifold of the simpler black hole, the Schwarzschild black hole. During the 1960s the black hole became well understood as a classical object, mainly due to the works of Penrose [4] and then Hawking [5] and Carter [6]. But, then in the 1970s due to Bekenstein [7] and Hawking [8], the whole field was revolutionized, the black hole concept entered the quantum arena. Of course, quantum dynamics is the underlying dynamics of the world and black holes have to be understood in this context. Conversely, the simple structure of a black hole, can be used to probe and learn about the quantum structure of gravitation. As such, a black hole is considered by many [9] as the gravitational equivalent of the hydrogen atom in mechanics, in the sense that this atom was used by

Bohr, Sommerfeld and others to touch and grasp the novel ideas of quantum mechanics [10].

Hawking's monumental discovery of 1974, perhaps the most important discovery in theoretical physics of the second half of the twentieth century, that a black hole radiates quantum mechanically, was followed by some interesting developments, but, perhaps, there was no ostensible growth, after that. Then, many physicists from different fields, like particle physics and field theory, moved to string theory. String theory, although a theory subject to criticisms on several grounds, can tackle important problems related to black hole physics. String theorists in taking the black hole problem into their hands back in the 1990s [11,12], out from the general relativists alone, opened the subject to the physics community overall, and revolutionized it into myriads of new directions. Before string theory attacked the problem, one should obey the general relativity bible, which was strict, allowing one to venture into other fields, such as non asymptotically flat spacetimes or naked singularities, only with extreme care and perhaps permission. String theory opened up the book and the theoretical discussion on black holes and their problems grew exponentially. Of course, general relativity itself benefited from it. For instance, new black hole solutions in general relativity, now called toroidal black holes, living in asymptotically anti-de Sitter spacetimes were found [13,14,15,16], and many other connections were made.

Thus, suddenly in the 1990s, the field of black holes was again lively and growing fast. Five topics of the utmost interest are: (a) the thermodynamics of black holes, (b) the black hole entropy and its degrees of freedom, (c) the information paradox, (d) the holographic principle and its connection to the generalized second law and to the covariant entropy bound, and (e) the inside of a black hole and its singularity. I will report on the first two topics, the others require reviews on their owns, and will be left for other opportunities. Due to a large bibliography on these two first topics I cannot be complete in listing references, the ones that are not mentioned will be left to another larger review. I have benefited tremendously from the reviews of Bekenstein [17,18] and Fursaev [19], see also the recent thorough reviews by Page [20] and Padmanabhan [21]. Unless otherwise stated we use units in which $G = c = \hbar = 1$.

2. Black hole thermodynamics and Hawking radiation

2.1. *Preliminaries*

It is now certain that a black hole can form from the collapse of an old massive star, or from the collapse of a cluster of stars. Many x-ray sources observed in our galaxy contain a black hole of about $10\,M_\odot$. Quasars, the most powerful distant objects, belong to a class which is one of the representatives of active galactic nuclei that are powered by a very massive black hole, with masses as high as $10^{10}M_\odot$. Our own Galaxy harbors a dead quasar with a mass of $10^6 M_\odot$ in its core. Mini black holes, with masses lying within a wide range, a typical one could have $10^{15}\,\text{gm} \sim 10^{-18}M_\odot$ (with radius $10^{-13}\,\text{cm}$), may have formed in the early universe. Finally, Planck black holes of mass $10^{-5}\,\text{gm}$ and radius $10^{-33}\,\text{cm}$ may form in an astronomical collider which could provide a center of mass energy of $10^{19}\,\text{GeV}$, or perhaps less if the idea of extra large dimensions is correct [22].

A black hole is a gravitational object whose interior region is invisible for the outside spacetime world. The boundary of this region is the event horizon of the black hole. To the outside world the black hole is like a tear in the spacetime, which interacts with its environment by attracting and scattering particles and waves in its neighborhood.

Black holes appear naturally, as exact solutions, in the theory of general relativity. The most simple is the Schwarzschild black hole which has only one parameter, the mass, and has a spherical horizon. By adding charge one obtains a black hole with more structure, the Reissner-Nordström black hole [23]. The theory of black holes received a tremendous boost after Kerr [24] found that a rotating black hole is also an exact solution of general relativity, a totally unexpected result at the time, that continues to flabbergast many people up to now. The Kerr black hole provides extra non-trivial dynamics to the spacetime, from which novel ideas sprang. Kerr black holes with charge are called Kerr-Newman black holes, a name also used to designate the whole family. There are now other important families of black holes, such as the family of anti-de Sitter spacetimes, with negative cosmological constant, whose horizons have topologies other than spherical [16], or other families in a variety of different theories of gravitation [25]. The Kerr-Neman family was the first to be thoroughly investigated classically. Some important properties, generically valid for other families, have been worked out in detail.

First, the event horizon acts as a one way membrane. Due to the strong gravitational field near the black hole, the light cones of the spacetime get tilted, so much so that their exterior boundary lies tangent to the horizon. The horizon thus acts as a one way membrane, i.e., no object, not even a light ray, that crosses it inwards can ever cross it back outwards. As a result, any physical quantity, such as energy, entropy or information, that is damped into the black hole remains permanently trapped inside, classically.

Second, a black hole has no hair. What does this mean? For an exterior observer, placed outside the horizon, the black hole forgets everything that it has swallowed. The black hole can have been formed from baryons alone, or from leptons alone, or from both, or anything else, the exterior observer cannot have access to what formed the black hole. The only thing the observer probes is the black hole mass M, electromagnetic charge Q, and angular momentum J. This is referred to as the baldness of the black hole, or as a black hole has no hair, in the language of Wheeler. In fact, it has three hairs, M, Q, J, but the nomenclature is still correct, one usually associates to someone that has three hairs that he is bald, has no hair!

Third, a black hole absorbs and scatters particles and waves. These properties involving scattering and absorption of particles and waves by black holes, specially by rotating black holes, were very important in the later developments. The whole subject started with the Penrose process [4], which branched into superradiance on one hand and the irreducible area concept on the other, and culminated with Hawking's theorem on area growth. Let us comment on these features briefly, first with special emphasis in superradiation. When a particle is scattered by a Kerr black hole and broken in two pieces in the process, energy can be extracted from the black hole rotation into the outgoing particle, using the existence of an ergosphere (a region just outside the event horizon) [4], a kind of relativistic sling shot phenomenom. The wave analog of the Penrose process, whereby an incoming wave (scalar, electromagnetic, gravitational or plasma) with positive energy that impinges on the rotating black hole splits up into an absorbed wave with negative energy and a reflected wave with enhanced positive energy [26,27,28], is called superradiance. Consider a wave of the form $e^{-i\omega t+im\phi}$, where ω is the frequency of the wave, t the time parameter, m the azimuthal angular wavenumber around the axis of rotation of the black hole, and ϕ the angular coordinate. Considering then that such a wave collides with the black hole, one concludes that if the frequency ω of the incident wave satisfies the superradiant condition $\omega < m\Omega$, where Ω is the angular velocity of the black hole, then the scattered wave is amplified [26,27,28], One simple way to get an idea of what is happening is by resorting to the inverse of the characteristic frequencies, i.e., the period of the wave

$\tau = 2\pi/\omega$, and the rotation period of the cylinder $T = 2\pi/\Omega$. Then the superradiant condition is now $m\,\tau > T$, which means that for $m = 1$ say, the wave suffers superradiant scattering if it takes a longer time in the neighborhood of the black hole than the time the black hole takes to make one revolution, so that there is enough time for the black hole to transfer part of its rotating energy to the wave. This way of seeing superradiance corresponds to giving a necessary condition, i.e., to exist superradiance there should exist enough time so that the black hole can transmit part of its energy to the wave.

Fourth, the area of a black hole always grows in any physical process. The Penrose process also led to the concept of irreducible mass [29], which in turn led Hawking [30] to prove a theorem stating that the black hole area always grows in any physical process, classically. This theorem proved to be decisive for further developments. In turn, and in passing, one can use this theorem to prove superradiance. Indeed, following the lines of Zel'dovich [26] one roughly finds that the scattering of a wave by a black hole obeys $\frac{\kappa}{8\pi}\frac{dA}{dt} = (P_i - P_r)\left(1 - \frac{m\Omega}{\omega}\right)$, where A is the area of the event horizon, κ is its surface gravity, and P_i and P_r are the incident and reflected power of the wave, respectively. From the area law for black holes, which states that the area of the event horizon never decreases, i.e., $d\,A \geq 0$ [30], one finds that if the frequency of the incident wave satisfies the superradiant condition, the second factor in the right hand side of the equation is negative. In order to guarantee that the area does not decrease during the scattering process, one must have $P_r > P_i$. Thus, the energy of the wave that is reflected is higher than the energy of the incident wave, as long has the superradiant condition is satisfied. On other developments on superradiance and how it can be used, along with a mirror, to build a black hole bomb see Ref. [31,32].

With these four ingredients, i.e., one-way membrane, no hair, scattering properties, and area law, all is set to put the black hole in a thermodynamic context.

2.2. *Thermodynamics and Hawking radiation*

A Kerr-Newman black hole, say, can form from the collapse of an extremely complex distribution of ions, electrons and radiation. But once it has formed the only parameters we need to specify the system are the parameters that characterize the Kerr-Newman black holes, the mass M, the charge Q and the angular momentum J. Thus we have a system specified by three parameters only, which hide lots of other parameters. In physics there is another instance of this kind of situation, whereby a system is specified and usefully described by few parameters, but on a closer look there are many more other parameters that are not accounted for in the compact description. This is the case in thermodynamics. For thermodynamical systems one gives the energy E, the volume V, and the number of particles N, say, and one can describe the system in a usefully manner, although the system encloses, and the description hides, a huge number of molecules.

Connected to this, was the question Wheeler was raising in the corridors of Princeton University [3], that in the vicinity of a black hole entropy can be dumped onto it, thus disappearing from the outside world, and grossly violating the second law of thermodynamics. Bekenstein, a Ph.D. student in Princeton at the time, solved part of the problem in one stroke. He postulated, entropy is area, more precisely [7], $S_{\rm BH} = \eta \frac{A}{l_{\rm pl}^2}\, k_{\rm B}$, where one is using full units, η is a number of the order of unity or so, that could not be determined, $l_{\rm pl} \equiv \sqrt{\frac{G\hbar}{c^2}}$ is the Planck length, of the order of 10^{-33} cm, and $k_{\rm B}$ is the Bolztmann constant. This is, of course, aligned with the area's law of Hawking, and the Penrose and superradiance processes. Bekenstein invoked several physical arguments to why the entropy S should go with A and not with $\sqrt{A}$ or A^2. For instance, it cannot go with $\sqrt{A}$ (A itself goes with $\sim M^2$) because when two black holes merge the final mass obeys $M < M_1 + M_2$ since there is emission of gravitational radiation. But if $S_{\rm BH} \propto M < M_1 + M_2 \propto S_{\rm BH1} + S_{\rm BH2}$ the entropy could decrease, so such a law is no good. The correct option turns out to be $S \propto A$, the one that Bekenstein took. Also correct, it seems, is to understand that this is a manifestation of quantum gravity, so that one should divide the area by the Planck area, and multiply by the Boltzmann constant to convert from the usual area units into the usual entropy units. There is thus a link between black holes and thermodynamics.

One can then wonder whether there is a relation obeyed by black hole dynamics equivalent to the first law of thermodynamics. For a Schwarzschild black hole one has that the area of the event horizon is given by $A = 4\pi r_+^2$. Since $r_+ = 2M$ one has $A = 16\pi M^2$. Then one finds $dM = 1/(32\,\pi\, M)\, dA$, which can be written as.

$$dM = \frac{\kappa}{8\,\pi} dA\,, \tag{1}$$

which is the first law of black hole dynamics [33]. The surface gravity of the event horizon of the Schwarzschild black hole is $\kappa = 1/4\,M$. Equation (1) can be compared with

$$dE = TdS\,, \tag{2}$$

which is the first law of thermodynamics. Note that, a priori, the analogy between S and A, and T and κ, is merely mathematical, whereas the analogy between E and M, is physical, they are the same quantity [34]. For a generic Kerr-Newman black hole one has the relation $dM = \kappa dA + \Phi dQ + \Omega dJ$, where Φ is the electric potential, and Ω the angular velocity of the black hole horizon. Comparing with the thermodynamical relation $dE = TdS + pdV + \mu dN$, where the symbols have their usual meanings, it further strengthens the analogy.

Following Bekenstein, this is no mere analogy though, the black hole system is indeed a thermodynamic system with the entropy of this system being proportional to the area. But what is η in the equation proposed by Bekenstein? Thermodynamic arguments alone were not sufficient to determine this number. Using quantum field theory methods in curved spacetime Hawking [8] showed that a Schwarzschild black hole radiates quantically as a black body at temperature

$$T = \frac{1}{8\pi M}\,. \tag{3}$$

Since $\kappa = 1/4\,M$, the temperature and the surface gravity are essentially the same physical quantity, with $T = \kappa/2\,\pi$. Moreover, from equation (2), one obtains $\eta = 1/4$, yielding finally

$$S = \frac{1}{4}\,A\,. \tag{4}$$

in geometrical units. Thus the Hawking radiation solved definitely the thermodynamic conundrum. However, it introduced several others puzzles.

The black hole is then a thermodynamic system. Thus, the second law of thermodynamics $\Delta\,S \geq 0$ should be obeyed. Since one does not know for sure the meaning of black hole entropy, it is useful to write the entropy as a sum of the black hole entropy $S_{\rm BH}$, and the usual matter entropy $S_{\rm matter}$, i.e., $S = S_{\rm BH} + S_{\rm matter}$, allowing one to write the second law as

$$\Delta\,S_{\rm BH} + \Delta S_{\rm matter} \geq 0\,, \tag{5}$$

commonly called the generalized second law [35]. The generalized second law proved important in many developments.

3. Statistical interpretation of black hole entropy

3.1. *Preliminaries*

In statistical mechanics, the entropy of an ordinary object is a measure of the number of states available to it, i.e., it is the logarithm of the number of quantum states that the object may access given its energy. This is the statistical meaning of the entropy. Since black holes have entropy, one can ask what does the black hole entropy represent? What is the statistical mechanics of a black hole as a thermodynamic object?

Retrieving full units to equation (4) one has

$$S_{\rm BH} = \frac{1}{4}\frac{A}{l_{\rm pl}^2}\, k_{\rm B}\,, \tag{6}$$

where again, $l_{\rm pl}^2 = \frac{c^2}{\hbar G}$ is the Planck area, and $k_{\rm B}$ is the Boltzmann constant. Judging from the four fundamental constants appearing in the formula, namely, $G, c, \hbar, k_{\rm B}$, one gets a system where relativistic gravitation, quantum mechanics and thermodynamics, are mixed together, indicating that a statistical interpretation should be in sight. Moreover, the number $\frac{A}{l_{\rm pl}^2}$ itself suggests that the sates of the black hole are some kind or another of quantum states. In addition, the factor 1/4 became a target for any theory that wants to explain black hole entropy from fundamental principles.

Note that black hole entropy is large. A neutron star with one solar mass has entropy of the order of $S \sim 10^{57}$ (in units where $k_{\rm B} = 1$) in a region within a radius of about 10 Km. A solar mass black hole has an entropy of 10^{79} in a region within a radius of 3 Km. There is a huge difference in entropy for these two objects of about the same size, suggesting, somehow, the black hole harnesses entropy that can be peeled away through the black hole's lifetime, i.e., the time the black hole takes to radiate its own mass via Hawking radiation.

3.2. *Entropy in the volume*

Bekenstein [36] tried first to connect the entropy of a black hole with the logarithm of the number of quantum configurations of any matter that could have served as the black hole origin, in perfect consonance with the no hair theorem. Now, the number of those quantum configurations can be associated to the number of internal states that a black hole can have, hinting in this way that the entropy of a black hole lies on the volume inside the black hole (see also Ref. [37]). This idea of bulk entropy, although interesting, has many drawbacks, see Refs. [38,39,40].

3.3. *Entropy in the area*

There are now many alternative interpretations that associate the black hole entropy with the its area, the area of the horizon. One can divide these interpretations into those that claim the degrees of freedom are on the quantum matter in the neighborhood of the horizon that gives rise to the Hawking radiation, those that claim that the degrees of freedom are on the gravitational field alone, and those that put the degrees of freedom on both, matter and gravitational fields, like string theory.

3.3.1. *Matter entropy*

(i) Entropy of quantum fields

One interpretation says that the black hole entropy comes from the entropy of quantum matter fields fluctuations in the vicinity of the horizon. This was advanced by Gerlach [41] who proposed that the entropy was related to the number of zero-point fluctuations that give rise to the Hawking radiation. So the entropy comes from the all time matter fields surrounding the horizon created by the Hawking process. Later Zurek and Thorne [42] proposed the quantum atmosphere picture and 't Hooft [43] developed the idea in the brick wall model. The advantage of these insights is that the linear dependence of $S_{\rm BH}$ on the horizon area, $S_{\rm BH} = \eta A$ comes automatically, as the matter that gives rise to the entropy is in a thin shell surrounding a surface, the horizon. One disadvantage, is that the coefficient η is infinite since arbitrarily small wavelengths also take part in the matter fields surrounding the horizon. One can cure this by imposing a cutoff at the Planck length, which, although sensible, is ad hock and incapable of giving the goal factor 1/4. Moreover, η is proportional to the number of fields existing in nature, making even harder to connect it with the coefficient 1/4.

(ii) Entropy of entanglement

Another, somehow connected, interpretation comes from Sorkin and collaborators [44], who suggested that the entropy is related to the entanglement entropy arising from tracing out the degrees of freedom existing beyond the horizon. In other words, the entropy is generated by dynamical degrees of freedom, excited at a certain time, associated to the matter in the black hole interior near the horizon through non-causal EPR correlations with the external matter. It has been used by many different authors, see e.g., Refs. [45,46]. This has the advantages and disadvantages of the above interpretation.

(iii) Entropy in induced gravity

Other interpretation that can be mentioned is the one that associates the degrees of freedom to heavy matter fields that when integrated out induce, naturally, general relativity. This way of seeing general relativity was envisaged by Sakharov [47] and the corresponding entropy interpertation was put forward in Ref. [48].

3.3.2. *Gravitational entropy*

(i) Entropy from boundary conditions

An improved interpretation, perhaps, is that of Solodukhin [49] and Carlip [50,51,52] who, independently, switched from matter field fluctuations to gravitational field fluctuations. They showed that the existence of a horizon, the surface where the fluctuations occur, makes the fluctuations themselves obey the laws of a conformal field theory in two spatial dimensions, this number two is related to the dimensionality of the horizon. Conformal field theory has been thoroughly investigated, yielding for the logarithm of number of states associated with the fluctuations, a value for the entropy that matches exactly the entropy formula for a black hole, with the coefficient 1/4 coming out perfect. The idea is to use the correct boundary conditions at a horizon so as to give rise to new degrees of freedom that do not exist in the bulk spacetime. However interesting it may be, see also Ref. [53], it lacks a direct physical interpretation, since the boundary conditions are too formal.

(ii) Heuristic interpretation for the degrees of freedom

A physical interpretation for the gravitational degrees of freedom comes from the intuitive idea of Bekenstein and Mukhanov [54,9] that the area of the horizon being an adiabatic invariant, should be quantized in Ehrenfest's way. Suppose, then, that the area of the horizon is quantized with uniformly spaced levels of order of the Planck length squared, i.e., $A = \alpha\, l_{\rm pl}^2\, n$ with α a pure number, and $n = 1, 2, \ldots.$ Thus a small black hole is constructed from a small number of Planck areas, one can build the next black hole putting an extra Planck area, and so on. The horizon, according with this view, can be thought of as a patchwork of patches with area $\alpha\, l_{\rm pl}^2$. If every Planck patch can have two distinct states, say, then a black hole with two Planck areas can be in four different states, a black hole with three Planck areas can be in eight different states, a large black hole with n Planck areas can be in 2^n different surface states. Now, degeneracy and entropy are connected in such a way that latter is the logarithm of the former, i.e., $S_{\rm BH} = \ln 2^n = (\ln 2)\, n = \frac{\ln 2}{\alpha}\frac{A}{l_{\rm pl}^2}$. The area law is then recovered, by default. Further, from Hawking's work we know that $\frac{\ln 2}{\alpha} = \frac{1}{4}$ so that the quantization law is $A = 4\,(\ln 2)\, l_{\rm pl}^2\, n$. We can instead think that every area patch has k distinct states instead of two. Then the same reasoning follows, and one has that a black hole with area $A = \alpha\, l_{\rm pl}^2\, n$ can be in any of k^n sates. The entropy is then $S_{\rm BH} = \ln k^n = (\ln k)\, n = \frac{\ln 2}{\alpha}\frac{A}{l_{\rm pl}^2}$, and the area quantization law is $A = 4\,(\ln k)\, l_{\rm pl}^2\, n$, and $\alpha = 4 \ln k$. The question is now, what is k? Hod [55] found a way to determine k. Inspired by Bohr's correspondence principle, that transition frequencies at large quantum numbers should equal classical oscillation frequencies, one should associate the classical oscillation frequencies of the black hole with the highly damped quasinormal frequencies, since these take no time, as quantum transitions take no time. So, for instance, the highly damped quasinormal frequencies of the Schwarzschild black hole are found to be $M\,\omega_n = \frac{\ln 3}{8\pi} - \frac{i}{4}\left(n + \frac{1}{2}\right)$, to leading order. The factor $\frac{\ln 3}{8\pi}$ was first found numerically [56], and much later analytically [57]. Then using $\Delta M = \omega$ and so $\Delta A = 32\,\pi\, M\, \Delta M = 32\,\pi\, M\, \omega = 4\,(\ln 3)\, l_{\rm pl}^2$, along with, from the very definition of A, $\Delta A = 4\,(\ln k)\, l_{\rm pl}^2 \Delta n$, constrained by $\Delta n = 1$ as it is required for a single simple area transition, one finds $k = 3$. Then the quantization of the area is given by $A_n = 4\,(\ln 3)\; l_{\rm pl}^2\, n$. This has been also used by people of loop quantum gravity, and received a boost as the whole idea of Hod fixes the Barbiero-Imirzi parameter, a loose parameter in the theory [58].

The spin-area parameter k was fixed in the case of a Schwarzschild black hole, $k = 3$. What can one say about the other black holes? The subject of quasinormal modes has been very active since Visheveshwara noticed that the signal from a perturbed black hole is, for most of the time, an exponentially decaying ringing signal, with the ringing frequency and damping timescales being characteristic of the black hole, depending only on its parameters like M, Q and J, and the cosmological constant Λ, say. Whereas for astrophysical black holes the most important quasinormal frequencies are the lowest ones, i.e., frequencies with small imaginary part, so that the signal can be detected, for black holes in fundamental physics the most important are the highly damped ones, since one is interested in the transition between classical and the quantum physics (see, e.g., Refs. [59,60]). Ultimately, one wants to understand whether the number $k = 3$ depends on the nature of the black hole (does a Kerr black hole give the Schwarzschild number), on the nature of spacetime (asymptotically flat, de Sitter, anti-de

Sitter), and on the dimension of spacetime or not. Different spacetimes yield different boundary conditions, and thus completely different behavior for quasinormal modes, whereas one might expect black hole area levels to depend only on local physics near the horizon, so that it is not obvious how to reconcile such locality with the quasinormal mode behavior. This makes it hard to argue that k is universal, as it should be. The study on other different black holes has not been conclusive.

(iii) York's interpretation

York [61] made a very interesting proposal where the entropy of the black hole comes from the statistical mechanics of zero-point quantum fluctuations of the metric, in the form of quasinormal modes, over the entire time of evaporation. The approach has thus a physical interpretation for the entropy, and gets the coefficients within the same range as the exact ones. York's idea is the translation of Gerlach's quantum matter fluctuations [41] to fluctuations in the gravitational field, and has been retaken in [62].

(iv) Other interpretations and methods

Other methods are Euclidean path integral [63], giving $S_{\mathrm{BH}} = \frac{1}{4}A$ directly, but it is flawed, since it uses a saddle point approximation at a point that is not a minimum. There is a method of surface fields and Euclidean conical singularities [64]. There is the Noether's charge method [65], a very useful one that has been frequently used. There are also hints that the entropy depends on the gravitational Einstein-Hilbert action alone, and like energy in general relativity, is a global concept [66].

There are other techniques that, although not constructed to yield an interpretation, corroborate that there should be a statistical interpretation. One of these is related to pair creation of black holes. In the Schwinger process of production of charged particles in a background electric field, the production rate grows as the number of particle species produced. If this is extrapolated to black hole production in a background field then the rate of the number of black hole pairs produced should go as the number of black hole states. Indeed, one can show that the factor $\Gamma \sim e^{\frac{1}{4}A_{\mathrm{BH}}} = e^{S_{\mathrm{BH}}}$ multiplies indeed the pair production amplitude, consistent with interpretation that the entropy counts black hole microstates. To work out these results one has to find the instanton solution, i.e., the solution that gives the transition rates, of the Euclidean C-metric, where the C-Metric is the solution for two black holes accelerating apart. This has been done for asymptotically flat spacetimes [67] and for de Sitter and anti-de Sitter spacetime [68,69,70,71].

The notion of black hole entropy has been extended to higher dimensional spacetimes where one can also have black p-branes. A black hole is a special case of a black p-brane, one with $p = 0$, a black string has $p = 1$, a black membrane has $p = 2$, and so on. These black branes suffer from a gravitational instability, the Gregory-Laflamme instability, and entropic arguments suggest that the fate of such a brane is a set of black holes [72].

3.3.3. *Entropy in string theory*

So far, I have not mentioned what is the contribution of string theory to the interpretation of black hole entropy. String theory has been extremely helpful in the advancement of black hole theory for several reasons. In relation to the calculation of black hole entropy it has given new methods, and many new and different black hole solutions on which one can apply these new methods. On the other hand, in relation to the interpretation of black hole entropy, to answer the question of where are the degrees of freedom, it has come short of a result. Let us see several developments in the context provided by string theory.

(i) Heuristics

First, heuristics [73]. String theory is a theory that provides many fields, which can be called matter fields, besides the gravitational field, and so the degrees of freedom for the entropy can come from both, the matter and the gravitational fields, now studied together in a coherent fashion. Since a string is matter, the entropy of a string goes with mass, and one can write $S_{\rm string} \simeq l_{\rm s} M$, where $l_{\rm s}$ is the fundamental string length (i.e., the string lengthscale), and M is the mass of the string (here we use string units). The entropy of a Schwarzschild black hole goes as $S_{\rm BH} \simeq G M^2 \simeq g^2 l_{\rm s}^2 M^2$, where now it was advisable to recover G, which in string units is equal to g the coupling of the string with the spacetime times the string lengthscale $l_{\rm s}$, both squared. The black hole radius goes as $r_{\rm BH} \simeq G M \simeq g^2 l_{\rm s}^2 M$. Now, the coupling g can be changed. Start from a black hole state, and assume one decreases the coupling reversibly and adiabatically, i.e., maintaining $S_{\rm BH}$ constant. Then $M \sim 1/g$ increases, and $r_{\rm BH} \sim g$ decreases. Thus as one puts less coupling, maintaining the entropy, the mass of the black hole increases so as to compensate in the number of states; on the other hand the radius decreases because there is much less gravity, a behavior that is similar to polytropic white dwarf stars. Now, one cannot go on decreasing the radius forever, the process has to stop when the radius of the black hole is of the order of the string scale $r_{\rm BH} \sim l_{\rm s}$. So, $l_{\rm s} \sim g_{\rm crit}^2\, l_{\rm s}^2\, M_{\rm crit}$ yielding from the black hole side $M_{\rm crit} \sim 1/(g_{\rm crit}^2\, l_{\rm s})$. This, in turn, implies $S_{\rm BH} \sim 1/g_{\rm crit}^2$, and from the string side $S_{\rm string} \sim 1/g_{\rm crit}^2$. Thus, this heuristic reasoning gives that there is a transition point from the black hole state to the string state, and vice versa, meaning that heavy string states form black holes, a not unexpected result. Unfortunately, there is no control as to where are the degrees of freedom when the black hole forms in this set up. One knows where are the degrees of freedom of the string, in the string itself, the way it curves, wiggles, vibrates, and so on, but then when it collapses and turns into a black hole at the transition point, it is a usual gravitational collapse, leaving us again in the dark. The nice thing about this calculation is that at the transition point the entropy is about the same for string and black hole, but how is the entropy transfered from the string to the black hole, or vice-versa, the calculation leaves us blind. See, however, the fuzzball proposal for black holes [74], where there is a retrieval of the intepretation that the entropy of a black hole lies on the volume inside the black hole, not in its area.

(ii) Exact calculations for extreme black holes

Extreme black holes allow an exact, tough tricky, calculation of the entropy, done for the first time by Strominger and Vafa [75]. A simple extreme black hole has mass M and charge Q that obey the relation $Q = (\sqrt{G})M$. The entropy is then $S_{\rm BH} = 4\pi G M^2 = 4\pi Q^2$. Since the entropy does not depend on the gravitational constant G, the entropy is a measure of the number of the elementary charges of the extreme black hole alone. As we now know, $G = g^2 l_{\rm s}^2$, and so the entropy does not depend on g. One can vary the string coupling g and obtain the same entropy. On the other hand, $r_{\rm BH} = G M = \sqrt{G}\, M = g\, l_{\rm s}\, Q$, so $r_{\rm BH}$ depends on g. For weak coupling one has $g << 1$ and so $r_{\rm BH} << l_{\rm s}$, the object is a condensed string in an almost flat spacetime, it is actually an intricate condensate of strings and branes, whereas for strong coupling one has $g >> 1$ and so $r_{\rm BH} >> l_{\rm s}$, the object is a black hole. Now, some extreme black holes have the property they are supersymmetric, i.e., supersymmetric transformations do not change the black hole, and there are some theorems that say that there are no quantum corrections when going from strong to weak coupling and vice-versa. So, one can calculate the entropy of the object at weak coupling, where one has an object in flat spacetime and then extrapolate directly and exactly this calculation to strong coupling. At weak coupling, one finds that the dual theory that governs the dynamics of the condensate of branes and strings is a conformal field theory. One can then use the machinery of conformal field theory, through the Cardy formula, and get the entropy. Amazingly, for certain black holes in string theory, with several different charges, it gives exactly the black hole entropy. This calculation is very interesting indeed, but again it leaves us blind to what are and where are the black hole degrees of freedom. Another snag of the calculation, is that it does not work out for general black holes, it works out only for extreme black holes, and even so not all extreme black holes.

(iii) What is conformal field theory?

We have been talking about conformal field theory, in various connections, namely in connection with the degrees of freedom of the horizon related to the method of Carlip [50] and Solodukhin [49], and in connection with the string theory methods. But what is conformal field theory? A way to see this [19] is to work with c massless scalar fields in one spatial dimension, i.e., in two spacetime dimensions. The Klein-Gordon equation for each field is

$$\left(\partial_t^2 - \partial_x^2\right) \phi_k(t,x) = 0 \,, \quad k = 1, ..., c \,, \tag{7}$$

valid in a one dimensional box of length b, i.e., with boundary conditions given by $\phi_k(t,0) = \phi_k(t,b) = 0$. This has a Planck radiation spectrum whose free energy is given by $F(T,b) = cT \, \Sigma_n \ln\left(1 - \mathrm{e}^{-\omega_n/T}\right)$, $\omega_n = \frac{\pi}{b}\, n \,, n = 1, 2, ...$, where the ω_n are the normal frequencies of the fields. In the thermodynamic limit $T\, b >> 1$, one can evaluate the sum to obtain $F(T,b) = -\frac{\pi c}{6}\, b T^2$, and thus

$$S(E,b) = 2\pi \sqrt{\frac{c}{6}\frac{b\, E}{\pi}} \,. \tag{8}$$

Now, how can one calculate this entropy using conformal field theory methods. First, one notes that the theory given in equation (7) is indeed conformal invariant. Using null coordinates $x_- = t - x$ and $x_+ = t + x$ the Klein-Gordon equation turns into $\partial_{x_-}\partial_{x_+}\phi_k = 0$. Indeed, this is invariant under conformal transformation $x_- \to x'_- = f(x_-)$ and $x_+ \to x'_+ = f(x_+)$. Now, when one has a symmetry, in this case conformal, one has an associated conserved charge. In turn these conserved charges are the generators of the corresponding symmetry transformation (for instance, the Hamiltonian is the generator of time translations, translations being included in conformal transformations). In two dimensions the generators, L_n and $\bar{L}_n$, of conformal transformations are infinite. They give the standard Virasoro algebra, $[L_n, L_m] = (n-m)\, L_{n+m}$, and the same for the complex conjugate, where the brackets are Poisson brackets. Interesting to note that the algebra of the generators is the same as the algebra of the Fourier components of the infinitesimal vector field that gives the coordinate transformations. The Hamiltonian generator is $\frac{2\pi}{b}(L_0 + \bar{L}_0)$. This is classical, and there is no entropy for the c scalar fields. However, when quantized the generators get an extra term, quantum mechanics yields always a scale which in turn produces an anomaly in the conformal field theory. This gives rise to an extra term for the algebra,

$$[L_n, L_m] = (n-m)\, L_{n+m} + \frac{c}{12}(n^3 - n)\, \delta_{n+m\,0}\,, \tag{9}$$

where the brackets should be viewed as a commutator now, and the generators as operators. The Hamiltonian operator is then $H = \frac{2\pi}{b}(L_0 + \bar{L}_0)$, which when applied in a state $|h, \bar{h} >$ gives the energy $E = \frac{2\pi}{b}(h + \bar{h})$. Now, the sate $|h, \bar{h} >$ can be constructed from vacuum, the box without fields, in many different ways, since $|h, \bar{h} >= \Pi_k (L_{-k})^{\alpha_k}\, \Pi_p (\bar{L}_{-p})^{\beta_p} |0 >$, with $\Sigma_k\, \alpha_k = h$, $\Sigma_p\, \alpha_p = \bar{h}$, and where $|0 >$ is the vacuum vector, and L_{-k} are creation operators. The state $|h, \bar{h} >$ is an eigenstate of L_0 and $\bar{L}_0$, sure. One can then find the degeneracy, as Cardy did [76], and show that $D = \mathrm{e}^{2\pi\sqrt{\frac{c}{6}\frac{bE}{\pi}}}$. Thus the entropy of the c conformal fields in a periodic box is $S = \ln D = 2\pi\sqrt{\frac{c}{6}\frac{bE}{\pi}}$, as in the thermodynamic result.

In possession of these ideas, we can better understand the Strominger-Vafa calculation. For low g one has a condensate of strings and branes instead of a black hole, which obey a conformal field theory. With the theory in hand one finds c, E and b, then one gets $S_{\rm CFT}$, and through supersymmetry arguments, extrapolates to high g, giving $S_{\rm BH}$, through $S_{\rm BH} = S_{\rm CFT}$. The entropy $S_{\rm BH}$ obtained in this way gives precisely $S_{\rm BH} = \frac{1}{4}A$. This is exact, but no interpretation for the entropy.

(iv) The BTZ black hole and the AdS/CFT conjecture

There is another place where these calculations are exact, it is the three dimensional BTZ black hole that lives in a cosmological constant Λ background, i.e., in an anti-de Sitter spacetime. The idea [77] came as follows. Brown and Henneaux [78] showed for the first time that the asymptotic group of three-dimensional anti-de Sitter spacetime is the conformal group in two dimensions, stating in addition that any quantum theory of such a type of spacetime should take this into account. At about the same time Cardy gave a formula, now famous, for the entropy of a two-dimensional conformal field theory with central charge [76]. Then, later, Strominger [77] applied the Cardy formula to Brown and Henneaux results [78] and showed that it gave the formula discovered by Bekenstein and Hawking, $S_{\rm BH} = \frac{1}{4}A$ [7,8]. More precisely, in this spacetime one has an intrinsic length scale, which is $l = 1/\sqrt{\Lambda}$. One also has the black hole radius $r_{\rm BH}$. Now, the black hole entropy can be calculated through gravitational methods to give $S_{\rm BH} = \frac{1}{4}A = \frac{1}{4}\frac{2\pi\, r_{\rm BH}}{G} = 2\pi\sqrt{\frac{l^2\, M}{2\, G}}$, with $M = \frac{r_{\rm BH}^2}{8\, l^2\, G}$. Now, compare with the Cardy formula $S_{\rm CFT} = 2\pi\sqrt{\frac{c\, b\, E}{6\pi}}$. For this put $b = 2\pi\, l$, forcing the conformal field theory to live on a cylinder of perimeter $2\pi\, l$, identify $M = E$, and then choose the central charge as $c = \frac{3}{2}\frac{l}{G}$ [77]. Then, with these choices $S_{\rm BH} = S_{\rm CFT}$. This equation relates classical and quantum quantities. The conformal theory is quantum lives on a flat spacetime M_2, one dimensional lower than the black hole, which lives in three dimensional spacetime. The metric on M_2 is a cylindrical flat metric, $ds^2 = -dt^2 + l^2\, d\varphi^2$. On the other hand, the metric for the black hole spacetime at constant large radius is $ds^2 = \frac{r^2}{l^2}(-dt^2 + l^2\, d\varphi^2)$. So M_2 can be seen, apart from a superfluous factor, as the asymptotic infinity of M_3, as its asymptotic boundary. Therefore, the black hole entropy (a semiclassical limit of quantum gravity), is determined by a quantum conformal field theory (CFT) defined at the asymptotic infinity of the bulk anti-de Sitter (AdS) spacetime. This is an example of the AdS/CFT conjecture of Maldacena [79], which was based in other spacetimes, and also works here. Since this type of computation for the black hole entropy is done at infinity, the infinity of anti-de Sitter spacetime, it does not see the details of the horizon. Thus, more than a direct computation of black hole entropy, this type of computation gives an upper bound for the entropy of anti-de Sitter spacetime in three dimensions. In this case, it is just as good, since the maximum of entropy in a region arises by inserting a black hole in it.

4. Conclusions

Thus we see that we are still far from a consensus [39]. Are the degrees of freedom located in the volume or in the area, or in both, or are they complementary descriptions? Are they realized in the matter or in the gravitational field or in both? The answer still lies ahead. The entropy puzzle does not exhaust the black hole. Other sources of fascinating problems and conundrums are the information paradox [80,81], the holographic principle [82,83] (for developments see Refs. [84,85,86,87]), and last but not the least the inside of a black hole and the problem of spacetimes singularities [88,89]. All of these are problems in fundamental physics whose solutions will help in a better understanding of the connections between quantum theory, statistical and information theory, and gravitation, and ultimately can lead us to the correct quantum theory of gravity.

Acknowledgments

I thank the organizing committee of the Fifth International Workshop on New Worlds in Astroparticle Physics, for providing a very stimulating atmosphere in the gracious city of Faro, in a meeting, held in January 2005, that opened up the celebrations of the first World Year of Physics. I thank conversations with my students Gonçalo Aprá Dias and Nuno Santos, as well as with Vitor Cardoso and Óscar Dias. I also thank Observatório Nacional do Rio de Janeiro for hospitality. This work was partially funded by FCT - Portugal through project POCTI/FNU/57552/2004.

References

1. J. P. S. Lemos, *astro-ph*/9612220 (1996).
2. J. A. Wheeler, in *Relativity, groups an topology*, eds. C. and B. de Witt (Gordon and Breach 1964), p. 315.
3. J. A. Wheeler, K. Ford, *Geons, black holes, and quantum foam: A life in physics*, (W. W. Norton & Company, 2000).
4. R. Penrose, *Nuov. Cimento* **1** (special number), 252 (1969).
5. S. W. Hawking, G. F. R. Ellis, *The large scale structure of space-time*, (Cambridge University Press 1973).
6. B. Carter, in *Black holes, Les astres occlus*, ed. C. and B. de Witt (Gordon and Breach 1973), p. 57.
7. J. D. Bekenstein, *Phys. Rev. D* **9**, 2333 (1973).
8. S. W. Hawking, *Nature* **248**, 30 (1974); *Commun. Math. Phys.* **43**, 199 (1975).
9. J. D. Bekenstein, *gr-qc*/9710076 (1997).
10. J. Mehra, H. Rechenberg, *The historical development of quantum theory* (6 Volumes), (Springer 2002).
11. E. Witten, *Phys. Rev. D* **44**, 314 (1991).
12. G. T. Horowitz, A. Strominger, *Nucl. Phys.* **B360**, 197 (1991).
13. J. P. S. Lemos, *Class. Quantum Grav.* **12**, 1081 (1995).
14. J. P. S. Lemos, *Phys. Lett.* **B353**, 46 (1995).
15. J. P. S. Lemos, V. T. Zanchin, *Phys. Rev. D* **54**, 3840 (1996).
16. J. P. S. Lemos, *gr-qc*/0011092 (2000).
17. J. D. Bekenstein, *gr-qc*/9808028 (1998).
18. J. D. Bekenstein, *gr-qc*/9409015 (1994).
19. D. V. Fursaev, *Phys. Elem. Part. Nucl.* **36**, 81 (2005).
20. D. N. Page, *hep-th*/0409024 (2004).
21. T. Padmanabhan, *Phys. Rep.* **406**, 49 (2005).
22. N. Arkani-Hamed, S. Dimopoulos, G. Dvali, *Phys. Lett.* **B429**, 263 (1998).

23. C. W. Misner, K. S. Thorne, J. A. Wheeler, *Gravitation*, (Freeman 1973).
24. R. Kerr, *Phys. Rev. Lett.* **11**, 237 (1963).
25. J. P. S. Lemos, *hep-th*/9701121 (1997).
26. Ya. B. Zel'dovich, *Sov. Phys. JETP* **35**, 1085 (1972).
27. W. G. Unruh, *Phys. Rev. D* **10**, 3194 (1974).
28. J. D. Bekenstein, M. Schiffer, *Phys. Rev. D* **58**, 064014 (1998).
29. D. Christodoulou, *Phys. Rev. Lett.* **25**, 1596 (1970).
30. S. W. Hawking, *Phys. Rev. Lett* **26**, 1344 (1971).
31. V. Cardoso, O. J. C. Dias, J. P. S. Lemos, S. Yoshida, *Phys. Rev. D* **70**, 044039 (2004).
32. O. J. C. Dias, J. P. S. Lemos, "Superradiant amplification in astrophysical black hole systems", unpublished (2004).
33. J. M. Bardeen, B. Carter, S. W. Hawking, *Commun. Math. Phys.* **31**, 161 (1973).
34. R. Wald, *Quantum field theory in curved spacetime and black hole thermodynamics* (University of Chicago Press 1994).
35. J. D. Bekenstein, *Phys. Rev. D* **9**, 3292 (1974).
36. J. D. Bekenstein, *Phys. Rev. D* **12**, 3077 (1975).
37. V. Frolov, I. Novikov, *Phys. Rev. D* **48**, 4545 (1993).
38. T. Jacobson, *gr-qc*/9908031 (1999).
39. T. Jacobson, D. Marolf, C. Rovelli, *hep-th*/0501103 (2005).
40. R. D. Sorkin, *hep-th*/0504037 (2005).
41. U. H. Gerlach, *Phys. Rev. D* **14**, 1479 (1976).
42. W. H. Zurek, K. S. Thorne, *Phys. Rev. Lett.* **54**, 2171 (1985).
43. G. 't Hooft, *Nucl. Phys.* **B256**,727 (1985).
44. L. Bombelli, R. K. Koul, J. Lee, R. D. Sorkin, *Phys. Rev. D* **34**, 373 (1986).
45. M. Srednicki, *Phys. Rev. Lett.* **71**, 666 (1993).
46. C. Adami, *quant-ph*/0405005 (2004).
47. A. D. Sakharov, Sov. Phys. Dokl. **12**, 1040 (1968); Gen. Rel. Grav. **32**, 365 (2000).
48. V. P. Frolov, D. V. Fursaev, *Phys. Rev. D* **56**, 2212 (1997).
49. S. N. Solodukhin, *Phys. Lett.* **B454**, 213 (1999).
50. S. Carlip, *Class. Quant. Grav.* **16**, 3327 (1999).
51. S. Carlip, *Phys. Rev. Lett.* **88**, 24130 (2002).
52. S. Carlip, *Class. Quant. Grav.* **22**, 1303 (2005).
53. G. A. S. Dias, J. P. S. Lemos, in preparation (2005).
54. J. D. Bekenstein, V. F. Mukhanov, *Phys. Lett.* **B360**, 7 (1995).
55. S. Hod, *Phys. Rev. Lett.* **81**, 4293 (1998).
56. H. P. Nollert, *Phys. Rev. D* **47**, 5253 (1993).
57. L. Motl, A. Neitzke, *Adv. Theor. Math. Phys.* **7**, 307 (2003).
58. J. Baez, in *gr-qc*/0303027 (2003).
59. V. Cardoso, J. P. S. Lemos, S. Yoshida, *Phys. Rev. D* **69**, 044004 (2004).
60. V. Cardoso, *gr-qc*/0404093 (2004).
61. J. W. York, *Phys. Rev. D* **28**, 2929 (1983).
62. N. Iizuka, D. Kabat, G. Lifschytz, D. A. Lowe, *Phys. Rev. D* **67**, 124001 (2003).

63. G. W. Gibbons, S. W. Hawking, *Phys. Rev. D* **15**, 2738 (1977).
64. C. Teitelboim, *Phys. Rev. D* **51**, 4315 (1995).
65. V. Iyer, R. M. Wald, *Phys. Rev. D* **50**, 846 (1994).
66. S. W. Hawking, C. J. Hunter, *Phys. Rev. D* **59**, 044025 (1999).
67. D. Garfinkle, S. B. Giddings, A. Strominger, *Phys. Rev. D* **49**, 958 (1994).
68. O. J. C. Dias, J. P. S. Lemos, *Phys. Rev. D* **69**, 084006 (2004).
69. O. J. C. Dias, *Phys. Rev. D* **70**, 024007 (2004).
70. O. J. C. Dias, J. P. S. Lemos, *Phys. Rev. D* **70**, 124023 (2004).
71. O. J. C. Dias, *hep-th*/0410294 (2004).
72. B. Kol, *hep-th*/0411240 (2004).
73. G. T. Horowitz, J. Polchinski, *Phys. Rev. D* **55**, 6189 (1997).
74. S. M. Mathur, *Fortschr. Phys.* **53**, 793 (2005).
75. A. Strominger, C. Vafa, *Phys. Lett.* **B379**, 99 (1996).
76. J. A. Cardy, *Nucl. Phys.* **B270**, 186 (1986).
77. A. Strominger, *JHEP* **9802**, 009 (1998).
78. J. D. Brown, M. Henneaux, *Comm. Math. Phys.* **104**, 207 (1986).
79. J. M. Maldacena, *Adv. Theor. Math. Phys.* **2**, 231 (1998).
80. D. N. Page, *hcp-th*/9305040 (1993).
81. S. W. Hawking, *hep-th*/0507171 (2005).
82. G. 't Hooft, *gr-qc*/9310026 (1993).
83. L. Susskind, *J. Math. Phys.* **36**, 6377 (1995).
84. R. Bousso, *Rev. Mod. Phys.* **74**, 825 (2004).
85. S. Gao, J. P. S. Lemos, *JHEP* **0404**, 017 (2004).
86. S. Gao, J. P. S. Lemos, *Phys. Rev. D* **71**, 084010 (2005).
87. S. Gao, J. P. S. Lemos, this volume (2005).
88. W. Israel, in *Black holes and relativistic stars*, ed. R. Wald (University of Chicago Press, 1998), p. 137.
89. R. Penrose, *The road to reality: A complete guide to the laws of the Universe*, (Knopf 2005).

Part 2 Contributions

Cosmic Ray Physics

PHENOMENOLOGY OF COSMIC RAY AIR SHOWERS

MARIA TERESA DOVA*

Departamento de Física,
Universidad Nacional de La Plata, CC67 La Plata (1900), Argentina
E-mail: dova@fisica.unlp.edu.ar

The properties of cosmic rays with energies above 10^6 GeV have to be deduced from the spacetime structure and particle content of the air showers which they initiate. In this review, a summary of the phenomenology of these giant air showers is presented. We describe the hadronic interaction models used to extrapolate results from collider data to ultra high energies, an also the main electromagnetic processes that govern the longitudinal shower evolution as well as the lateral spread of particles.

1. Introduction

For primary cosmic ray energies above 10^6 GeV, the flux becomes so low that direct detection of the primary particle using detectors in or above the upper atmosphere is not longer possible. In these cases the primary particle has enough energy to initiate an extensive air shower (EAS) in the atmosphere. If the primary cosmic ray particle is a nucleon or nucleus the cascade begins with a hadronic interaction. The number of hadrons increases through subsequent generations of particle interactions. In each generation about 20% of the energy is transferred to an electromagnetic cascade by rapid decays of neutral pions. Ultimately, the electromagnetic cascade dissipates roughly 90% of the primary particle's energy trough ionization. The remaining energy is carried by muons and neutrinos from charged pion decays. The electromagnetic and weak interactions are rather well understood. However, uncertainties in hadronic interactions at ultra high energies constitute one of the most problematic sources of systematic error in analysis of air showers. In what follows a brief report of the phenomenology of EAS is presented, focusing in those aspects which are the

*Work partially supported by CONICET, Argentina

main source of systematic uncertainties affecting somehow the determination of primary energy and mass composition. A complete review on the phenomenology of cosmic ray air showers can be found in Ref. 1.

2. Hadronic Processes

Soft multiparticle production with small transverse momenta with respect to the collision axis is a dominant feature of most hadronic events at center-of-mass energies 10 GeV $< \sqrt{s} <$ 50 GeV. Despite the fact that strict calculations based on ordinary QCD perturbation theory are not feasible, there are some phenomenological models that successfully take into account the main properties of the soft diffractive processes. These models, inspired by $1/N$ QCD expansion are also supplemented with generally accepted theoretical principles like duality, unitarity, Regge behavior, and parton structure. The interactions are no longer described by single particle exchange, but by highly complicated modes known as Reggeons. Up to about 50 GeV, the slow growth of the cross section with $\sqrt{s}$ is driven by a dominant contribution of a special Reggeon, the Pomeron.

At higher energies, semihard interactions arising from the hard scattering of partons that carry only a very small fraction of the momenta of their parent hadrons can compete successfully with soft processes. These semihard interactions lead to the minijet phenomenon, i.e., jets with transverse energy ($E_T = |p_T|$) much smaller than the total center-of-mass energy. Unlike soft processes, this low-p_T jet physics can be computed in perturbative QCD. The parton-parton minijet cross section is given by

$$\sigma_{\rm QCD}(s, p_T^{\rm cutoff}) = \sum_{i,j} \int \frac{dx_1}{x_1} \int \frac{dx_2}{x_2} \int_{Q^2_{\rm min}}^{\hat{s}/2} d|\hat{t}| \, \frac{d\hat{\sigma}_{ij}}{d|\hat{t}|} \, x_1 f_i(x_1, |\hat{t}|) \; x_2 f_j(x_2, |\hat{t}|) \;, \quad (1)$$

where x_1 and x_2 are the fractions of the momenta of the parent hadrons carried by the partons which collide, $d\hat{\sigma}_{ij}/d|\hat{t}|$ is the cross section for scattering of partons of types i and j according to elementary QCD diagrams, f_i and f_j are parton distribution functions (pdf's), $\hat{s} = x_1 x_2 s$ and $-\hat{t} = \hat{s}\,(1 - \cos\vartheta^*)/2 = Q^2$ are the Mandelstam variables for this parton-parton process, and the sum is over all parton species. The integration limits satisfy $Q^2_{\rm min} < |\hat{t}| < \hat{s}/2$, with $Q_{\rm min}$ the minimal momentum transfer.

A first source of uncertainty in modeling cosmic ray interactions at ultra high energy is encoded in the extrapolation of the measured parton densities several orders of magnitude down to low x. Primary protons that impact on the upper atmosphere with energy $\approx 10^{11}$ GeV, yield partons with $x \equiv 2p_{\parallel}^{*}/\sqrt{s} \approx m_\pi/\sqrt{s} \sim 10^{-7}$, whereas current data on quark and gluon densities are only available for $x \approx 10^{-4}$ to within an experimental accuracy of 3% for $Q^2 \approx 20$ GeV^2 [6]. Moreover, application of HERA data to baryonic cosmic rays assumes universality of the pdf's.

For large Q^2 and not too small x, the Dokshitzer-Gribov-Lipatov-Altarelli-Parisi (DGLAP) equations successfully predict the Q^2 dependence of the quark and gluon densities (q and g, respectively). In the double–leading–logarithmic approximation the DGLAP equations predict a steeply rising gluon density, $xg \sim x^{-0.4}$, which dominates the quark density at low x, in agreement with experimental results obtained with the HERA collider [7]. Specifically, HERA data are found to be consistent with a power law, $xg(x,Q^2) \sim x^{-\Delta_{\rm H}}$, with an exponent $\Delta_{\rm H}$ between 0.3 and 0.4 [8].

The high energy minijet cross section is then determined by the small-x behavior of the parton distributions or, rather, by that of the dominant gluon distribution (via the lower limits of the x_1, x_2 integrations) which gives [9]:

$$\sigma_{\rm QCD}(s) \propto \int_{2\,Q^2_{\rm min}/s}^{1} \frac{dx_1}{x_1}\, x_1^{-\Delta_{\rm H}} \int_{2\,Q^2_{\rm min}/s}^{1} \frac{dx_2}{x_2}\, x_2^{-\Delta_{\rm H}} \sim s^{\Delta_{\rm H}} \ln(s/s_0)\,, \quad (2)$$

where s_0 is a normalization constant. One caveat is that the inclusive QCD cross section given in Eq. (2) is a Born approximation, and therefore automatically violates unitarity.

The procedure of calculating the inelastic cross section from inclusive cross sections is known as unitarization. In the eikonal model [11] of high energy hadron-hadron scattering,the inelastic cross section, assuming a real eikonal function, is given by

$$\sigma_{\rm inel} = \int d^2\vec{b}\, \left\{1 - \exp\left[-2\chi_{\rm soft}(s,\vec{b}) - 2\chi_{\rm hard}(s,\vec{b})\right]\right\}\,, \quad (3)$$

where the scattering is compounded as a sum of QCD ladders via hard and soft processes through the eikonals $\chi_{\rm hard}$ and $\chi_{\rm soft}$. It should be noted that we have ignored spin-dependent effects and the small real part of the scattering amplitude, both good approximations at high energies. Now, if the eikonal function, $\chi(s,\vec{b}) \equiv \chi_{\rm soft}(s,\vec{b}) + \chi_{\rm hard}(s,\vec{b}) = \lambda/2$, indicates the mean number of partonic interaction pairs at impact parameter $\vec{b}$, the probability p_n for having n independent partonic interactions using Poisson statistics reads, $p_n = (\lambda^n/n!)\, e^{-\lambda}$. Therefore, the factor $1 - e^{-2\chi} = \sum_{n=1}^{\infty} p_n$ in Eq. (3) can be interpreted semiclassically as the probability that at least 1

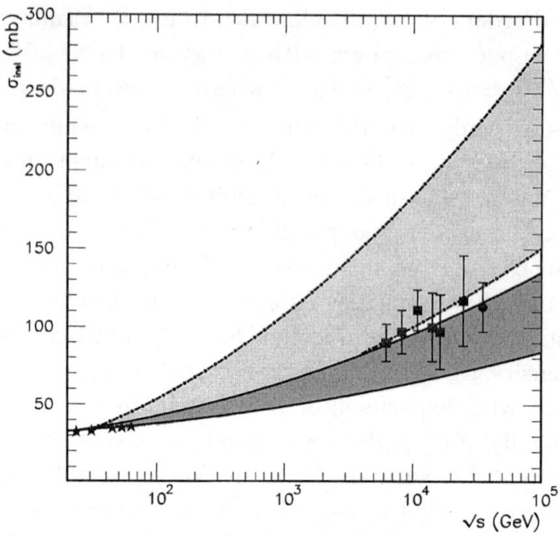

Figure 1. Energy dependence of the pp inelastic cross section as predicted by Eqs. (5) and (6) with $0.3 < \Delta_H < 0.4$. The darkly shaded region between the solid lines corresponds to the model with Gaussian parton distribution in $\vec{b}$. The region between the dashed-dotted lines corresponds to the model with exponential fall-off of the parton density in $\vec{b}$.

of the 2 protons is broken up in a collision at impact parameter $\vec{b}$. With this in mind, the inelastic cross section is simply the integral over all collision impact parameters of the probability of having at least 1 interaction, yielding a mean minijet multiplicity of $\langle n_{\rm jet}\rangle \approx \sigma_{\rm QCD}/\sigma_{\rm inel}$ [12]. The leading contenders to approximate the (unknown) cross sections at cosmic ray energies, SIBYLL [13] and QGSJET [14], share the eikonal approximation but differ in their *ansätse* for the eikonals. In both cases, the core of dominant scattering at very high energies is the parton-parton minijet cross section given in Eq. (1),

$$\chi_{\rm hard} = \frac{1}{2}\,\sigma_{\rm QCD}(s, p_T^{\rm cutoff})\; A(s,\vec{b})\,, \tag{4}$$

where the normalized profile function, $\int d^2\vec{b}\; A(s,\vec{b}) = 1$, indicates the distribution of partons in the plane transverse to the collision axis.

In the QGSJET-like models, the core of the hard eikonal is dressed with a soft-pomeron pre-evolution factor. This amounts to taking a parton distribution which is Gaussian in the transverse coordinate distance $|\vec{b}|$.

In SIBYLL-like models, the transverse density distribution is taken as

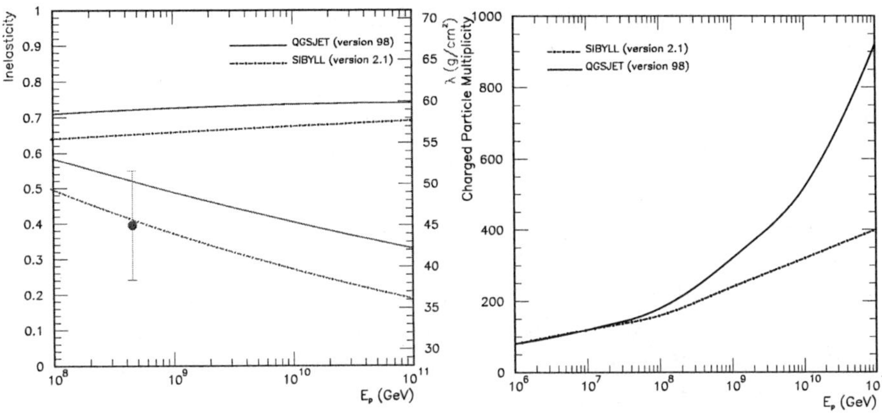

Figure 2. Left panel: The slowly rising curves indicate the mean inelasticity in proton air collisions as predicted by QGSJET and SIBYLL. The falling curves indicate the proton mean free path in the atmosphere. Right panel: Mean multiplicity of charged secondary particles produced in inelastic proton-air collisions processed with QGSJET and SIBYLL.

the Fourier transform of the proton electric form factor, resulting in an energy-independent exponential (rather than Gaussian) fall-off of the parton density profile with $|\vec{b}|$. The main characteristics of the pp cascade spectrum resulting from these choices are readily predictable: the harder form of the SIBYLL form factor allows a greater retention of energy by the leading particle, and hence less available for the ensuing shower. Consequently, on average SIBYLL-like models predict a smaller multiplicity than QGSJET-like models (see *e.g.* Refs. 16, 17, 18, 19).

At high energy, $\chi_{\rm soft} \ll \chi_{\rm hard}$, and so the inelastic cross section is dominated by the hard eikonal. With the appropriate choice of normalization, the cross section in Eq. (2) can be well–approximated by a power law. This implies that the growth of the inelastic cross section according to QGSJET-like models is given by

$$\sigma_{\rm inel} \sim \int d^2\vec{b}\ \Theta(b_s - |\vec{b}|) = \pi b_s^2 \sim 4\pi\,\alpha'_{\rm eff}\ \Delta_{\rm H}\ \ln^2(s/s_0)$$
$$\sim 0.52\ \Delta_{\rm H}\ \ln^2(s/s_0)\ \text{mb}\ . \qquad (5)$$

For SIBYLL-like models, the growth of the inelastic cross section also saturates the $\ln^2 s$ Froissart bound, but with a multiplicative constant which is larger than the one in QGSJET-like models [18]. Namely,

$$\sigma_{\rm inel} \sim 3.2\ \Delta_{\rm H}^2 \ln^2(s/s_0)\ \text{mb}\ . \qquad (6)$$

Figure 1 illustrates the large range of predictions for *pp* inelastic cross section which remain consistent with HERA data. When the two leading order approximations discussed above are extrapolated to higher energies, both are consistent with existing cosmic ray data. Note, however, that in both cases the range of allowed cross-sections at high energy varies by a factor of about 2 to 3. The points in Fig. 1 correspond to the most up-to-date estimate of the *pp* cross section from cosmic ray shower data [21].

There are three event generators, SIBYLL [13], QGSJET [14], and DPMJET [24] which are tailored specifically for simulation of hadronic interactions up to the highest cosmic ray energies. The latest versions of these packages are SIBYLL 2.1 [25], QGSJET 01 [26], and DPMJET III [27]; respectively. In QGSJET, both the soft and hard processes are formulated in terms of Pomeron exchanges. To describe the minijets, the soft Pomeron mutates into a "semi-hard Pomeron", an ordinary soft Pomeron with the middle piece replaced by a QCD parton ladder, as sketched in the previous paragraph. This is generally referred to as the "quasi-eikonal" model. In contrast, SIBYLL and DPMJET follow a "two channel" eikonal model, where the soft and the semi-hard regimes are demarcated by a sharp cut in the transverse momentum: SIBYLL 2.1 uses a cutoff parametrization inspired in the double leading logarithmic approximation of the DGLAP equations, whereas DPMJET III uses an *ad hoc* parametrization for the transverse momentum cutoff 8.

The transition process from asymptotically free partons to colour-neutral hadrons is described in all codes by string fragmentation models [28]. Different choices of fragmentation functions can lead to some differences in the hadron multiplicities. However, the main difference in the predictions of QGSJET-like and SIBYLL-like models arises from different assumptions in extrapolation of the parton distribution function to low energy.

Now we turn to nucleus-nucleus interactions, which cause additional headaches for event generators which must somehow extrapolate *pp* interactions in order to simulate the proton-air collisions of interest. All the event generators described above adopt the Glauber formalism [10].

Since the codes described above are still being refined, the disparity between them can vary even from version to version. At the end of the day, however, the relevant parameters boil down to two: the mean free path, $\lambda = (\widetilde{\sigma}_{\rm prod}\, n)^{-1}$, and the inelasticity, $K = 1 - E_{\rm lead}/E_{\rm proj}$, where n is the number density of atmospheric target nucleons, $E_{\rm lead}$ is the energy of the most energetic hadron with a long lifetime, and $E_{\rm proj}$ is the energy of the projectile particle. Overall, SIBYLL has a shorter mean free path and a smaller inelasticity than QGSJET, as indicated in Fig. 2. Since a shorter mean free path tends to compensate a smaller inelasticity, the two codes generate similar predictions for an air shower which has lived through several generations. The different predictions for the mean charged particle multiplicity in proton-air collisions are shown in Fig. 2. Both models predict the same multiplicity below about 10^7 GeV, but the predictions diverge above that energy. Such a divergence readily increases with rising energy. As it is extremely difficult to observe the first interactions experimentally, it is not straightforward to determine which model is closer to reality.

3. Electromagnetic Component

The evolution of an extensive air shower is dominated by electromagnetic processes. The interaction of a baryonic cosmic ray with an air nucleus high in the atmosphere leads to a cascade of secondary mesons and nucleons. The first few generations of charged pions interact again, producing a hadronic core, which continues to feed the electromagnetic and muonic components of the showers. Up to about 50 km above sea level, the density of atmospheric target nucleons is $n \sim 10^{20}$ cm^{-3}, and so even for relatively low energies, say $E_{\pi^\pm} \approx 1$ TeV, the probability of decay before interaction falls below 10%. Ultimately, the electromagnetic cascade dissipates around 90% of the primary particle's energy, and hence the total number of electromagnetic particles is very nearly proportional to the shower energy.

By the time a vertically incident 10^{11} GeV proton shower reaches the ground, there are about 10^{11} secondaries with energy above 90 keV in the the annular region extending 8 m to 8 km from the shower core. Of these, 99% are photons, electrons, and positrons, with a typical ratio of γ to e^+e^- of 9 to 1. Their mean energy is around 10 MeV and they transport 85% of the total energy at ground level. Of course, photon-induced showers are even more dominated by the electromagnetic channel, as the only significant muon generation mechanism in this case is the decay of charged pions and kaons produced in γ-air interactions.

It is worth mentioning that these figures dramatically change for the case of very inclined showers. For a primary zenith angle, $\theta > 70°$, the electromagnetic component becomes attenuated exponentially with atmospheric depth, being almost completely absorbed at ground level. We remind the reader that the vertical atmosphere is ≈ 1000 g/cm^2, and is about 36 times deeper for completely horizontal showers.

In contrast to hadronic collisions, the electromagnetic interactions of shower particles can be calculated very accurately from quantum electrodynamics. Electromagnetic interactions are thus not a major source of systematic errors in shower simulations. The first comprehensive treatment of electromagnetic showers was elaborated by Rossi and Greissen [29]. This treatment was recently cast in a more pedagogical form by Gaisser [30] and a summary is presented in Ref. 1.

The generation of the electromagnetic component is driven by electron bremsstrahlung and pair production [31]. Eventually the average energy per particle drops below a critical energy, ϵ_0, at which point ionization takes over from bremsstrahlung and pair production as the dominant energy loss mechanism. The $e^{\pm}$ energy loss rate due to bremsstrahlung radiation is nearly proportional to their energy, whereas the ionization loss rate varies only logarithmically with the $e^{\pm}$ energy. The changeover from radiation losses to ionization losses depopulates the shower. One can thus categorize the shower development in three phases: the growth phase, in which all the particles have energy $> \epsilon_0$; the shower maximum, $X_{\rm max}$; and the shower tail, where the particles only lose energy, get absorbed or decay.

The relevant quantities participating in the development of the electromagnetic cascade are the probability for an electron of energy E to radiate a photon of energy $k = yE$ and the probability for a photon to produce a pair e^+e^- in which one of the particles (hereafter e^-) has energy $E = xk$. These probabilities are determined by the properties of the air and the cross sections of the two processes.

In the energy range of interest, the impact parameter of the electron or photon is larger than an atomic radius, so the nuclear field is screened by its electron cloud. In the case of complete screening, where the momentum transfer is small, the cross section for bremsstrahlung can be approximated by [32]

$$\frac{d\sigma_{e\to\gamma}}{dk} \approx \frac{A_{\rm eff}}{X_0 N_A k}\left(\frac{4}{3} - \frac{4}{3}y + y^2\right) , \qquad (7)$$

where $A_{\rm eff}$ is the effective mass number of the air, X_0 is a constant, and N_A is Avogadro's number. In the infrared limit (*i.e.*, $y \ll 1$) this approximation is inaccurate at the level of about 2.5%, which is small compared to typical experimental errors associated with cosmic air shower detectors. Of course, the approximation fails as $y \to 1$, when nuclear screening becomes incomplete, and as $y \to 0$, at which point the LPM and dielectric suppression effects become important.This infrared divergence is eliminated by the interference of bremsstrahlung amplitudes from multiple scattering centers. This collective effect of the electric potential of several atoms is known as

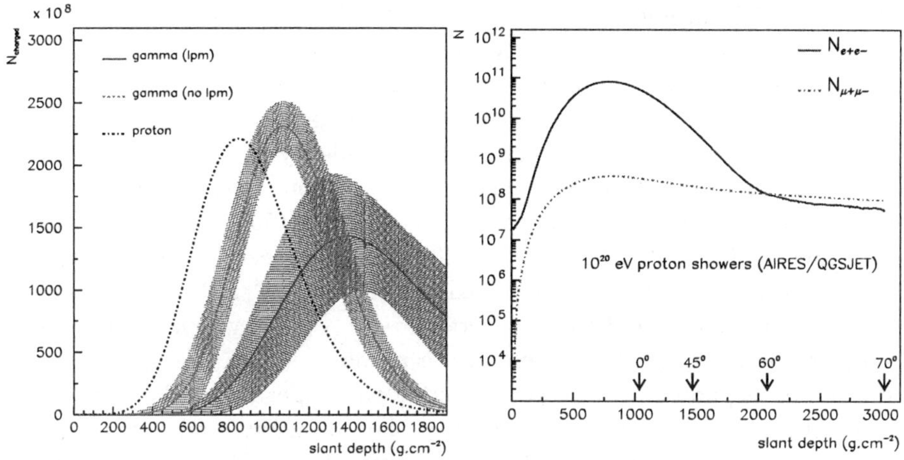

Figure 3. Left panel:Average longitudinal shower developments of 10^{11} GeV proton (dashed-dotted line) and γ-rays with and without the LPM effect (solid and dotted lines, respectively). The primary zenith angle was set to $\theta = 60°$. Right panel: Longitudinal development of muons and electrons as a function of the slant depth for 10^{11} GeV proton-induced showers.

the Landau-Pomeranchuk-Migdal (LPM) effect [33,34]. Using similar approximations, the cross section for pair production can be obtained [32].

The LPM suppression of the cross section results in an effective increase of the mean free path of electrons and photons. This effectively retards the development of the electromagnetic component of the shower. It is natural to introduce an energy scale, $E_{\rm LPM}$, at which the inelasticity is low enough that the LPM effect becomes significant [35].

The experimental confirmation of the LPM effect at Stanford Linear Accelerator Center (SLAC) [36] has motivated new analyses of its consequences in cosmic ray physics [37,38,39,40]. The most evident signatures of the LPM effect on shower development are a shift in the position of the shower maximum $X_{\max}$ and larger fluctuations in the shower development.

Since the upper atmosphere is very thin the LPM effect becomes noticeable only for photons and electrons with energies above $E_{\rm LPM} \sim 10^{10}$ GeV. For baryonic primaries the LPM effect does not become important until the primary energy exceeds 10^{12}GeV. To give a visual impression of how the LPM effect slows down the initial growth of high energy photon-induced showers, we show the average longitudinal shower development of 10^{10} GeV proton and γ-ray showers (generated using AIRES 2.6.0 [41]) with and without the LPM effect in Fig. 3.

At energies at which the LPM effect is important (*viz.*, $E > E_{\rm LPM}$), γ-ray showers will have already commenced in the geomagnetic field at almost all latitudes. This reduces the energies of the primaries that reach the atmosphere, and thereby compensates the tendency of the LPM effect to retard the shower development. The first description of photon interactions in the geomagnetic field dates back at least as far as 1966 [42], with a punctuated revival of activity in the early 1980's [43]. More recently, a rekindling of interest in the topic has led to refined calculations [44,45]. Primary photons with energies above 10^{10} GeV convert into e^+e^- pairs, which in turn emit synchrotron photons. Regardless of the primary energy, the spectrum of the resulting photon "preshower" entering the upper atmosphere extends over several decades below the primary photon energy, and is peaked at energies below 10^{10} GeV [44]. The geomagnetic cooling thus switches on at about the same energy at which the LPM effect does, and thereby preempts the LPM-related observables which would otherwise be evident.

The relevant parameter to determine both conversion probability and synchrotron emission is $E \times B_\perp$, where E is the γ-ray energy and $B_\perp$ the transverse magnetic field. This leads to a large directional and geographical dependence of shower observables. Thus, each experiment has its own preferred direction for identifying primary gamma rays.

3.1. *Electron lateral distribution function*

The transverse development of electromagnetic showers is dominated by Coulomb scattering of charged particles off the nuclei in the atmosphere. The lateral development in electromagnetic cascades in different materials scales well with the Molière radius $r_{\rm M} = E_s X_0/\epsilon_0$, which varies inversely with the density of the medium,$r_{\rm M} = r_{\rm M}(h_{\rm OL}) \frac{\rho_{\rm atm}(h_{\rm OL})}{\rho_{\rm atm}(h)} \simeq \frac{9.0 \text{ g/cm}^2}{\rho_{\rm atm}(h)}$, where $E_s \approx 21$ MeV and the subscript OL indicates a quantity taken at a given observation level.

Approximate calculations of cascade equations in three dimensions to derive the lateral structure function for a pure electromagnetic cascade in vertical showers were obtained by Nishimura and Kamata [51], and later worked out by Greisen [52] in the well-known NKG formula,

$$\rho(r) = \frac{N_e}{r_{\rm M}^2} C \left(\frac{r}{r_{\rm M}}\right)^{s_{\rm NKG}-2} \left(1 + \frac{r}{r_{\rm M}}\right)^{s_{\rm NKG}-4.5}, \tag{8}$$

where N_e is the total number of electrons, r is the distance from the shower axis. For a primary of energy E_0, the so-called "age parameter",

$s_{\rm NKG} = 3/(1 + \frac{2\ln(E_0/\epsilon_0)}{t})$, characterizes the stage of the shower development in terms of the depth of the shower in radiation lengths, *i.e.*, $t = \int_z^\infty \rho_{\rm atm}(z)\, dz/X_0$.

The NKG formula may also be extended to describe showers initiated by baryons [53]. In such an extension, one finds a deviation of behavior of the Molière radius when using a value of the age parameter which is derived from theoretical predictions for pure electromagnetic cascades. It is possible to generalize the NKG formula for the electromagnetic component of baryon-induced showers by modifying the exponents in Eq. (8) [53]. The derived NKG formula provides a good description of the e^+e^- lateral distribution at all stages of shower development for values of r sufficiently far from the hadronic core. Fortunately, this is the experimentally interesting region, since typical ground arrays can only measure densities at $r > 100$ m from the shower axis, where detectors are not saturated. It should be mentioned that an NKG-like formula can be used to parametrize the total particle's density observed in baryon-induced showers [54].

In the case of inclined showers, one normally analyzes particle densities in the plane perpendicular to the shower axis. Simply projecting distributions measured at the ground into this plane is a reasonable approach for near-vertical showers, but is not sufficient for inclined showers. In the latter case, additionally asymmetry is introduced because of both unequal attenuation of the electromagnetic components arriving at the ground earlier than and later than the core [53]. Moreover, deflections on the geomagnetic field become important for showers inclined by more than about 70°.

In the framework of cascade theory, any effect coming from the influence of the atmosphere should be accounted as a function of the slant depth t [51]. Following this idea, a LDF valid at all zenith angles $\theta < 70°$ can be determined by considering

$$t'(\theta, \zeta) = t \sec\theta \, (1 + K \cos\zeta)^{-1} , \tag{9}$$

where ζ is the azimuthal angle in the shower plane, $K = K_0 \tan\theta$, and K_0 is a constant extracted from the fit [53,55]. Then, the particle lateral distributions for inclined showers $\rho(r, t')$ are given by the corresponding vertical LDF $\rho(r, t)$ but evaluated at slant depth $t'(\theta, \zeta)$ where the dependence on the azimuthal angle is evident.

For zenith angles $\theta > 70°$, the surviving electromagnetic component at ground is mainly due to muon decay and, to a much smaller extent, hadronic interactions, pair production and bremsstrahlung. As a result the lateral distribution follows that of the muon rather closely. In Fig. 3

the longitudinal development of the muon and electron components are shown. It is evident from the figure that for very inclined showers the electromagnetic development is due mostly to muon decay [57,56].

4. The Muon Component

The muonic component of EAS differs from the electromagnetic component for two main reasons. First, muons are generated through the decay of cooled charged pions, and thus the muon content is sensitive to the initial baryonic content of the primary particle. Furthermore, since there is no "muonic cascade", the number of muons reaching the ground is much smaller than the number of electrons. Specifically, there are about 5×10^8 muons above 10 MeV at ground level for a vertical 10^{11} GeV proton induced shower. Second, the muon has a much smaller cross section for radiation and pair production than the electron, and so the muonic component of EAS develops differently than does the electromagnetic component. The smaller multiple scattering suffered by muons leads to earlier arrival times at the ground for muons than for the electromagnetic component.

The ratio of electrons to muons depends strongly on the distance from the core; for example, the e^+e^- to $\mu^+\mu^-$ ratio for a 10^{11} GeV vertical proton shower varies from 17 to 1 at 200 m from the core to 1 to 1 at 2000 m. The ratio between the electromagnetic and muonic shower components behaves somewhat differently in the case of inclined showers. For zenith angles greater than 60°, the $e^+e^-/\mu^+\mu^-$ ratio remains roughly constant at a given distance from the core. As the zenith angle grows beyond 60°, this ratio decreases, until at $\theta = 75°$, it is 400 times smaller than for a vertical shower. Another difference between inclined and vertical showers is that the average muon energy at ground changes dramatically. For horizontal showers, the lower energy muons are filtered out by a combination of energy loss mechanisms and the finite muon lifetime: for vertical showers, the average muon energy is 1 GeV, while for horizontal showers it is about 2 orders of magnitude greater.

High energy muons lose energy through e^+e^- pair production, muon-nucleus interaction, bremsstrahlung, and knock-on electron (δ-ray) production [58]. The first three processes are discrete in the sense that they are characterized by high inelasticity and a large mean free path. On the other hand, because of its short mean free path and its small inelasticity, knock-on electron production can be considered a continuous process. The muon bremsstrahlung cross section is suppressed by a factor of $(m_e/m_\mu)^2$

with respect to electron bremsstrahlung, see Eq. (7). Since the radiation length for air is about 36.7 g/cm^2, and the vertical atmospheric depth is 1000 g/cm^2, muon bremsstrahlung is of negligible importance for vertical air shower development. Energy loss due to muon-nucleus interactions is somewhat smaller than muon bremsstrahlung. Energy loss by pair production is slightly more important than bremsstrahlung at about 1 GeV, and becomes increasingly dominant with energy. Finally, knock-on electrons have a very small mean free path, but also a very small inelasticity, so that this contribution to the energy loss is comparable to that from the hard processes.

In addition to muon production through charged pion decay, photons can directly generate muon pairs, or produce hadron pairs which in turn decay to muons. In the case of direct pair production, the large muon mass leads to a higher threshold for this process than for electron pair production. Furthermore, QED predicts that $\mu^+\mu^-$ production is suppressed by a factor $(m_e/m_\mu)^2$ compared the Bethe-Heitler cross section. The cross section for hadron production by photons is much less certain, since it involves the *hadronic structure* of the photon. This has been measured at HERA for photon energies corresponding to $E_{\rm lab} = 2 \times 10^4$ GeV. This energy is still well below the energies of the highest energy cosmic rays, but nonetheless, these data do constrain the extrapolation of the cross sections to high energies.

The muon content of the shower tail is quite sensitive to unknown details of hadronic physics. This implies that attempts to extract composition information from measurements of muon content at ground level tend to be systematics dominated. The muon LDF is mostly determined by the distribution in phase space of the parent pions. However, the pionization process together with muon deflection in the geomagnetic field obscures the distribution of the first generation of pions. A combination of detailed simulations, high statistics measurements of the muon LDF and identification of the primary species using uncorrelated observables could shed light on hadronic interaction models.

Acknowledgments

I would like to thank the organizers for the financial support and warn hospitality. I am grateful to L. Anchordoqui, A. Mariazzi, T. McCauley, T. Paul, S. Reucroft and J. Swain for providing a very productive and agreeable working atmosphere in order to write our review article.

References

1. L. Anchordoqui, M. T. Dova, A. Mariazzi, T. McCauley, T. Paul, S. Reucroft and J. Swain. *Ann. Phys.* **314**, 145 (2004) [arXiv:hep-ph/0407020].
2. A. Capella, U. Sukhatme, C. I. Tan and J. Tran Thanh Van, *Phys. Rept.* **236**, 225 (1994).
3. T. K. Gaisser and F. Halzen, *Phys. Rev. Lett.* **54**, 1754 (1985).
4. G. Pancheri and C. Rubbia, *Nucl. Phys. A* **418**, 117C (1984).
5. C. Albajar *et al.* [UA1 Collaboration], *Nucl. Phys. B* **309**, 405 (1988).
6. C. Adloff *et al.* [H1 Collaboration], *Eur. Phys. J. C* **21**, 33 (2001), *Phys. Rev. D* **59**, 074022 (1999) [arXiv:hep-ph/9805268].
7. H. Abramowicz and A. Caldwell, *Rev. Mod. Phys.* **71**, 1275 (1999) [arXiv:hep-ex/9903037].
8. R. Engel, *Nucl. Phys. Proc. Suppl.* **122**, 40 (2003).
9. J. Kwiecinski and A. D. Martin, *Phys. Rev. D* **43**, 1560 (1991).
10. R. J. Glauber and G. Matthiae, *Nucl. Phys. B* **21**, 135 (1970).
11. L. Durand and H. Pi, *Phys. Rev. D* **38**, 78 (1988).
12. T. K. Gaisser and T. Stanev, *Phys. Lett. B* **219**, 375 (1989).
13. R. S. Fletcher, T. K. Gaisser, P. Lipari and T. Stanev, *Phys. Rev. D* **50**, 5710 (1994).
14. N. N. Kalmykov, S. S. Ostapchenko and A. I. Pavlov, *Nucl. Phys. Proc. Suppl.* **52B**, 17 (1997).
15. F. Abe *et al.* [CDF Collaboration], *Phys. Rev. D* **56**, 3811 (1997).
16. L. A. Anchordoqui, M. T. Dova, L. N. Epele and S. J. Sciutto, *Phys. Rev. D* **59**, 094003 (1999) [arXiv:hep-ph/9810384].
17. L. A. Anchordoqui, M. T. Dova and S. J. Sciutto, Proc. 26th International Cosmic Ray Conference, Salt Lake City **1**, 147 (1999) [arXiv:hep-ph/9905248].
18. J. Alvarez-Muniz, R. Engel, T. K. Gaisser, J. A. Ortiz and T. Stanev, *Phys. Rev. D* **66**, 033011 (2002) [arXiv:astro-ph/0205302].
19. R. Engel and H. Rebel, *Acta Phys. Polon. B* **35**, 321 (2004).
20. U. Amaldi and K. R. Schubert, Nucl. Phys. B **166**, 301 (1980).
21. M. M. Block, F. Halzen and T. Stanev, *Phys. Rev. D* **62**, 077501 (2000) [arXiv:hep-ph/0004232].
22. M. Honda *et al.*, *Phys. Rev. Lett.* **70**, 525 (1993).
23. R. M. Baltrusaitis *et al.*, *Phys. Rev. Lett.* **52**, 1380 (1984).
24. J. Ranft, *Phys. Rev. D* **51**, 64 (1995).
25. R. Engel, T. K. Gaisser, T. Stanev and P. Lipari, Proc. 26th International Cosmic Ray Conference, Salt Lake City **1**, 415 (1999).
26. D. Heck, Proc. 27th International Cosmic Ray Conference, Hamburg, 233 (2001).
27. S. Roesler, R. Engel and J. Ranft, arXiv:hep-ph/0012252.
28. T. Sjostrand, *Int. J. Mod. Phys. A* **3**, 751 (1988).
29. B. Rossi and K. Greisen, *Rev. Mod. Phys.* **13**, 240 (1941).
30. T. K. Gaisser, *Cosmic Rays and Particle Physics,* (Cambridge University Press, 1990).

31. H. Bethe and W. Heitler, *Proc. Roy. Soc. Lond. A* **146**, 83 (1934).
32. Y. S. Tsai, *Rev. Mod. Phys.* **46**, 815 (1974) [Erratum *Rev. Mod. Phys.* **49**, 421 (1977)].
33. L. D. Landau and I. Pomeranchuk, *Dokl. Akad. Nauk Ser. Fiz.* **92**, 535 (1953).
34. A. B. Migdal, *Phys. Rev.* **103**, 1811 (1956).
35. T. Stanev, C. Vankov, R. E. Streitmatter, R. W. Ellsworth and T. Bowen, *Phys. Rev. D* **25**, 1291 (1982).
36. S. Klein, *Rev. Mod. Phys.* **71**, 1501 (1999) [arXiv:hep-ph/9802442].
37. J. Alvarez-Muniz and E. Zas, *Phys. Lett. B* **434**, 396 (1998) [arXiv:astro-ph/9806098].
38. A. N. Cillis, H. Fanchiotti, C. A. Garcia Canal and S. J. Sciutto, *Phys. Rev. D* **59**, 113012 (1999) [arXiv:astro-ph/9809334].
39. J. N. Capdevielle, C. Le Gall and K. N. Sanosian, *Astropart. Phys.* **13**, 259 (2000).
40. A. V. Plyasheshnikov and F. A. Aharonian, *J. Phys. G* **28**, 267 (2002) [arXiv:astro-ph/0107592].
41. S. J. Sciutto, arXiv:astro-ph/9911331.
42. T. Erber, *Rev. Mod. Phys.* **38**, 626 (1966).
43. B. Mcbreen and C. J. Lambert, *Phys. Rev. D* **24**, 2536 (1981).
44. X. Bertou, P. Billoir and S. Dagoret-Campagne, Astropart. Phys. **14**, 121 (2000).
45. P. Homola *et al.*, arXiv:astro-ph/0311442.
46. W. Heitler. *The Quantum Theory of Radiation*, 2nd. Edition, (Oxford University Press, London, 1944).
47. D. Heck, G. Schatz, T. Thouw, J. Knapp and J. N. Capdevielle, FZKA-6019 (1998).
48. J. Knapp, D. Heck, S. J. Sciutto, M. T. Dova and M. Risse, *Astropart. Phys.* **19**, 77 (2003) [arXiv:astro-ph/0206414].
49. H. J. Drescher and G. R. Farrar, *Phys. Rev. D* **67**, 116001 (2003) [arXiv:astro-ph/0212018].
50. H. J. Drescher and G. R. Farrar, *Astropart. Phys.* **19**, 235 (2003) [arXiv:hep-ph/0206112].
51. K. Kamata and J. Nishimura,*Progr. Theor. Phys. Suppl.* **6**, 93 (1958).
52. K. Greisen, *Ann. Rev. Nucl. Sci.* **10**, 63 (1960).
53. M. T. Dova, L. N. Epele and A. G. Mariazzi, *Astropart. Phys.* **18**, 351 (2003) [arXiv:astro-ph/0110237].
54. M. Roth [AUGER Collaboration], arXiv:astro-ph/0308392.
55. M. T. Dova, L. N. Epele and A. Mariazzi, *Nuovo Cim.* **24C**, 745 (2001).
56. A. Cillis and S. J. Sciutto, *J. Phys. G* **26**, 309 (2000).
57. M. Ave, R. A. Vazquez, E. Zas, J. A. Hinton and A. A. Watson, *Astropart. Phys.* **14**, 109 (2000) [arXiv:astro-ph/0003011].
58. A. N. Cillis and S. J. Sciutto, *Phys. Rev. D* **64**, 013010 (2001) [arXiv:astro-ph/0010488].

FIRST RESULTS FROM THE MAGIC EXPERIMENT

ALESSANDRO DE ANGELIS*
Dipartimento di Fisica dell'Università di Udine and INFN
Via delle Scienze 208, I-33100 Udine (Italia)
E-mail: deangelis@fisica.uniud.it

AND THE MAGIC COLLABORATION

With its diameter of 17m, the MAGIC telescope is the largest Cherenkov detector for gamma ray astrophysics, and it has the lowest energy threshold (well below 100 GeV). MAGIC started operations in October 2003 and is currently taking data. This report summarizes its main characteristics, its first results and its potential for physics.

1. Introduction

The MAGIC (Major Atmospheric Gamma Imaging Cherenkov) telescope has been designed [1] with the aim to achieve the lowest possible gamma energy threshold with a ground-based gamma IACT (Imaging Air Cherenkov Telescope). A clear-case physics motivation was the search for gamma ray sources in the unexplored energy range between 30 GeV and 300 GeV [2] to study the apparent discrepancy between the well populated sky-map of sources below 10 GeV observed by EGRET (more than half of them still unidentified) and the only handful of sources discovered by IACTs above 300 GeV. The observation program of MAGIC includes several galactic and extragalactic types of sources such as Supernova Remnants (SNR), pulsars, microquasars and Active Galactic Nuclei (AGNs). The low energy threshold allows MAGIC to extend the observation of extragalactic sources up to z∼1 and beyond, which is impossible at higher energies due to the interaction of gammas with the extragalactic background light. Another unique feature of MAGIC is the fast repositioning time of the telescope that allows to observe gamma ray bursts (GRBs) within 25 seconds after their detection by satellite detectors. Besides, MAGIC has a huge potential for

*Also at IST, Lisboa.

studies related to fundamental physics (search for dark matter, study of anomalous dispersion relations for photons such as prediced by quantum gravity inspired models and by models in which the Lorentz symmetry is violated).

MAGIC incorporates many technological innovations in order to fulfill the requirements imposed by the physics goals [3]. Cost and prototyping considerations led to the decision to construct, as a first step, a single large telescope incorporating the latest technological developments.

Figure 1. *The MAGIC telescope in August 2004.*

2. Description and status of the telescope

MAGIC is installed at 2200 m a.s.l. on the Canary Island of La Palma. The construction was finished in October 2003, with a few modifications during the commissioning phase in August 2004. With a 17m diameter high reflectivity mirror dish and a high quantum efficiency PMT camera MAGIC is designed for an energy threshold as low as 30 GeV.

2.1. *The technological novelties*

In order to make the construction of a very large telescope realistic and to meet the price-performance requirements, one needs to develop new techniques and technologies. Listed below are the most important technological novelties developed by the collaboration, extensively tested over the past years and finally used in the construction of MAGIC.

- The square shaped mirrors of MAGIC with a side length of 49.5 cm are entirely made of Al and include internal heating (with a power selectable between 12 and 50 W). The mirror front surface is polished by using a diamond cutter providing a beam divergence of $1'$. A single mirror tile weighs $\sim$4.2 kg, and a quartz layer deposited on its front surface is protecting it from weathering.
- Every four mirror tiles are fixed on a carrying panel of $\sim 1\text{m}^2$ size and are optically adjusted to act as a single piece 1m^2 mirror. The reflector of MAGIC includes 241 panels of the above-mentioned type. The panels are attached to stepping motor-based electromechanical actuators (steered by local micro-controllers) that allow one to always provide their optimal orientation in spite of deformations of the reflector structure (gravitational loads are changing) when rotating. In the current design of MAGIC we are using semiconductor laser pointers working at a wavelength of about 650 nm in the active mirror control (AMC) system. These lasers are fixed at the mirror panel centers, and their light spots are adjusted to show the location of the reflected light spots of the corresponding mirrors in the focal plane. A CCD camera is fixed close to the centre of the reflector that measures the differences in the orientation of different laser pointers from their ideal positions and sends correction signals to the corresponding panel steering micro-controllers that will drive the motors.
- In order to provide low weight, inertia and to limit thermal expansion, the three-layer reflector space frame is made from reinforced carbon-fiber tubes and universal Al spherical joints.
- 6-dynode ultra-fast hemispherical PMTs with enhanced quantum efficiency have been especially designed for the needs of MAGIC.
- Light guides of a special form, well matching the hemispherical shape of the PMT input window, are designed to maximize the double hit chance of the semi-transparent photo cathode by incident light. These provide an increase in the photon detection

efficiency.

- A special "milky" wavelength shifting coating of PMT windows provides a substantial increase in quantum efficiency.
- Ultra-fast analog signal transmission by using VCSEL diodes and optical fibers helps, in spite of the 162 m fiber length from the camera to the counting house, to preserve the original shape of the ultra-fast pulse for further processing. The entire fiber bundle has significantly less weight than the solution based on coaxial cables and is immune against electromagnetic pick-up and lightening and grounding problems.
- A 10-layer printed circuit motherboard in the camera is used to distribute the control and supply voltages, and simultaneously serves as a mechanical support for holding the pixels.
- The heat released inside the camera by the PMTs and electronics (~600 W) is transported away by circulating water through its case. A closed loop regulated thermostat located on one of the bogeys of the telescope allows one to control the temperature inside the camera.
- The 3 level trigger provides tight time coincidence (equivalent to a 5 ns gate) and strong rejection of different backgrounds on hardware level. While the 1st level does simple discrimination, the 2nd level provides a programmable next neighbour trigger logic and the 3rd level allows one to trigger on given patterns.

A short summary of the main parameters of the telescope is shown in the following table:

Mount type:	alt-azimuth
Total weight:	65 tons
Re-positioning time to an arbitrary Direction in the sky:	22 s
Reflector diameter, area, shape:	17 m, 239 m², parabolic
Reflector optics:	F/D = 1
Reflector frame structure:	3-layer reinforced carbon-fibre space frame
Number of mirror tiles, Mean reflectivity, angular resolution:	940 85 % 1'
Camera field of view:	3.9°
Pixel sizes:	397 central pixels of 0.10° 180 outer pixels of 0.20°
PMTs:	6 dynode enhanced bialkali hemispherical ET 9116 (1") and ET 9117 (1.5")
DAQ:	300 MSample/s FADC readout
DAQ dynamic range:	effective 10 bits
DAQ event rate capability:	~1 kHz-sustained
Average trigger rate at current threshold:	~ (250-300) Hz near zenith from a dark region

The sensitivity of MAGIC as calculated from Monte Carlo is shown together with the expected sensitivity from other gamma-ray detectors in the GeV-TeV range in Fig. 2.

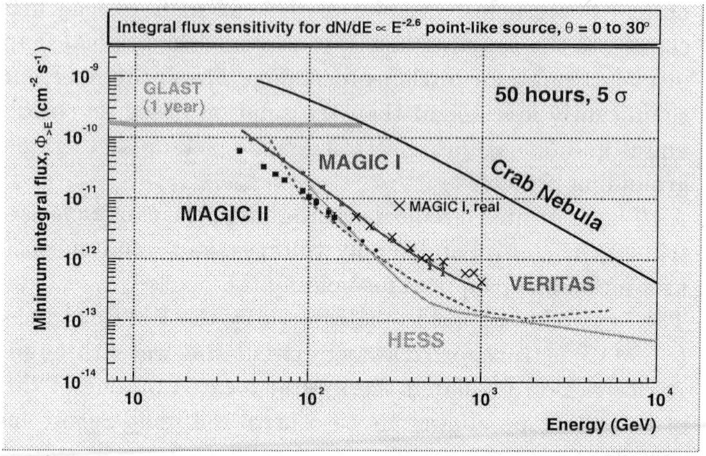

Figure 2. *Sensitivities for some operating and proposed gamma detectors.*

3. First observations of sources and performance on data

3.1. *Simple gamma/hadron separation applied to the data from Crab Nebula*

Below we show the currently achieved sensitivity of the telescope with the example of the analysis of a small sample of data, in total 110 min observation time, taken from Crab Nebula in September 2004. The variables used for the discrimination are the classical Hillas parameters [4]. The Random Forest method [5], after short training on Monte Carlo gammas and on the measured OFF data, is applied to the Crab data. An image cleaning of 3 and 2 sigma is applied correspondingly to the core and the boundary pixels; at least 5 core pixels are selected. Even this very simple, very straightforward analysis reveals the strong potential of the telescope, with an efficiency of around 20σ/hour (Fig. 3).

A 5 sigma signal, not yet optimized, is present even in the lowest SIZE bin Corresponding to energies between 50 and 75 GeV, calculated from a calibration on data at the 400 GeV range (Fig. 4). The total rate of the excess gamma rays is $\sim$ 0.5Hz after the analysis. Our Monte Carlo simulations confirm a threshold energy of the telescope of 50 GeV after analysis.

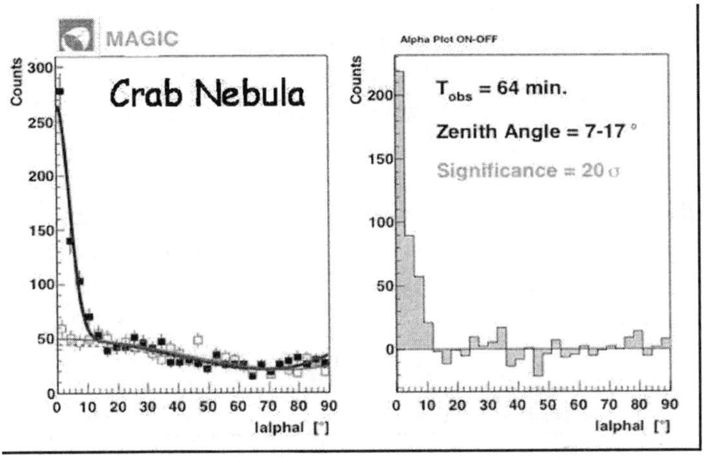

Figure 3. *Results of an analysis of the data from the Crab Nebula. It is shown that MAGIC achieves a sensitivity of around* 20σ *in an hour (3 to 4 minutes are thus needed for a* 5σ *sensitivity).*

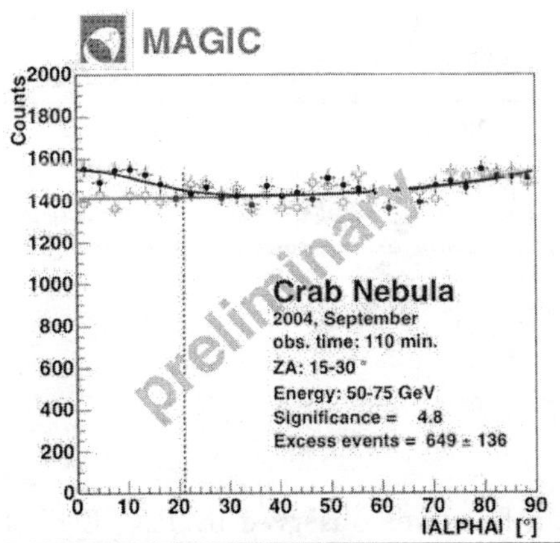

Figure 4. *ALPHA plot for Crab in the energy region between 50 GeV and 75 GeV. A signal is seen.*

The energy spectrum between 50 GeV and 3 TeV is shown in Fig. 5. This is the first time that the low part of the GeV spectrum is measured from the Crab Nebula. The hardening of the spectrum of Crab Nebula when lowering the energies, predicted some 12 years ago by theoreticians (Harding, De Jager, 1992) for the Synchrotron Self-Compton (SSC) mechanism of gamma production, can contribute to degradation of the signal to noise ratio.

A 2-d image of the source is shown in Fig. 6.

The results of the above shown straightforward analysis could be further improved by using other analysis methods (work is in progress) as well as by using more parameters and better cuts for a better signal to noise ratio. Especially in the very low energy region (low SIZE cut) one may anticipate a strong improvement. Our first reconstruction of the energy spectrum in the high energy part is in a reasonably good agreement with the results of other experiments.

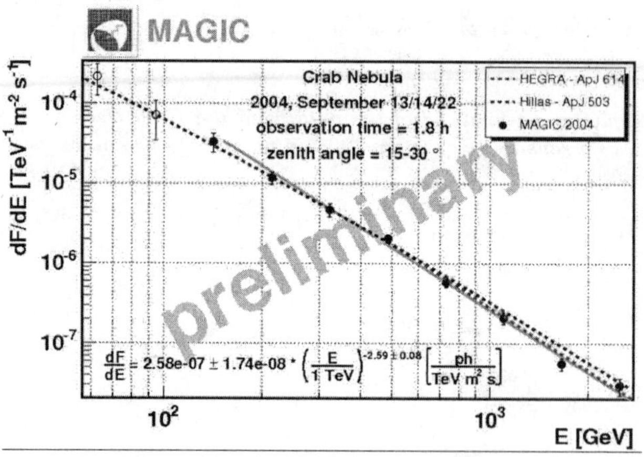

Figure 5. *The energy spectrum of Crab Nebula measured by MAGIC. Errors are statistical only. We are progressing on the calculation of the efficiency related to the two lowest energy points.*

3.2. *Mkr421*

In December 2004, a flare was observed by RXTE in the AGN Markarian 421. The source was also observed simultaneously by MAGIC and HESS; such simultaneous observations will allow in the future, when systematically done, cross-calibration of the telescopes.

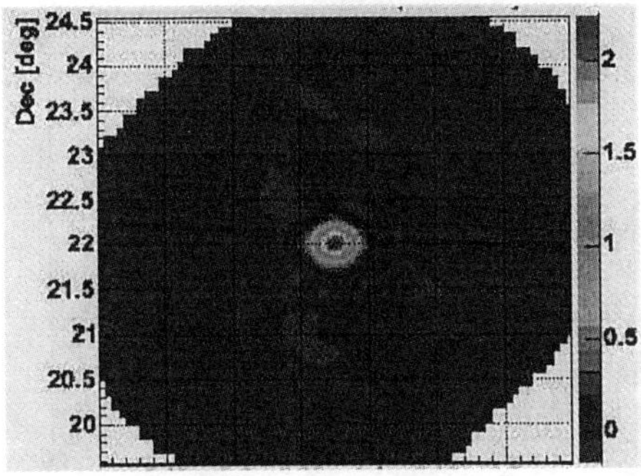

Figure 6. *Two-dimensional plot of the signal from the Crab nebula, showing an angular RMS resolution of* 0.1° *on the position of a point source.*

The signal seen by MAGIC simultaneously with HESS is shown in Fig. 7. A light curve can be studied, by slicing the signal in intervals as small as 2 min.

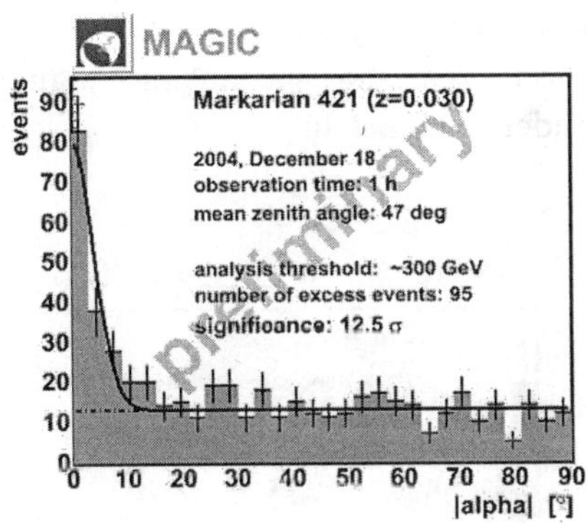

Figure 7. *Signal from Markarian 421 (see text).*

3.3. *1ES1959+650*

1ES1959 is an AGN at $z \sim 0.05$ previously observed by HEGRA in 150 h, and by Whipple. It is as faint as 0.1 Crab. A 7σ detection was possible with 6.5 hours of observation (Fig. 8).

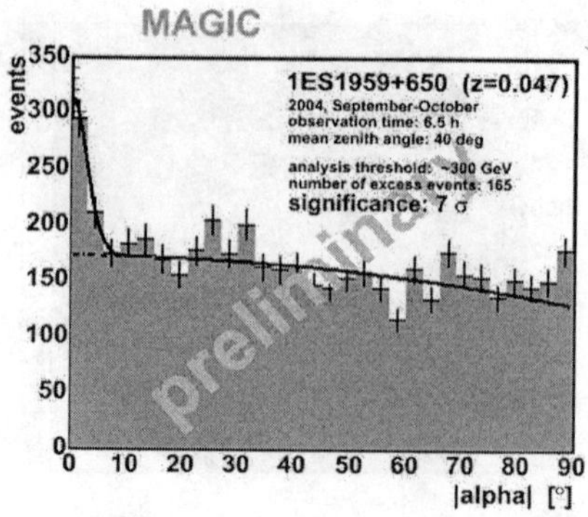

Figure 8. *Results of an analysis of the data from the AGN 1ES1959+650.*

3.4. *The galactic center*

The conditions for observing the Galactic Center from La Palma are difficult, being the zenith angle larger than 58°. We estimate that this reflects on an effective threshold of the order of 1 TeV.

Nevertheless, a 3 hours test shows a hint of a signal (Fig. 9), for which the systematics are under evaluation.

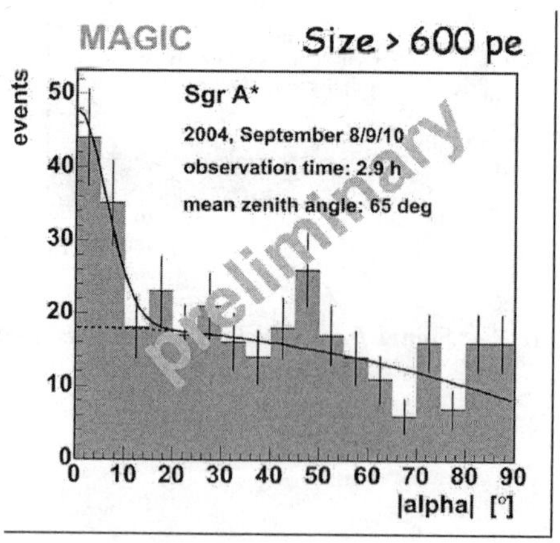

Figure 9. *Results of an analysis of the data from the Galactic Center.*

4. Expected improvements in the performance

Although we are working on improving many parameters of the telescope, we think that the performance of MAGIC I is essentially understood.

4.1. *Better sensitivity and precision with two telescopes*

As it is now established in the TeV energy regime adding up of more telescopes (the so-called stereo regime) is increasing the sensitivity of the installation roughly as $\sim \sqrt{N}$ where N is the number of telescopes, when pointing to the same source. In the sub-100 GeV energy regime one shall probably expect that some other aspect of the multi-telescope approach, namely the coincidence character, will play a more dominant role, allowing one to lower the threshold setting in spite of strong backgrounds.

As a next step the MAGIC collaboration is constructing the second MAGIC telescope, at some 80m distance from the first MAGIC telescope. The construction has already been started, and the second telescope shall be completed in 2007. The coincident / stereo operation of the two telescopes will provide two times higher sensitivity.

5. Conclusions

The MAGIC telescope has started its regular operation. Gamma ray signals from strong sources have first been measured (outbursts of Mrk421, Crab Nebula); a signal from the Galactic Center has been observed; a source of an intensity of 0.1 Crab, 1ES1959, has been recorded. Currently we are calibrating our threshold setting and the sensitivity by observing the Crab Nebula. The comparison of the measured data with the Monte Carlo simulations indicates that we are working currently at a threshold setting of $\simeq$50 GeV.

The construction of a second detector (MAGIC II) to be put at 80 m from MAGIC has started, and completion is expected in 2006.

References

1. Barrio J. et al., "The MAGIC Telescope Design report", MPI Institute Report MPI-PhE/98-5 (March 1998)
2. For reviews, see for example Hoffman C. M., C. Sinnis, P. Fleury and M. Punch, Rev. Mod. Phys. 71 (1999) 897; Aharonian F.A., "Very High Energy Cosmic Gamma Radiation", World Scientific 2004
3. Baixeras C. et al., NIM A518 (2004) 188
4. Weekes T. C. et al., Astrophys. J. 342 (1989) 379
5. Bock R. et al., NIM A516 (2004) 511

HOW TO SELECT UHECR IN EUSO – THE TRIGGER SYSTEM

PEDRO ASSIS†

LIP

Av. Elias Garcia, 14-1,

1000-149 Lisboa, Portugal

EUSO is a space mission to detect Ultra High Energy Cosmic Rays (UHECR) by collecting the Ultra-Violet light generated by UHECR in its interaction with Earth's atmosphere. The online discrimination of events from the background requires several triggering levels and sophisticated algorithms. In this paper the trigger system is described.

1. The EUSO Mission

EUSO Extreme Universe Space Observatory [1] is a space based fluorescence detector for UHECR detection that will observe an area of ~150000 km^2 and is expected to collect ~1000 events with an energy greater than 10^{20}eV. The EUSO detector (Fig. 1) consists of an UV telescope to be installed at the International Space Station (ISS) pointing at nadir. Its detection principle (Fig. 2) relies on the detection of the fluorescence light and the reflected Čerenkov light generated by an EAS traversing Earth atmosphere.

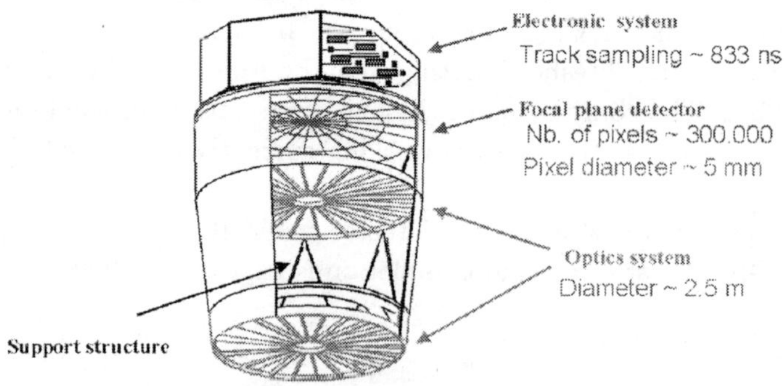

Figure 1. The EUSO instrument

† Work supported by FCT grant SFRH//BD/14226/2003.

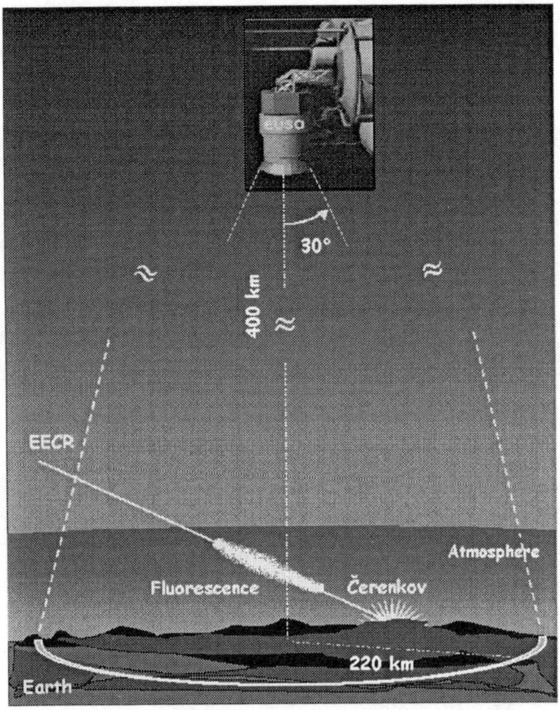

Figure 2. EUSO detection principle

2. Trigger System Architecture

The EUSO trigger has been designed to provide a good discrimination of shower events from the background for energies as low as possible (~$5x10^{19}$eV). High energetic shower events are clearly distinguishable from the background and a simple signal persistency trigger can be used. For lower energetic events (<10^{20}eV) the signal to noise ratio is reduced and the detection of the signal must rely on the identification of its properties, namely the space-time patterns, that differ a lot for events and for background.

The Euso trigger system relies basically in three trigger levels. The two lower level triggers analyse the persistency, with time, of the signal at pixel level and at cell level* using low thresholds. When these levels issue a trigger alert the third level trigger is asked to validate the event.

* In the EUSO detector the pixels are grouped in n x n pixel cells. Cells are independent from each other.

The third level trigger uses an online, real time, pattern recognition algorithm (Track Find Algorithm) which is discussed in the following section.

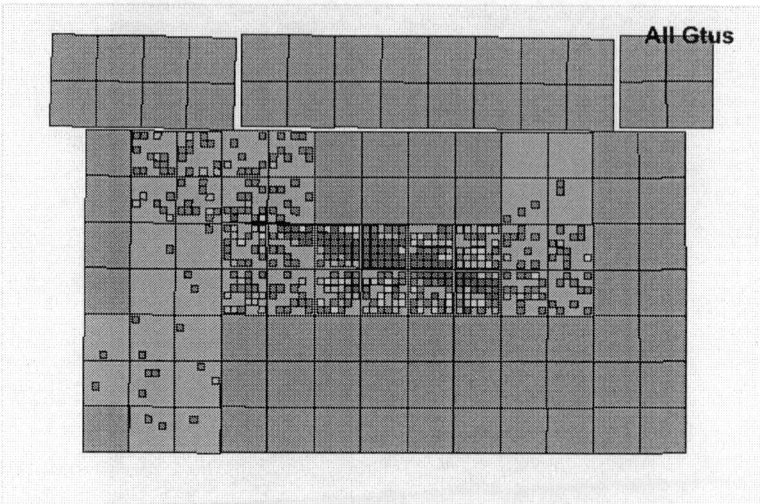

Figure 3. A shower event registered in the EUSO focal surface

3. Track Find Algorithm

The "track finding" algorithm has been conceived for minimizing the fake rate triggers in the EUSO experiment. In EUSO the pixels are grouped in independent cells and there are two registers that contain the projection of the signals in the X and Y directions (X wired-OR and Y wired-OR). The pixel signal is integrated in a Gate Time Unit (GTU) of ~2μs. The algorithm is based on a very simple approach and uses the available information residing in the X and Y wired-or memories and performing simple Boolean operations (AND, OR and logic shift) so that it can be easily implemented in a FPGA-like device.

At the reception of an alert trigger, the first location of the X memory is ANDed with the second location of the memory, i.e. X wired-or information for GTU=1 and GTU=2, and the result stored in a register C with the same length in bits (36 bits). The third memory location will be ANDed with C and the result stored in the C register. In case register C is zero the second location is ANDed with the third and the result put on C and so on until reaching GTU=m.

In parallel the first location is ANDed with the shifted, one position on the left, value of the second location and the result stored in a register L, as like as before the operation will continue iteratively storing the result always in the register L.

The same procedure is applied right-shifting the registers, storing the result in register R.

Every time any of the C, L, R (ORed) registers are not equal to zero, the position n of the corresponding location (n=0; to 19, if 20 locations are used for the memory) is stored in the corresponding bit of a XT register.

The same operations are done in parallel for the Y Memory. At the end also for the Y memory there will be a register YT holding the track of the position n of the corresponding location where the AND conditions were fulfilled.

As final operation the XT and YT registers are ANDed and the result counted for the number of consecutive bits set to 1, that give the condition of detected track-like if the number of consecutive "1" is greater or equal to the set persitency-1.

4. Trigger Simulation

For the EUSO mission simulation it was developed a C++ based software: ESAF - EUSO Simulation and Analysis Framework [2]. It is a software framework that will provide the framework for the whole process of data-simulations and data-analysis, from the simulation of the primary particle interaction in atmosphere, to the transport of light to the EUSO optical pupil, to the detector response simulation and, finally, to the reconstruction and the physical analysis.

ESAF is designed using Object Oriented (OO) technology to achieve the high degree of modularity and flexibility needed.

The Trigger algorithm has been already implemented in ESAF and validation tests have started.

5. Summary

The EUSO triggering scheme is presented. It consists of two signal persistency levels and a pattern recognition algorithm – Track Find algorithm. The track find algorithm has been designed to use simple logic operations so that it can be implemented in an FPGA-like device. The trigger system has been coded in the general EUSO Simulation and Analysis Framework – ESAF.

References

1. L. Scarsi, *Il Nuovo Cimento* **C24**, 471 (2001).
2. M. Pallavicini and A. Thea, EUSO-SDA-REP-014-1 (2004).

PRESSURE AND TEMPERATURE DEPENDENCE OF THE PRIMARY SCINTILLATION IN AIR*

M. FRAGA[1], A. ONOFRE[1], N. F. CASTRO[1], R. FERREIRA MARQUES[1], S. FETAL[1,3], F. FRAGA[1], M. PIMENTA[4], A. POLICARPO[1], F. VELOSO[1]

[1] *LIP- Coimbra, Dep. Física, Univ. Coimbra, 3004-516 Coimbra, Portugal*

[2] *UCP, R. Dr. Mendes Pinheiro, 24, 3080 Figueira da Foz, Portugal*

[3] *ISEC, Quinta da Nora, 3030-199 Coimbra, Portugal*

[4] *LIP-IST, Av. Elias Garcia 14, 1100-149 Lisboa, Portugal*

A study of the temperature dependence of nitrogen fluorescence is presented. Preliminary measurements of the light yield for the 0-0 band of 2nd positive system (centred at 337.1 nm) at different gas densities were performed using alpha particles as excitation source.

1. Introduction

Charged particles, produced in an ultra high energy cosmic ray shower, move downwards in the lower atmosphere and loose their energy in collisions with the N_2 and O_2 molecules. A small fraction of the energy loss (~5×10^{-5}) appears as UV light (between 300 and 420 nm) which results from the radioactive de-excitation of the N_2 $C^3\Pi_u$ molecular state (2nd positive system) [1]. This light is assumed to be emitted isotropically and proportional to the energy loss by ionization. The detection of the nitrogen scintillation (air fluorescence) is a well established technique that has been used in the study of ultra high energy cosmic rays [2,3]. An accurate knowledge of the scintillation efficiency as a function of altitude is of utmost importance for the absolute calibration of the detectors based on this technique. In this work, we present preliminary measurements of the temperature dependence of the nitrogen fluorescence.

2. Experimental set-up

The experimental set-up is represented schematically in Fig. 1. A chamber, made of stainless steel, is placed inside a cooling unit, which allows measurements

---

* This work is supported by contract POCTI/FP/FNU/50340/2003 with the Portuguese FCT.

between room temperature and -25°C. The light produced inside the chamber is collected from each side by a photomultiplier (Photonis XP2020Q), through two ½" thick fused silica windows (one directly coupled to the chamber and the other separating the PM chamber from the cooling unit, see Fig.1). One of the optical channels includes an interference filter (with a central wavelength of 340 nm and a FWHM of 10 nm) that selects the 0-0 molecular band of the 2nd positive system of N_2 [1]. The PMs operate in the counting mode and the signals from both of them are set into coincidence. The chamber is evacuated, at room temperature, with a diffusion pump for several hours prior to each gas filling. Once the working pressure is reached the entrance valve is closed. The pressure and the temperature in the gas chamber are measured, respectively, with a SETRA pressure transducer model 216 and a Pt-100 sensor from Minco EC AG. The temperature at the photocathode of the PM is also recorded with another temperature sensor. When thermal equilibrium is reached the temperature at the photocathode of the PM is about 10 degrees higher than the temperature inside the chamber. A LABVIEW program was developed for the data acquisition and the temperature control of the cooling unit. As excitation source the alpha particles from ^{241}Am are used. The source is placed outside the chamber and the radiation enters the chamber through a 12.5 μm mylar window. The distance between the ^{241}Am source and the mylar window is 6 mm.

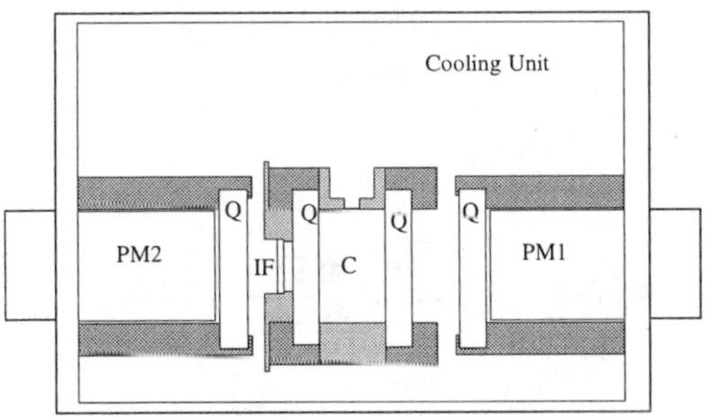

Fig. 1. Experimental set-up. C – Gas chamber; Q – fused silica window; IF – Interference filter.

3. Theoretical model

In pure nitrogen, the vibrational level v' of the N_2 $C^3\Pi_u$ molecular state can de-excite either radioactively to one of the vibrational levels of the intermediate electronic state $B^3\Pi_g$ (v´´=0,1,2,..) or by collisions with N_2 molecules on their ground state. The lifetime $\tau_{v'}$ of the vibrational level v´ is, then, given by

$$\frac{1}{\tau_{v'}}=\frac{1}{\tau_0}+k_{v'}[N_2], \qquad (1)$$

where τ_0 is the natural lifetime, $[N_2]$ is the concentration of nitrogen molecules and $k_{v'}$ is the rate coefficient for the collisions with N_2, respectively. The rate constant is related with the cross section σ for the process by

$$k_{v'}=\left(\frac{8k_BT}{\mu\pi}\right)^{1/2}\sigma \qquad (2)$$

where k_B is the Boltzmann constant, μ is the reduced mass of the molecule and T is the absolute temperature of the gas. Assuming a constant gas density (ρ), the temperature dependence of $\tau_{v'}$ can be described by

$$\tau_{v'}=\left(A_{v'}+B\sqrt{T}\right)^{-1}, \qquad (3)$$

where $A_{v'}=\frac{1}{\tau_0}$ is the Einstein coefficient for the v' level and

$$B=\left(\frac{8k_B}{\mu\pi}\right)^{1/2}\sigma\,[N_2] \qquad (4)$$

The intensity of the molecular band (v´-v´´) is given by

$$Y_{v'v''}=A_{v'v''}\tau_{v'}(\rho,T)\;\varepsilon(\lambda)\int_0^R\frac{\Omega(x)}{4\pi}T(x)N_{v'}^{exc}(\rho)\frac{dE}{dx}dx \qquad (5)$$

where $A_{v'v''}$ is the Einstein coefficient for the v´-v´´ transition, $\varepsilon(\lambda)$ is a factor that takes into account the transmission of the fused silica windows and the detection efficiency of the photomultiplier, $\Omega(x)$ is the solid angle of photon collection, $T(x)$ is the transmission of the interference filter (that depends on the incidence angle), $N_{v'}^{exc}(\rho)$ is the number of excitations of the v' vibrational state, per alpha particle, per unit energy and dE/dx and R are the stopping power and range of the alpha particles, respectively. Since the gas density is kept constant during a cooling cycle, the light yield is expected to increase as the temperature decreases according with Eq. 3.

4. Results and analysis

Measurements of light yields were performed, as a function of temperature, for the 0-0 band, centered at 337.1 nm, for different gas densities. Preliminary results are shown in Fig. 2. The represented data are only corrected for the accidental coincidences during the counting period. The decrease of the light yield with the temperature, for the higher gas densities, can be attributed to the fact that the alpha particle source is outside the chamber: as the temperature goes down, the density of air inside the cooling unit increases (P=constant) and, therefore, the energy loss of the alpha particles before entering the mylar window also increases. Consequently, the alpha particles enter the gas with a lower energy (around 4% lower for $\rho=0.8\rho_o$, ρ_o being the density for atmospheric pressure and room temperature) and since the alpha particle ranges are shorter, the solid angle of photon collection also decreases. The simulation of the chamber, based the GEANT4 tool packet, is under way, in order to estimate these corrections, for the different gas densities and for the different temperatures. Preliminary results seem to indicate that, once the geometrical corrections are introduced, all data exhibit a behaviour similar to the one measured for $\rho=0.43\rho_o$ (see Fig.2). However, other effects such as the variation of the gain and/or the efficiencies of the photocathodes of the PMs with temperature, may also be important and are under study.

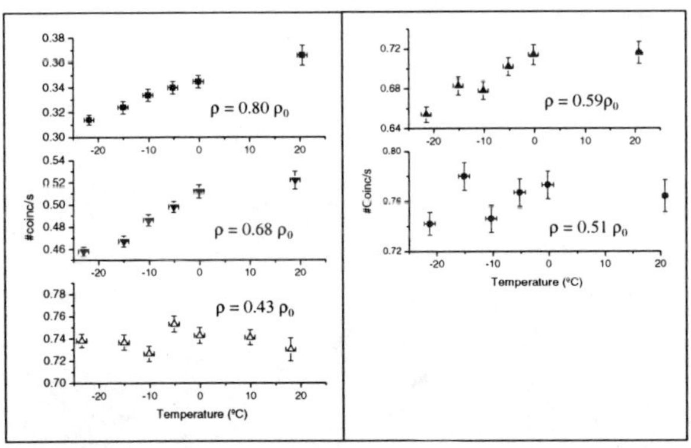

Fig. 2. Variation of the coincidence rate (for the 0-0 band of the N_2 2nd positive system) with the temperature for different gas densities (ρ_o is the density at atmospheric pressure and at room temperature).

5. Conclusions

The simulation of the geometry of the set-up is of utmost importance for the analysis of the present data. Preliminary calculations seem to indicate that once the geometrical corrections are introduced all data exhibit a similar behaviour of the light yield dependence on temperature. However, the uncertainties are still very large and have to be reduced. Further studies on the excitation and de-excitation mechanisms of the emitting state are also needed in order to check the accuracy of the present theoretical model.

References

1. M. Fraga, A. Onofre, M. Pimenta and A. Policarpo, Air Light 03, Bad Liebenzell, Germany, Dec. 11 - 14, 2003, http://www.auger.de/events/air-light-03/#wp.
2. C.C. Jui et al. (HiRes Collaboration), in: Invited Rapporteur and Highlight Papers, in: Proc. 26th Int. Cos. Ray Conf., Salt Lake City, 2000, 370.
3. J. Abraham et al. (Pierre Auger Collaboration) , Nucl. Instr. Meth. Phys. Res. A 523 (2004) 50.

OVERVIEW OF THE GLAST PHYSICS

N. GIGLIETTO*, M. BRIGIDA, A. CALIANDRO, C. FAVUZZI, P. FUSCO,
F. GARGANO, F. GIORDANO, F. LOPARCO, M.N. MAZZIOTTA,
S. RAINÒ, P. SPINELLI

Dipartimento Interateneo di Fisica,
Politecnico di Bari e Università di Bari,
via E. Orabona 4,
70124 Bari, Italy

We present an overview of the physics that GLAST satellite can explore. The improvement of the detector quality both in sensitivity and in angular resolution respect to previous satellite experiments, let GLAST detector to have big opportunities to understand γ-ray sky and discover new sources.

1. The GLAST mission

The Universe is almost transparent in the energy range between 20 MeV to few hundred GeVs, therefore the photon observations in this energy band is extremely useful to give answers to many of the high energy astrophysics questions. The GLAST satellite is equipped with two different instruments: a Gamma Ray Burst Monitor[1], working in the energy range from 20 KeV to 20 MeV, mainly dedicated to the detection of GRBs, and the Large Area Telescope[2] (LAT), able to reconstruct γ-ray directions and energies in the range 20 MeV up to at least 300 GeV. High energy measurements are based on the γ-ray pair conversions in the silicon tracker. The GLAST satellite will be launched in august 2007 and will be set on a low-Earth orbit at 550 Km altitude and about 28° inclination with the possibility to operate both in rocking zenith mode or pointed mode. In the rocking mode the instrument will be pointed on the zenith and rocked periodically by $\pm 35°$ north-south on alternate orbits, in order to cover the poles, and east-west to fill gaps around the South Atlantic Anomaly.

*e-mail: giglietto@ba.infn.it

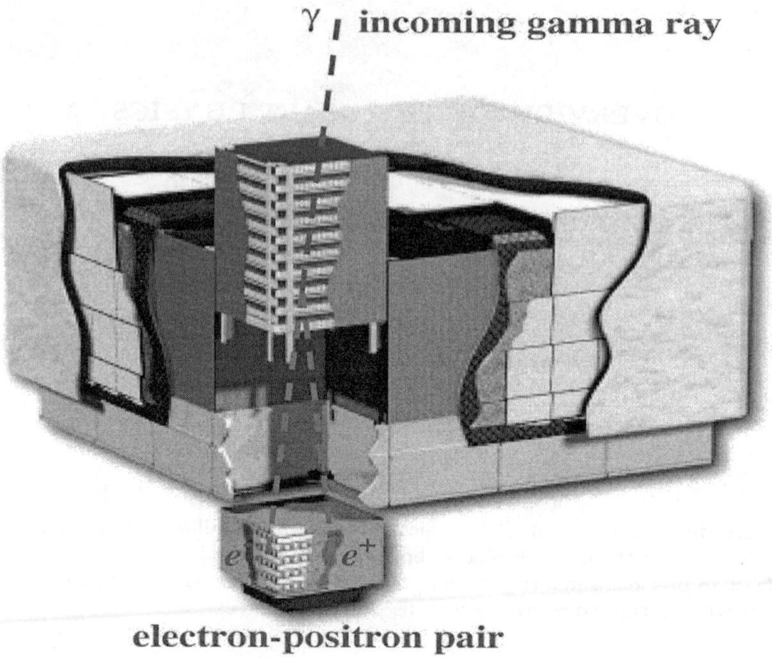

Figure 1. The GLAST-LAT detector

2. The Large Area Telescope (LAT)

The LAT detector is a sophisticated detector based on silicon strip sensors. It has a modular structure consisting of 4×4 elements called "tower". Each tower is composed of 18 xy tracking planes with single-sided silicon strip detectors, 228μm pitch. This array is then shielded by a segmented detector composed of plastic scintillator tiles in order to perform the anticoincidence for charged cosmic rays. Finally energy measurements are obtained by the LAT calorimeter, based on segmented CsI (Tl) crystals, giving both longitudinal and transverse information about the energy deposits, and able to measure γ-ray energies up to about 500 GeV with an energy resolution of less than 10% and located in the lower part of the detector as shown in Fig.1. The dimensions of each tower result then to be 40×40 cm^2 giving a total peak effective area larger than 8000 cm^2, taking into account also the cuts to reject background. The overall area, larger than a factor 10 respect to EGRET, provides to this detector the right sensitivity to discover thousand of new sources and understand γ-ray production mechanisms for many of them. Moreover the fast response of the detector reduces the deadtime to less than 100 μs, a characteristic that is particularly important to study short time variations in γ-ray fluxes. Another fundamental behaviour is the angular resolution of the LAT apparatus that is better than 0.15° for γ-ray energies larger than 10 GeV with a field of view larger than 2 sr. Therefore the sensitivity to pointlike sources should result $< 6 \times 10^{-9}\text{cm}^{-2}\text{s}^{-1}$.

3. Main scientific goals of the GLAST mission

Gamma ray studies can address answers to many astrophysical questions, in particular GLAST observations can explain the origin of the cosmic rays, give indications for the presence of Dark Matter and understand many astrophysical sources like Gamma Ray Bursts, pulsars and AGNs. Due to the excellent angular resolution of GLAST-LAT, the identification of pointlike sources will be done with great richness of details. A detailed exploration of SNRs regions should permit a clear identification of the cosmic ray interactions with the interstellar medium at the shell of SNR, well resolved from the pulsar core. In this case the energy spectra of the region showing the cosmic ray interactions with the interstellar medium, should be also characterized by the presence of the typical bump due to the π_0 decays, not present in the pulsar core spectrum, confirming in this way that these regions constitute the main engine for the accelerations of HE CR protons. AGNs studies will be mainly performed by a multiwavelength campaign of observations together Earth-based γ-ray observatories. The signature of AGNs is a fast energy release, of the order of 10^{49} erg/s in a compact central volume of the object. Therefore these sources should be characterized by a daily scale luminosity variability and a typical energy spectrum covering different energy scales, with a two peak behaviour, the most energetic often at TeV energies. Due to the large sensitivity of the detector, GLAST can produce a large catalogue containing at least 3000 AGNs, not yet observed in the γ energy band. A detailed study of the energy spectrum moreover can help to distinguish leptonic from hadronic emission in AGNs, as proposed in different models [3,4]. A different measurable effect should be the observation of a redshift dependence of the AGN spectra cutoff due to the interactions between extra galactic background light (EBL) and TeV γ-rays. The $\gamma - \gamma$ interaction between TeV and UV photons, these latter heavily produced at the time of galaxy formation, should produce a cut off in the observed AGN spectrum particularly evident above 10 GeV, for AGNs having a $z > 2$ redshift[5].

The good energy resolution and the pointing abilities of LAT detector permit also an indirect search for the Dark Matter that can be realized by looking for anomalies in the observed γ-ray spectrum due to the eventual $\chi - \overline{\chi}$ annihilations producing γs in the final state[6]. Decay rates are strongly dependent from the SUSY models but also from the DM distribution in the galaxies. Therefore a possible experimental strategy can be obtained by a comparison of the spectrum looking toward the Galactic Center respect to the spectrum from the galactic halo or nearby dwarfs[7]. GLAST will give

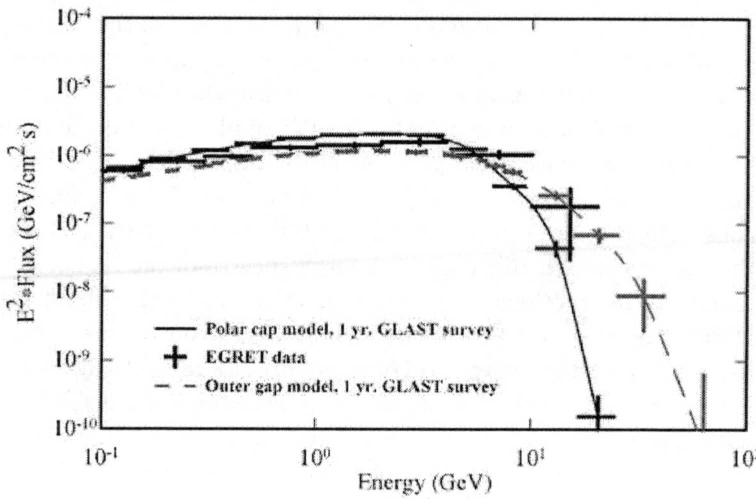

Figure 2. The pulsar energy spectrum predicted for GLAST 1 year sky survey, for the outer gap and polar cap models.

many contribution to pulsar studies in the γ-ray observations. Due to the sensitivity and resolution GLAST will observe probably at least 100 new pulsars emitting in the γ band[8] and giving detailed spectra to test pulsar emission physics and comparing different emission models. GLAST should be able to discriminate the polar-cap[9] and outer-gap[10] emission models by a comparison of the pulsars energy spectrum collected during at least one year sky survey, as shown in Fig.2.

Gamma Ray Bursts (GRB) are one of the most interesting phenomena observed in the last years. Their emissions generally consist in a short period, from milliseconds to thousand of seconds, of very intense γ-ray emission followed by a afterglow stage observed at several wavelength bands (radio,optical, X-ray and γ-ray) and lasting for a longer timescale (from hours to days). One of the most widely accepted models for GRBs postulates that they are powered by a relativistically expanding fireball. Electrons accelerated by the shocks produced by the colliding shells of the material inside the fireball, called internal shocks, produce radiation during the prompt phase via synchrotron emission. The afterglow emission is instead due to the radiation by non-thermal particles when the external shock is formed when the fireball blastwave sweeps up the external medium. LAT together the GBM will measure the energy spectra of GRBs from keV to GeV during their flares and therefore will contribute to the GRB understanding producing a large catalogue with much more detailed and finer lightcurves. Due to the fast response of the apparatus, GLAST will furnish a prompt alert for other observers. In this case the detector reduced deadtime and the good energy resolution, should be particularly important to fully understand the physics of the central engine of these objects.

4. Conclusions

GLAST observations will contribute to a full exploration of the γ-ray universe, giving the opportunity to understand better known sources and finding new sources. The γ-ray emission of our Galaxy can also give indications about the Dark Matter presence and its density profile. Moreover the possibility to establish multiwavelength campaign of observations with other detectors, both ground-based and satellite, will permit a detailed study of astrophysical sources.

References

1. A. von Kienlin et al., The GLAST Burst Monitor for GLAST, in Proc of the SPIE-Conference,Glasgow 2004.
2. E. Bloom, G. GodFrey, S. Ritz Proposal for the Gamma-ray Large Area Telescope, SLAC-R-22.February, (1998).
3. K. Katarzyński, H. Sol and A. Kus, *Astron. Astrophys.* **367**,809 (2001).
4. P. Padovani and P. Giommi, *Astrophys. J.* **444**, 567 (1995).
5. A. Chen, L. C. Reyes and S. Ritz, *The Astrophys. J.***608** 686 (2004).
6. S. Peirani, R Mohayaee, J. A. de Freitas Pacheco,*Phys.Rev.D***70**, 043503 (2004).
7. A. Cesarini, F. Fucito, A. Lionetto, A. Morselli and P. Ullio, *Astropart. Phys.* **21**, 267 (2004).
8. D. J. Thompson,*astro-ph***0312272** (2003).
9. J. K. Daugherty, A. K. Harding,*Astrophys. J.***458**, 278 (1996).
10. L. Zhang, K. S. Cheng, *Astrophys. J.***487**,370 (1997).

VELOCITY AND CHARGE RECONSTRUCTION WITH THE AMS/RICH DETECTOR

LUÍSA ARRUDA, F.BARÃO, J.BORGES, F.CARMO,
P.GONÇALVES, R.PEREIRA M.PIMENTA
LIP/IST
Av. Elias Garcia, 14, 1º andar
1000-149 Lisboa, Portugal
e-mail: luisa@lip.pt

The Alpha Magnetic Spectrometer (AMS), to be installed on the International Space Station (ISS) in 2008, will be equipped with a proximity focusing Ring Imaging Čerenkov detector (RICH). This detector will be equipped with a dual radiator (aerogel+NaF), a lateral conical mirror and a detection plane made of 680 photomultipliers and light-guides, enabling measurements of particle electric charge and velocity. A likelihood method for the Čerenkov angle reconstruction was applied leading to a velocity determination for protons with a resolution around 0.1%. The electric charge reconstruction is based on the counting of the number of photoelectrons and on an overall efficiency estimation on an event-by-event basis. Results from the application of both methods are presented.

1. The AMS02 detector

AMS [1] (Alpha Magnetic Spectrometer) is a precision spectrometer designed to search for cosmic antimatter, dark matter and to study the relative abundance of elements and isotopic composition of the primary cosmic rays. It will be installed in the International Space Station (ISS), in 2008, where it will operate, at least, for a period of three years.

The spectrometer will be capable of measuring the rigidity ($R \equiv pc/|Z|e$), the charge (Z), the velocity (β) and the energy (E) of cosmic rays within a geometrical acceptance of $\sim 0.5\,\mathrm{m}^2$.sr. Fig. 1 shows a schematic view of the AMS spectrometer. On top, a Transition Radiation Detector (TRD) will discriminate between leptons and hadrons. It will be followed by the first of the four Time-of-Flight (TOF) system scintillator planes. The TOF will provide a fast trigger, charge and velocity measurements for charged particles, as well as information on their direction of incidence. The tracking system will be surrounded by Veto Counters and embedded in a magnetic field of about 0.9 Tesla produced by a superconducting magnet.

It will consist on a Silicon Tracker, constituted of 8 double sided silicon planes, providing both charge and rigidity measurements with an accuracy better than 2% up to 20 GV. The maximum detectable rigidity is around 1 TV. The Ring Imaging Čerenkov Detector (RICH), described in the next section, will be located right after the last TOF plane and before the Electromagnetic Calorimeter (ECAL) which will enable e/p separation and will measure the energy of the detected photons.

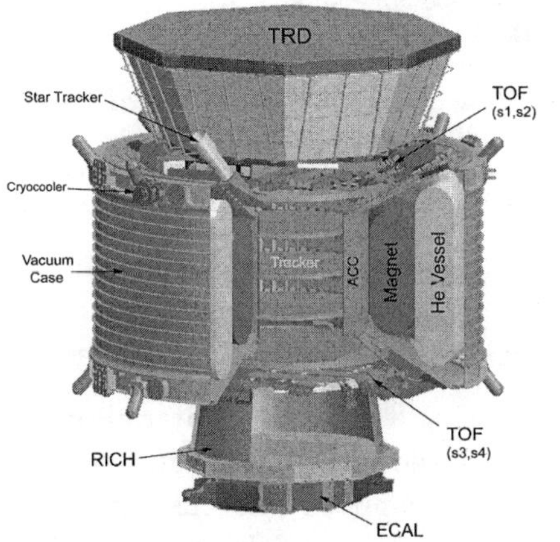

Figure 1. A whole view of the AMS Spectrometer.

1.1. *The RICH detector*

The RICH is a proximity focusing device with a dual radiator configuration on the top (low refractive index aerogel 1.050, 3 cm thick and a central square of sodium fluoride (NaF), 0.5 cm thick); a lateral conical mirror of high reflectivity increasing the reconstruction efficiency and a detection matrix with 680 photomultipliers and light guides. The active pixel size of the PMTs is planned to be of 8.5 mm with a spectral response ranging from 300 to 650 nm with a maximum at $\lambda \sim 420$ nm. There will be a large non-active area at the centre of the detection area due to the insertion of the ECAL. For a more detailed description of the RICH detector see Ref. 2. The RICH detector of AMS was designed to measure the velocity of charged particles with a resolution $\Delta\beta/\beta$ of 0.1%, to extend the electric charge separation capability up to Z$\sim$26, to provide more information on the albedo rejection and to contribute in e/p separation. Its acceptance is of $\sim 0.35\, m^2.sr$. Figure 2 shows a view of the RICH and a beryllium event display with a detailed view of the PMT matrix.

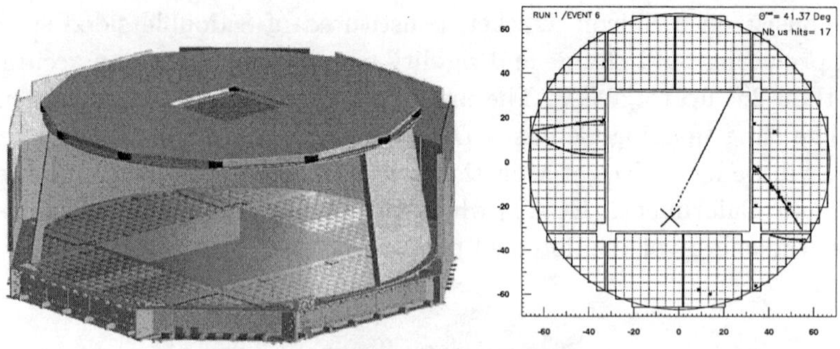

Figure 2. *On the left:* View of the RICH detector. *On the right:* Beryllium event display generated in a NaF radiator. The reconstructed photon pattern (full line) includes both reflected and non-reflected branches. The outer circular line corresponds to the lower boundary of the conical mirror. The square is the limit of the non-active region.

2. Velocity reconstruction

A charged particle crossing a dielectric material of refractive index n with a velocity β, greater than the speed of light in that medium, emits photons. The aperture angle of the emitted photons with respect to the radiating particle is known as the Čerenkov angle, θ_c, and it is given by (see Ref. 3).

$$\cos\theta_c = \frac{1}{\beta\, n} \tag{1}$$

It follows that the velocity of the particle, β, is straightforward derived from the Čerenkov angle reconstruction, which is based on a fit to the pattern of the detected photons. Complex photon patterns can occur at the detector plane due to mirror reflected photons, as can be seen on right display of Fig. 2. The event displayed is generated by a simulated beryllium nucleus in a NaF radiator.

The Čerenkov angle reconstruction procedure relies on the information of the particle direction provided by the Tracker. The tagging of the hits signaling the passage of the particle through the solid light guides in the detection plane provides an additional track element, however, those hits are excluded from the reconstruction. The best value of θ_c will result from the maximization of a Likelihood function, built as the product of the probabilities, p_i, that the detected hits belong to a given (hypothetical) Čerenkov photon pattern ring,

$$L(\theta_c) = \prod_{i=1}^{nhits} p_i^{n_i}\,[r_i(\theta_c)]\,. \tag{2}$$

Here r_i is the closest distance of the hit to the Čerenkov pattern and n_i the signal strength. For a more complete description of the method see Ref. 4. The resolution achieved for singly charged particles crossing the aerogel radiator with $\beta \sim 1$ is $\sim$4 mrad and for those crossing the NaF radiator the resolution is $\sim$8 mrad. The evolution of the relative resolution of β with the charge can be observed on the left plot of Fig. 3. It was extracted from reconstructed events generated in a test beam at CERN in October 2003 with fragments of an Indium beam of 158 GeV/nuc, in a prototype of the RICH detector.

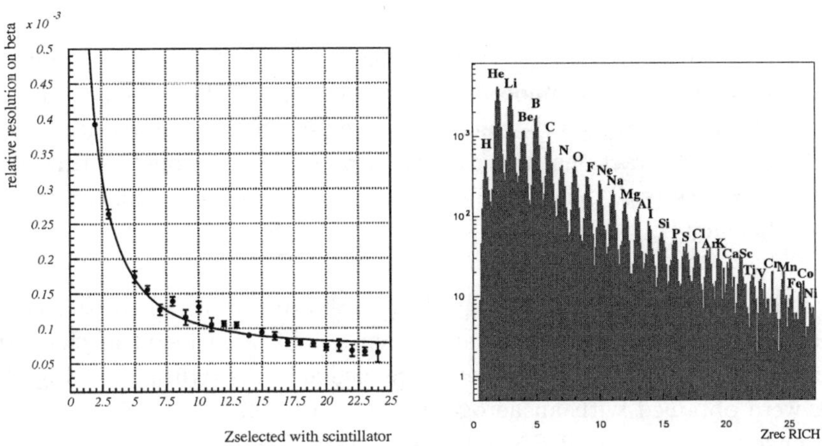

$$\sqrt{(\tfrac{A}{Z})^2 + B^2}$$

$$A = 7.77 \times 10^{-4} \pm 5 \times 10^{-6}$$

$$B = 7.3 \times 10^{-5} \pm 2 \times 10^{-6}$$

$$\chi^2 = 2.76 \qquad npfit = 22$$

Figure 3. At left evolution of the relative resolution on β with the charge and at right the reconstructed charge peaks. Both are reconstructions with data from a test beam at CERN in October 2003, using an Indium beam of 158 GeV/nuc.

3. Charge reconstruction

The Čerenkov photons produced in the radiator are uniformly emitted along the particle path inside the dielectric medium, L, and their number per unit of energy depends on the particle's charge, Z, and velocity, β, and on the refractive index, n, according to the expression:

$$\frac{dN_\gamma}{dE} \propto Z^2 L \left(1 - \frac{1}{\beta^2 n^2}\right) = Z^2 L \sin^2 \theta_c \tag{3}$$

So to reconstruct the charge the following procedure is required:

- Čerenkov angle reconstruction.
- Estimation of the particle path, L, which relies on the information of the particle direction provided by the Tracker.
- Counting the number of photoelectrons. The number of photoelectrons related to the Čerenkov ring has to be counted within a fiducial area, in order to exclude the uncorrelated background. Therefore, photons which are scattered in the radiator are excluded. A distance of 13 mm to the ring was defined as the limit for photoelectron counting, corresponding to a ring width of ~5 pixels.
- Evaluation of the photon detection efficiency. The number of radiated photons (N_γ) which will be detected ($n_{p.e}$) is reduced due to the interactions with the radiator (ε_{rad}), the photon ring acceptance (ε_{geo}), light guide efficiency (ε_{lg}) and photomultiplier efficiency (ε_{pmt}).

$$n_{p.e.} \sim N_\gamma \, \varepsilon_{rad} \, \varepsilon_{geo} \, \varepsilon_{lg} \, \varepsilon_{pmt} \tag{4}$$

The charge is then calculated according to expression 3, where the normalization constant can be evaluated from a calibrated beam of charged particles. Reconstructed charge peaks are visible in the right plot of Fig. 3. Data were obtained with an aerogel radiator 1.05, 2.5 cm thick from the mentioned test beam at CERN in October 2003. The charge resolution obtained for helium is ~0.2 and it is possible to separate charges up to Z=27. For a more complete description of the charge reconstruction method see Ref. 4.

4. Conclusions

AMS is a spectrometer designed for antimatter and dark matter searches and for measuring relative abundances of nuclei and isotopes. The instrument will be equipped with a proximity focusing RICH detector based on a mixed radiator of aerogel and sodium fluoride, enabling velocity measurements with a resolution of about 0.1% and extending the charge measurements up to the Iron element. Velocity reconstruction is made with a Likelihood method. Charge reconstruction is made in an event-by-event basis. Both algorithms were successfully applied to simulated data samples with flight configuration. Evaluation of the algorithms on real data taken with the RICH prototype was performed at the LPSC, Grenoble in 2001 and in the test beam at CERN, in October 2002 and 2003.

References

1. S. P. Ahlen et al., Nucl. Instrum. Methods A **350**,34 (1994).
   V. M. Balebanov et al., *AMS proposal to DOE*, approved April 1995.
2. M.Buenerd. Proceedings of the Fourth Workshop on Rich Detectors (RICH02) June 5-10, 2002, Pylos, Greece.
3. T.Ypsilantis and J.Seguinot, Nucl. Instrum. Methods A **343**, 30 (1994).
4. F. Barao,L. Arruda et al., Čerenkov angle and charge reconstruction with the RICH detector of the AMS experiment, Nucl. Instrum. Methods A **502**,310 (2003).

ISOTOPE SEPARATION WITH THE RICH DETECTOR OF THE AMS EXPERIMENT

L. ARRUDA, F. BARÃO, J. BORGES, F. CARMO, P. GONÇALVES, RUI PEREIRA, M. PIMENTA

LIP/IST
Av. Elias Garcia, 14, 1º andar
1000-149 Lisboa, Portugal
e-mail: pereira@lip.pt

The Alpha Magnetic Spectrometer (AMS), to be installed on the International Space Station (ISS) in 2008, is a cosmic ray detector with several subsystems, one of which is a proximity focusing Ring Imaging Čerenkov (RICH) detector. This detector will be equipped with a dual radiator (aerogel+NaF), a lateral conical mirror and a detection plane made of 680 photomultipliers and light guides, enabling precise measurements of particle electric charge and velocity. Combining velocity measurements with data on particle rigidity from the AMS Tracker it is possible to obtain a measurement for particle mass, allowing the separation of isotopes.

A Monte Carlo simulation of the RICH detector, based on realistic properties measured at ion beam tests, was performed to evaluate isotope separation capabilities. Results for three elements — H (Z=1), He (Z=2) and Be (Z=4) — are presented.

1. The AMS02 experiment

Alpha Magnetic Spectrometer (AMS)[1] is an experiment designed to study the cosmic ray flux by direct detection of particles above the Earth's atmosphere. The deployment of the final detector (AMS-02) to the ISS is scheduled for 2008, for a minimum operating period of 3 years. A preliminary version of the detector (AMS-01) was successfully flown aboard the US space shuttle Discovery in June 1998.

On the ISS, orbiting at an average altitude of 400 km, AMS will collect an extremely large number of cosmic ray particles. Its main goals are *(i)* a detailed study of cosmic ray composition and energy spectrum through the collection of an unprecedented volume of data, *(ii)* a search for heavy antinuclei ($Z \geq 2$) which if discovered would signal the existence of antimatter domains in the Universe, and *(iii)* a search for dark matter constituents by examining possible signatures of their presence in the cosmic ray spectrum.

AMS is a spectrometer equipped with a superconducting magnet. It is composed of several subdetectors: a Transition Radiation Detector (TRD), a Time-of-Flight (TOF) detector, a Silicon Tracker, Anticoincidence Counters (ACC), a Ring Imaging Čerenkov (RICH) detector and an Electromagnetic Calorimeter (ECAL). Fig. 1 shows a schematic view of the full AMS detector. The present work evaluates reconstruction capabilities of the RICH detector of AMS.

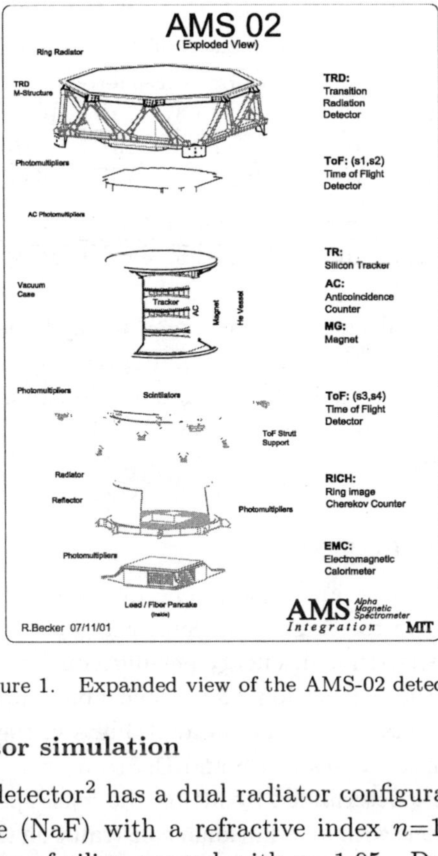

Figure 1. Expanded view of the AMS-02 detector

2. RICH detector simulation

The AMS RICH detector[2] has a dual radiator configuration with a square of sodium fluoride (NaF) with a refractive index n=1.334 at the centre surrounded by tiles of silica aerogel with n=1.05. Detector efficiency is increased by the presence of a highly reflective ($\approx$ 85%) conical mirror surrounding the expansion volume. For additional information on the RICH detector capabilities see also ref. 3.

A full-scale Monte Carlo simulation of the RICH detector was performed to evaluate isotope separation capabilities using the GEANT 3 software package. Data on particle rigidity, which in the experimental setup are expected to come from the AMS silicon tracker, were created by adding a random smearing to the simulated rigidity. The function giving smearing magnitude was adjusted to match real tracker performance.

Table 1 shows the total number of events generated for each simulation. The total number of simulated events corresponds to approximately one day of data, in the cases of H and He, and one year in the case of Be. Simulated distributions were based respectively on ref. 4 for H, ref. 5 for He and ref. 6 for Be and adjusted to the AMS detector acceptance.

Table 1. Statistics for the AMS RICH simulations

Simulations for NaF+aerogel		
H total : $\mathbf{1.63 \times 10^7}$	^{1}H (p) : 1.61×10^7	^{2}H (d) : 1.39×10^5
He total : $\mathbf{2.02 \times 10^6}$	^{3}He : 3.39×10^5	^{4}He : 1.68×10^6
Be total : $\mathbf{8.47 \times 10^5}$	^{9}Be : 6.97×10^5	^{10}Be : 1.49×10^5

Simulation for NaF only		
H total : $\mathbf{1.53 \times 10^7}$	^{1}H (p) : 1.52×10^7	^{2}H (d) : 1.31×10^5

An additional simulation was performed for hydrogen events radiating in NaF, with a total statistics corresponding to approximately one week of data taking. This was due to the relatively low number of NaF events produced by generic simulations (NaF events correspond to only $\sim 10\%$ of the total RICH data).

3. Reconstruction procedure

For each particle, charge and kinetic energy were determined using the procedure described in ref. 7. Since isotopic ratios are a function of energy, the reconstructed spectrum in energy-per-nucleon was divided in narrow regions for which the calculation of isotopic abundances was performed separately. Only events with a minimum of 3 hits in the Čerenkov pattern were considered. In the cases of He and Be, total isotopic abundances for each energy bin were determined by fitting the mass spectrum to a sum of two gaussian functions with an additional constraint on the mass resolutions $(\frac{\sigma_1}{\sigma_2} = \frac{m_1}{m_2})$.

Hydrogen was a special case due to low masses and the small d/p ratio. An inverse mass $(1/m)$ spectrum was used, with separate fits being performed for the two mass peaks. A gaussian fit was used for p, while for d a sum of a gaussian and a constant was used to account for proton background.

For all elements, final isotopic ratios were calculated from the gaussian fit integrals corresponding to each peak.

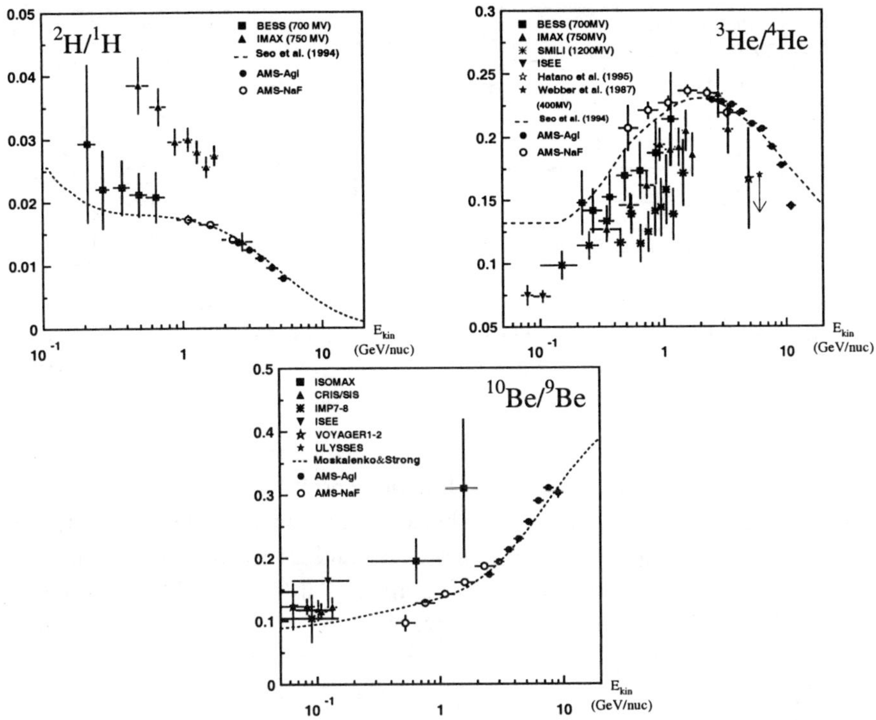

Figure 2. Reconstruction of simulated isotopic ratios in AMS compared with data from other experiments

4. Reconstruction results

Fig. 2 shows the results obtained for isotopic ratios compared with the simulated distributions. Data from previous experiments are also shown for comparison. In the cases of He and Be, satisfactory fits were obtained for the energy regions from the Čerenkov thresholds up to ~ 3 GeV/nucleon (NaF) and ~ 10 GeV/nucleon (aerogel). In the case of H good fits were only obtained for the regions between 0.9 and 3 GeV/nucleon in NaF and from the Čerenkov threshold up to ~ 6 GeV/nucleon in aerogel. These figures clearly show that even a small fraction of the expected AMS statistics will represent a major improvement on existing results for any of the three elements.

For each element, data on mass resolution and separation power for different energies were obtained from fit results. Separation power was defined as the ratio $\frac{\Delta m}{\sigma_m}$. Fig. 3 shows mass resolution and separation power as functions of energy for both radiators. Optimal mass resolutions were reached around 1 GeV/nucleon in NaF and 3 GeV/nucleon in aerogel.

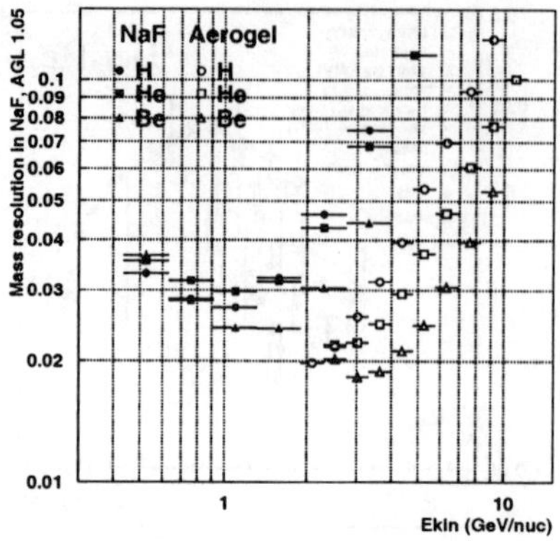

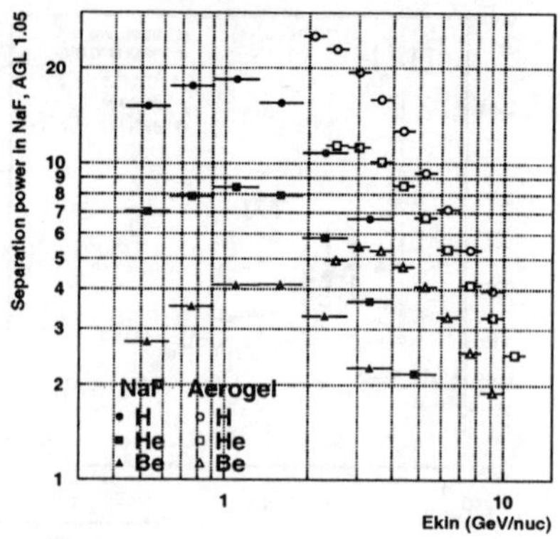

Figure 3. Simulation results for mass resolution *(left)* and separation power *(right)*

Separation power is higher for lighter elements, suggesting isotope separation should be possible up to higher energies in the case of hydrogen. However, the greater difference between proton and deuteron statistics ($d/p \sim 10^{-2}$) compared to the cases of He and Be isotopes eventually leads to the separation being only possible up to $\sim$ 6 GeV/nucleon compared to $\sim$ 10 GeV/nucleon for the other elements.

5. Conclusions

AMS will provide a major improvement on existing data for isotopic abundances in cosmic rays. Simulation results indicate that the separation of light isotopes using the combination of RICH data and tracker rigidity measurements is feasible. The dual radiator configuration of NaF and aerogel makes isotope separation of light elements possible for energies in the range from 0.5 to 10 GeV/nucleon, approximately. Best mass resolutions are $\sim$ 2% at 3 GeV/nucleon for aerogel, and $\sim$ 3% at 1 GeV/nucleon for NaF. Techniques presented here may also be applied in the separation of antimatter isotopes which is of great importance in dark matter studies.

References

1. S. P. Ahlen *et al.*, *Nucl. Instrum. Methods A* **350**, 34 (1994).
   V. M. Balebanov *et al.*, *AMS proposal to DOE*, approved April 1995.
2. M. Buenerd, Proceedings of the Fourth Workshop on Rich Detectors (RICH02) June 5-10, 2002, Pylos, Greece.
3. L. Arruda, These proceedings.
4. E. S. Seo *et al.*, *Astrophys. J.* **432**, 656 (1994).
5. E. S. Seo *et al.*, *Astrophys. J.* **431**, 705 (1994).
6. A. W. Strong and I. V. Moskalenko, *Adv. Space Res.* **27**, 717 (2001).
7. F. Barao, L. Arruda *et al.*, *Nucl. Instrum. Methods A* **502**, 310 (2003).

Gravitational Waves and Compact Sources

GRAVITATIONAL RADIATION FROM 3D COLLAPSE TO ROTATING BLACK HOLES

LUCA BAIOTTI

SISSA, International School for Advanced Studies, Trieste, Italy and Max-Planck-Institut für Gravitationsphysik, Golm, Germany

IAN HAWKE

Max-Planck-Institut für Gravitationsphysik, Golm, Germany and School of Mathematics, University of Southampton, UK

LUCIANO REZZOLLA

SISSA, International School for Advanced Studies and INFN, Trieste, Italy and Department of Physics, Louisiana State University, Baton Rouge, LA USA

ERIK SCHNETTER

Max-Planck-Institut für Gravitationsphysik, Golm, Germany

We discuss the application of our new three-dimensional fully general relativistic hydrodynamics code which uses high-resolution shock-capturing techniques and a conformal traceless formulation of the Einstein equations, to the study of the gravitational collapse of uniformly rotating neutron stars to Kerr black holes. We investigate the dynamics of the matter and of the trapped surfaces and provide precise measurements of the black-hole mass and spin. We also present the first calculation of the gravitational wave emission produced in fully three-dimensional simulations. An essential aspect of these simulations is the use of progressive mesh-refinement techniques which allow to move the outer boundaries of the computational domain to regions where gravitational radiation attains its asymptotic form. The waveforms have been extracted using a gauge-invariant approach in which the numerical spacetime is matched with the non-spherical perturbations of a Schwarzschild spacetime. Overall, the results indicate that the waveforms have features related to the properties of the initial stellar models (in terms of their w-mode oscillations) and of the newly produced rotating black holes (in terms of their quasi-normal modes). While our waveforms are in good qualitative agreement with those computed in two-dimensional simulations, our amplitudes are about one order of magnitude smaller and this difference is most likely due to our less severe pressure reduction. For a neutron star rotating uniformly near mass-shedding and collapsing at 10 kpc, the signal-to-noise ratio computed uniquely from the burst is $S/N \sim 0.25$ for LIGO/VIRGO, but this grows to be $S/N \lesssim 4$ for LIGO II.

1. Introduction

The numerical investigation of gravitational collapse of rotating stellar configurations leading to black-hole formation is a long standing problem. However, it is through numerical simulations in general relativity that one can hope to improve our knowledge of fundamental aspects of Einstein theory such as the cosmic censorship hypothesis and black-hole no-hair theorems, along with that of current open issues in relativistic astrophysics research, such as the mechanism responsible for gamma-ray bursts. Furthermore, numerical simulations of stellar gravitational collapse to black holes provide a unique means of computing the gravitational waveforms emitted in such events, believed to be among the most important sources of detectable gravitational radiation.

The modelling of black-hole spacetimes with collapsing matter sources in multidimensions is very difficult. This is due, on the one hand, to the inherent difficulties and complexities of the system of equations which is to be integrated (*i.e.* the Einstein field equations coupled to the general-relativistic hydrodynamics equations) and, on the other hand, to the immense computational resources needed to integrate the equations in the case of three-dimensional (3D) evolutions. In addition to the practical difficulties encountered in the accurate treatment of the hydrodynamics involved in the gravitational collapse of a rotating neutron star to a black hole, the precise numerical computation of the gravitational radiation emitted in the process is particularly challenging as the energy released in gravitational waves is much smaller than the total rest-mass energy of the system. Even though in recent years many studies have extended to three spatial dimensions the investigation of gravitational collapse to black holes[26,27,16], these have not included calculations of the emitted gravitational radiation.

We have recently developed and carefully tested `Whisky`, a 3D parallel code for the solution of the relativistic hydrodynami equations over a generic curved background spacetime[5]. The details of the formulation we use for the Einstein and hydrodynamics equations can be found in Baiotti et al.[6] but we here stress that an important feature of our approach to the solution of the hydrodynamics equations is that it extends to a general relativistic context the powerful numerical methods developed in classical hydrodynamics, in particular the high-resolution shock-capturing (HRSC) schemes based on linearized Riemann solvers. Such schemes are essential for a correct representation of shocks, whose presence is expected in several astrophysical scenarios.

Another fundamental improvement we have implemented is the ability to excise from the evolved grid regions of spacetime contained within horizons. Such regions are causally disconnected from the rest of the spacetime and so do not have any influence on the exterior evolution. The field and hydrodynamical variables in these regions, though, could reach very large gradients and compromise the evolution. More details on how the hydrodynamical excision is applied in practice, as well as tests showing that this method is stable, consistent and converges to the expected order are published in Hawke et al.[18]. For all the details about each of the following sections, which contain only brief summaries of our work on this subject, the reader is referred to Baiotti et al.[6,7].

2. Initial stellar models

The initial data for our simulations are constructed using a 2D numerical code, that computes accurate stationary equilibrium solutions for axisymmetric and rapidly rotating relativistic stars in polar coordinates[29]. For simplicity, we have focused on initial models constructed assuming a polytropic EOS with polytropic exponent $\Gamma = 2$ and polytropic constant $K = 100$. These stellar models are, at least qualitatively, representative of what is expected from observations of neutron stars. In the following we report in particular about two of the models we have studied, one slowly rotating (which we refer to as D1) and one rapidly rotating (D4). Both models are dynamically unstable and the collapse is triggered by a small (*i.e.* $\lesssim 2\%$) reduction of the pressure.

3. Dynamics of the matter

We present in Fig. 1 a representative snapshot of the final stages of the evolution of a slowly rotating initial model in which we show the isocontours of the rest-mass density and velocity field in the (x, y) plane (left column) and in the (x, z) plane (right column), respectively. Soon after an apparent horizon is found and when this has grown to a sufficiently large area, the portion of the computational domain containing the singularity is excised. The excised region of the computational domain is indicated as an area filled with squares. Also shown with a thick dashed line is the coordinate location of the apparent horizon and it should be remarked that, because of rotation, this surface is not a coordinate two-sphere, although such deviations cannot be appreciated in Fig. 1 (see also Section 5). At $t = 0.57$ ms, the time which Fig. 1 refers to, most of the matter has already fallen within the apparent

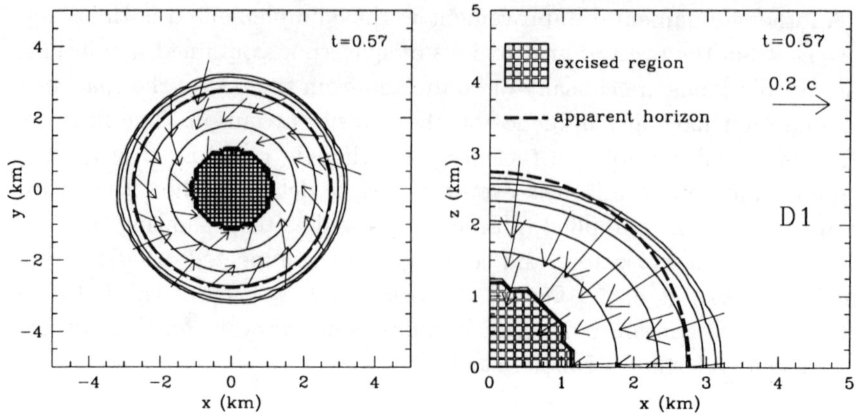

Figure 1. View of the final stages of the collapse of model D1 where we show the isocontours of the rest-mass density and velocity field in the (x, y) plane (left column) and in the (x, z) plane (right column), respectively.

horizon and has assumed an oblate shape.

Overall, confirming what was already discussed by several authors in the past, the gravitational collapse of the slowly rotating stellar model D1 takes place in an almost spherical manner and we have found no evidence of shock formation which could prevent the prompt collapse to a black hole, nor appreciable deviations from axisymmetry.

In Fig. 2 we show a representative snapshot of the evolution of the

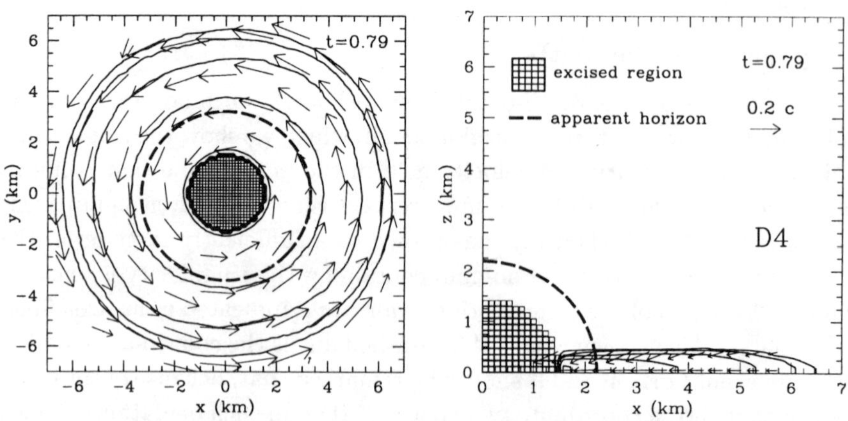

Figure 2. The same as in Fig. 1 but for model D4.

rapidly rotating model D4. The isocontour levels shown for the rest-mass density are the same as the ones used in Fig. 1. The dynamics is very similar to the one discussed for model D1 up to a time $t \sim 0.49$ ms. However, as the collapse proceeds, significant differences between the two models start to emerge and in the case of model D4 the large angular velocity of the progenitor stellar model produces significant deviations from a spherical infall. Indeed, the parts of the star around the rotation axis that are experiencing smaller centrifugal forces collapse more promptly and, as a result, the configuration increases its oblateness. In this case, by the time the apparent horizon has been found, the star has flattened considerably, all the matter near the rotation axis has fallen inside the apparent horizon, but a disc of low-density matter remains near the equatorial plane orbiting at relativistic velocities $\gtrsim 0.2\ c$. The centrifugal barrier preventing a purely radial infall of matter far from the rotational axis is the consequence of the larger initial angular momentum and of the pressure wave originating from the collapse along the rotational axis.

Note that the disc formed outside the apparent horizon is *not* dynamically stable and, in fact, it rapidly accretes onto the newly formed black hole. This is shown in Fig. 2, where one can notice that the disc is considerably flattened but also has large radial inward velocities which induce it to be accreted rapidly onto the black hole. Note that as the area of the apparent horizon increases, so does the excised region.

4. Dynamics of the horizons

In order to investigate the formation of a black hole and measure its properties, we have used the horizon finders available within the `Cactus` framework[12], which compute both the *apparent*[30] and the *event*[14] horizon. Using these tools it is possible to implement several different methods for measuring the black-hole mass and spin. The first and simplest method of these methods for computing the black-hole mass is to use the formula $M = C_{\rm eq}/4\pi$, where $C_{\rm eq} \equiv \int_0^{2\pi} \sqrt{g_{\phi\phi}} d\phi$ is the proper equatorial circumference. But this would only apply exactly to the Kerr (or Schwarzschild) solution. Note that this procedure is strictly valid in the case of a Kerr black hole only and is therefore expected to be accurate but not exact for the not-yet stationary black hole produced in our simulations.

A second method for estimating the horizon mass is to look at the distortion of the horizon using the ratio of polar and equatorial proper circumferences, $C_r \equiv C_{\rm pol}/C_{\rm eq}$, as in Brandt et al.[11]. For a perturbed Kerr black hole this is expected to oscillate around the asymptotic Kerr value with the form of a quasi-normal mode. By fitting to this mode we extract an estimate of the angular momentum parameter a/M_{hor} as[10]

$$\frac{a}{M_{hor}} = \sqrt{1 - (-1.55 + 2.55 C_r)^2} \,, \tag{1}$$

where M_{hor} coincides with the black-hole mass M only if the spacetime has become axisymmetric and stationary.

A third method for calculating instead J and hence measuring M is to use the *isolated* and *dynamical*-horizon frameworks of Ashtekar and collaborators[2,15]. This method assumes the existence of a Killing vector field intrinsic to a marginally trapped surface such as an apparent horizon and provides an unambiguous definition of the spin of the black hole and hence of its total mass. If there is an energy flux across the horizon, the isolated-horizon framework needs to be extended to the *dynamical*-horizon formalism[3]. The advantage of the dynamical-horizon framework is that it gives a measure of mass and angular momentum which is computed locally, without a global reconstruction of the spacetime. One possible disadvantage is that the horizon is required to have an (approximate) Killing vector field, so that no information can be found with this approach in the case in which the horizon deviates largely from axial symmetry. However, arbitrarily large distortions are allowed as long as they are axisymmetric in the sense of having a Killing vector field; the coordinate representation of the horizon does not need to have any symmetries. As expected, since our simulations do not deviate significantly from axisymmetry, we have not encountered problems in applying the dynamical horizon framework to the horizons found in our simulations.

A fourth method for computing J only applies if an event horizon is found and if its angular velocity has been measured. In a Kerr background, in fact, the generators of the event horizon rotate with a constant angular velocity $\omega \equiv -g_{t\phi}/g_{\phi\phi} = \sqrt{g_{tt}/g_{\phi\phi}}$ such that

$$\frac{a}{M} = \frac{J}{M^2} = \left[\frac{A\omega^2}{\pi}\left(1 - \frac{A\omega^2}{4\pi}\right)\right]^{1/2} \,, \tag{2}$$

where A is the event-horizon proper area.

The direct comparison of different methods employed allowed to measure the mass of the black hole with a precision of 0.5% in the case of model D1 and of $\sim 1\%$ in the case of model D4[a].

---

[a]The precision is based on the initial ADM mass computed on a compactified domain.

5. Reconstructing the global spacetime

All of the results presented and discussed in the previous Sections describe only a small portion of the spacetime which has been solved during the collapse. It is also interesting and instructive to collect all of these pieces of information into a *global* description of the spacetime and look for those features which mark the difference between the collapse of slowly and rapidly rotating stellar models. As we discuss below, these features emerge in a very transparent way within a global view of the spacetime.

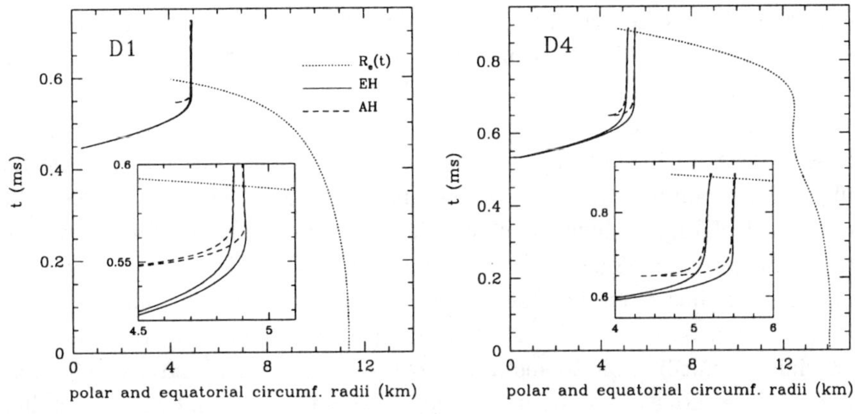

Figure 3. Evolution of the most relevant surfaces during the collapse for the cases D1 and D4.

To construct this view, we use the worldlines of the most representative surfaces during the collapse, namely those of the equatorial stellar surface, of the apparent horizon and of the event horizon. For all of them we need to use quantities measured by proper observers and, in particular, circumferential radii. The results of this spacetime reconstruction are shown in Fig. 3, whose left and right panels refer to the collapse of models D1 and D4, respectively. The different lines indicate the worldlines of the circumferential radius of the stellar surface (dotted line), as well as that of the apparent horizon (dashed line) and of the event horizon (solid line). Note that for the horizons we show both the equatorial and the polar circumferential radii, with the latter being always smaller than the former. For the stellar surface, on the other hand, we show the equatorial circumferential radius only. Note that in both panels of Fig. 3 the event horizon grows from an essentially zero size to its asymptotic value. In contrast, the apparent horizon grows from an initially non-zero size and, as it should, is always contained within the event horizon. At late times, the worldlines merge to the precision at which we can compute them. A quick look at the two panels of Fig. 3 is sufficient to appreciate the different properties in the dynamics of the collapse of slowly and rapidly rotating models.

The worldlines of the stellar equatorial circumferential radius are very different in the two cases. In the slowly rotating model D1, in particular, the star collapses smoothly and the worldline always has a negative slope, thus reaching progressively smaller radii as the evolution proceeds (*cf.* left panel of Fig. 3). By time $t \simeq 0.59$ ms, the stellar equatorial circumferential radius has shrunk below the corresponding value of the event horizon. In the case of the rapidly rotating model D4, on the other hand, this is no longer true and after an initial phase which is similar to the one described for D1, the worldline does not reach smaller radii. Rather, the stellar surface slows its inward motion and, at around $t \sim 0.6$ ms, the stellar equatorial circumferential radius does not vary appreciably. Indeed, the right panel of Fig. 3 shows that at this stage the stellar surface moves to slightly larger radii. This behaviour marks the phase in which a flattened configuration has been produced and the material at the outer edge of the disc experiences a stall. As the collapse proceeds, however, also this material will not be able to sustain its orbital motion and, after $t \sim 0.7$ ms, the worldline moves to smaller radii again. By time $t \simeq 0.9$ ms, the stellar equatorial circumferential radius has shrunk below the corresponding value of the event horizon.

6. Gravitational waves

In addition to the technical difficulties due to the accurate treatment of the hydrodynamics involved in the collapse and reported above, the precise calculation of the gravitational radiation emitted in the process is particularly challenging as the energy released in gravitational waves is much smaller than the total rest-mass energy of the system. Indications of the difficulties inherent to the problem of calculating the gravitational-wave emission in rotating gravitational collapse have emerged in the first and only work, which dates back almost 20 years[28]. In 1985, in a landmark work in numerical relativity, Stark and Piran used their axisymmetric general-relativistic code to evolve rotating configurations and to compute the gravitational radiation produced by their collapse to black holes. The results referred to initial configurations consisting of polytropic stars which underwent collapse after the pressure was reduced by a factor ranging from 60% up to 99% for the rapidly rotating models. The initial data effectively consisted of spherically-symmetric solutions with a uniform rotation simply "added" on. Not being stationary solutions of the Einstein equations, these stars could reach dimensionless spins up to $a \equiv J/M^2 = 0.94$, with J and M the angular momentum and mass of the star, respectively.

Overall, their investigation revealed that while the nature of the collapse depended on the parameter a, the form of the waves remained roughly the same over the entire range of the values of a, with the amplitude increasing with a. Particularly important was evidence that, despite the complex matter dynamics during the collapse, the gravitational-wave emission could essentially be related to the oscillations of a perturbed black hole spacetime.

The lack of published works about the gravitational-wave emission from 3D simulations of this process can be justified in part by the difficulties in measuring a signal often below the truncation error of the 3D simulations, but most importantly by the fact that all of the above calculations made use of Cartesian grids with uniform spacing. With the computational resources currently available, this choice, in fact, places the outer boundaries too close to the source to detect gravitational radiation.

The progressive mesh-refinement (PMR) techniques we have used have removed these restrictions, enabling us to place the outer boundaries of the computational domain at very large distances from the collapsing star. This conceptually simple but technologically challenging improvement has two important physical consequences. Firstly, it reduces the influence of inaccurate boundary conditions at the outer boundaries of the domain whilst retaining the required accuracy in the region where the black hole forms. Secondly, it allows the wave-zone to be included in the computational domain and thus to extract the gravitational waves produced in the collapse.

In practice, we have adopted a Berger-Oliger prescription for the refinement of meshes on different levels[9] and used the numerical infrastructure described in Schnetter et al.[25]. In addition to this, we have also implemented a simplified form of adaptivity in which new refined levels are added at predefined positions during the evolution. More specifically, given an initial stellar model of mass M and equatorial coordinate radius R, our initial grid consists of four levels of refinement, with the innermost one covering the star with a typical resolution of $\Delta x \sim 0.17\,M$ and with the outermost having a typical resolution of $\Delta x \sim 1.38\,M$ and extending up to $\sim 82.5\,M \simeq 20.9\ R$. As the collapse proceeds and the star occupies smaller portions of the computational domain, three more refined levels are added one by one, nested in the four original ones. By the time the simulation is terminated at $\sim 81.5\,M$, the finest typical spatial resolution is $\Delta x \sim 0.02\,M$. A detailed discussion of the grid and of its evolution will be given in Baiotti et al.[8]. Also, hereafter we will restrict the discussion to the collapse of the most rapidly rotating dynamically unstable model, namely model D4. The discussion of the emission from stellar models rotating at smaller velocities will be presented in Baiotti et al.[8].

For the gravitational-wave extraction, we have adopted a gauge-invariant approach in which the spacetime is matched with the non-spherical perturbations of a Schwarzschild black hole[1,24,13,23,4,17]. In practice, a set of "observers" is placed on 2-spheres of fixed coordinate radius $r_{\rm ex}$, where they extract the gauge-invariant, odd $Q^{(\rm o)}_{\ell m}$ and even-parity $\Psi^{(\rm e)}_{\ell m}$ metric perturbations[21]. Here ℓ, m are the indices of the angular decomposition and we usually compute modes up to $\ell = 5$ with $m = 0$; modes with $m \neq 0$ are essentially zero because of the high degree of axisymmetry in the collapse. Although the position of such observers is arbitrary and the information they record must be the same for waves extracted in the wave-zone, we place our observers between $40\,M$ and $50\,M$ from the centre of the grid so as to maximize the length of the extracted waveform. We note that while a similar choice was made by Stark and Piran[28], it still provides only an approximate description of what would be observed at spatial infinity.

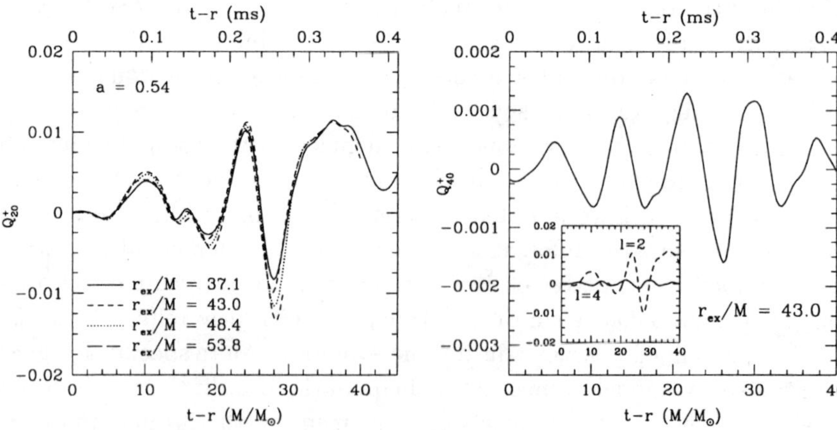

Figure 4. LEFT: The $\ell = 2$, even-parity perturbation as extracted by observers at different positions $r_{\rm ex}$ expressed in retarded time. RIGHT: The $\ell = 4$ mode, with the inset offering a comparison in the amplitudes of the two modes.

Using the odd and even-parity perturbations $Q^{\times}_{\ell m} = \lambda Q^{(\rm o)}_{\ell m}$ and $Q^{+}_{\ell m} =$

$\lambda\Psi^{(e)}_{\ell m}$, where $\lambda \equiv \sqrt{2(\ell+2)!/(\ell-2)!}$, we report in Fig. 4 and for model D4 the lowest-order multipoles for $Q^+_{\ell m}$ with the offset produced by the stellar quadrupole removed[8]. The left panel, in particular, refers to the $\ell = 2$ mode as extracted by four different observers at increasing distances and expressed in retarded time. The right panel instead refers to the $\ell = 4$ mode, with the inset giving a comparison between the two modes and showing that the gravitational-wave signal is essentially quadrupolar, with the $\ell = 2$ mode being about an order of magnitude larger than the $\ell = 4$ mode.

The very good overlap of the waveforms measured at different positions is important evidence that the extraction has been performed in the wave-zone, since the invariance under a retarded-time scaling is a property of the solutions of a wave equation. The overlap disappears if the outer boundary is too close or when the waves are extracted at smaller radii. A similar overlap is seen also for the $\ell = 4$ mode (not shown).

Another indication that the waveforms in Fig. 4 are an accurate description of the gravitational radiation produced by the collapse comes by analysing their power spectra. The collapse can be viewed as the rapid transition between the spacetime of the initial equilibrium star and the spacetime of the produced rotating black hole. It is natural to expect that the waveforms produced in this process will reflect the basic properties of both spacetimes and in particular the fundamental frequencies of oscillation. We validate this in Fig. 5 (left), where we show the power spectral densities (PSD) of the waveforms of the metric perturbations Q^+_{20} and Q^+_{40} reported in Fig. 4 (the units on the y-axis are arbitrary). The upper panel of Fig. 5 (left), in particular, shows the PSD of Q^+_{20} and compares it with the frequencies of the $\ell = 2, m = 0$ quasi-normal mode (QNM) of a Kerr black hole with $M = 1.861\, M_\odot$ and $a = 0.6$[20] (dashed line at 6.7 kHz) as well as with the first $w_{\rm II}$ "interface" mode for a typical compact star with $M = 1.27\, M_\odot$ and $R = 8.86$ km (dotted line at 8.8 kHz)[19]. Similarly, the lower panel of Fig. 5 (left), shows the PSD of Q^+_{40} comparing it with the $\ell = 4, m = 0$ QNM of a Schwarzschild black hole[19] (dashed line at 14.0 kHz) and the first w_1 "curvature" mode[19] (dotted line at 12.8 kHz). The frequencies for the QNMs of the black hole and for the w-modes of the star are those most easily obtained from tabulated values in the literature and serve here as indicative references. It is apparent that both peaks in the PSDs are rather far from the maximum sensitivity area of modern interferometric and bar detectors.

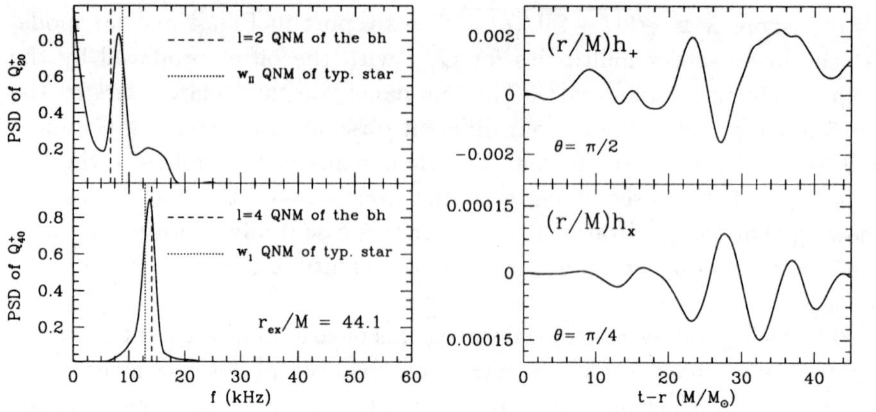

Figure 5. LEFT: Power spectra of the waveforms reported in Fig. 4. The dashed vertical lines indicate the frequencies of the QNMs of a black hole, while the dotted ones the $w_{\rm I}$ and $w_{\rm II}$ modes of a typical star. RIGHT: gravitational-wave amplitudes in the TT gauge.

Although the waveforms have very short duration with very broad PSDs, Fig. 5 (left) shows that these have strong and narrow peaks (a similar behaviour can be shown to be present also for the odd-parity modes[8]). Indeed, the excellent agreement between the position of these peaks and the fundamental frequencies of the vacuum and non-vacuum spacetimes is an important confirmation of the robustness of the results obtained.

Using the extracted gauge-invariant quantities it is also possible to calculate the transverse traceless (TT) gravitational-wave amplitudes in the two polarizations h_+ and $h_\times$ as

$$h_+ - {\rm i}h_\times = \frac{1}{2r}\sum_{\ell,m}\left(Q^+_{\ell m} - {\rm i}\int_{-\infty}^{t} Q^\times_{\ell m}(t')dt'\right)\ {}_{-2}Y^{\ell m}, \qquad (3)$$

where ${}_{-2}Y^{\ell m}$ is the $s = -2$ spin-weighted spherical harmonic. Because of the small amplitude of higher-order modes, the TT wave amplitudes can be simply expressed as $h_+ \simeq h_+(Q^+_{20}, Q^+_{40})$ and $h_\times \simeq h_\times(Q^\times_{30}, Q^\times_{50})$, where $Q^\times_{30} \gg Q^\times_{50}$. Their waveforms are shown in Fig. 5 (right) for the detector at $r_{\rm ex} = 37.1\,M$ and for two different inclination angles[8]. Note that the amplitudes in the cross polarization are about one order of magnitude smaller than those in the plus polarization, with the maximum amplitudes in a ratio $|(r/M)h_\times|_{\rm max}/|(r/M)h_+|_{\rm max} \simeq 0.06$. Here the odd-parity perturbations, which are zero in spacetimes with axial and equatorial symmetries, are just the result of the coupling, induced by the rotation, with the even-parity perturbations.

A precise comparison of the amplitudes in Fig. 5 (right) with the corresponding ones calculated by Stark and Piran[28] is made difficult by the differences in the choice of initial data and, in particular, by the impossibility of reaching $a \gtrsim 0.54$ when modelling consistently stationary polytropes in uniform rotation. However, when interpolating their results for the relevant values of a, we find a very good agreement in the form of the waves, but also that our estimates are about one order of magnitude smaller, with $|(r/M)h_+|_{\rm max} \simeq 0.00225$. Furthermore, we observe the amplitude of the gravitational waves to increase with the pressure reduction, suggesting that the origin of the difference is related mainly to this (see Baiotti et al.[8]).

Following Thorne[31] and considering the optimal sensitivity of VIRGO for the burst signal only, we set an upper limit for the characteristic amplitude produced in the collapse of a rapidly and uniformly rotating polytropic star at 10 kpc to be $h_c = 5.77 \times 10^{-22}(M/M_\odot)$ at a characteristic frequency $f_c = 931$ Hz. In the case of LIGO I, instead, we obtain $h_c = 5.46 \times 10^{-22}(M/M_\odot)$ at $f_c = 531$ Hz. In both cases, the signal-to-noise ratio is $S/N \sim 0.25$, but this can grow to be $\lesssim 4$ in the case of LIGO II. These ratios could be increased considerably with the detection of the black hole ringing following the initial burst (see Baiotti et al.[8]). Computing the emitted power as

$$\frac{dE}{dt} = \frac{1}{32\pi} \sum_{\ell,m} \left(\left| \frac{dQ^{+}_{\ell m}}{dt} \right|^2 + \left| Q^{\times}_{\ell m} \right|^2 \right) , \tag{4}$$

the total energy lost to gravitational radiation is $E = 1.45 \times 10^{-6}(M/M_\odot)$. This is about two orders of magnitude smaller than the estimate made by Stark and Piran[28] for a star with $a = 0.54$, but still larger than the energy losses computed recently in the collapse of rotating stellar cores to protoneutron stars by Müller et al.[22].

7. Conclusion

We have developed recently developed and tested `Whisky`, a 3D general-relativistic numerical evolution code that combines state-of-the-art numerical methods for the spacetime evolution with an accurate hydrodynamical evolution employing several high-order HRSC methods and hydrodynamical excision techniques. The evolution of the spacetime and of the hydrodynamics is coupled transparently through the method of lines, which allows for the straightforward implementation of various different time-integrators.

As a first astrophysical problem for this novel setup, we have here focused on the collapse of rapidly rotating relativistic stars to Kerr black holes. The stars are assumed to be in uniform rotation and dynamically unstable to axisymmetric perturbations. While the collapse of slowly rotating initial models proceeds with the matter remaining nearly uniformly rotating, the dynamics is shown to be very different in the case of initial models rotating near the mass-shedding limit. Although the stars become highly flattened during collapse, attaining a disc-like shape, the collapse cannot be prevented because the specific angular momentum is not sufficient for a stable disc to form. Instead, the matter in the disc spirals towards the black hole and angular momentum is transferred inward to produce a spinning black hole.

Several different approaches have been employed to compute the mass and angular momentum of the newly formed Kerr black hole. Besides more traditional methods involving the measure of the geometrical properties of the apparent *and* event horizons, we have fitted the oscillations of the perturbed Kerr black hole to specific quasi-normal modes obtained by linear perturbation theory. In addition, we have also considered the recently proposed isolated and dynamical horizon frameworks, finding it to be simple to implement and yielding estimates which are accurate and more robust than those of other methods. This variety of approaches has allowed for the determination of both the mass and angular momentum of the black hole with an accuracy unprecedented for a 3D simulation.

We have also presented the first waveforms from the gravitational collapse of rapidly rotating stars to black holes using 3D grids with Cartesian coordinates. The great potential shown by the PMR techniques employed here opens the way to a number of applications that would be otherwise intractable with uniform Cartesian grids. Work is in progress to consider initial models with realistic EOSs or in differential rotation, for which values $a \gtrsim 1$ can be reached and more intense gravitational radiation is expected.

Acknowledgments

It is a pleasure to thank T. Font, F. Löffler, P. Montero, E. Seidel, N. Stergioulas and O. Zanotti for useful discussions. We are also grateful to the `Cactus`-code team, for their efforts in producing an efficient infrastructure for numerical relativity. The computations were performed on the *Albert100* cluster at the University of Parma and on the *Peyote* cluster at the Albert-Einstein-Institut.

References

1. G. Allen, K. Camarda, E. Seidel, gr-qc/9806036 (1998)
2. A. Ashtekar, C. Beetle, S. Fairhurst, Class. Quantum Grav. **17**, 253 (2000)
3. A. Ashtekar, B. Krishnan, Phys. Rev. Lett. **89**, 261101 (2002)
4. J. Baker, S. Brandt, M. Campanelli, C. O. Lousto, E. Seidel, R. Takahashi, Phys. Rev. D **62**, 127701 (2000)
5. L. Baiotti and I. Hawke and P. Montero and L. Rezzolla, Mem. Soc. Astron. It., **1**, 327 (2003)
6. L. Baiotti, I. Hawke, P. J. Montero, F. Löffler, L. Rezzolla, N. Stergioulas, J.A. Font, E. Seidel, Phys. Rev. **D71**, 024035 (2005)
7. L. Baiotti, I. Hawke, L. Rezzolla, E. Schnetter, Phys. Rev. Lett., **94**, *in press*, (2005)
8. L. Baiotti *et al.*, *in preparation* (2005)
9. M.J. Berger, J. Oliger, J. Comput. Phys. **53**, 484 (1984)
10. S. Brandt, E. Seidel, Phys. Rev. **D52**, 856 (1995)
11. S. Brandt J. A. Font, J. M. Ibáñez, J. Massó, E. Seidel, Comp. Phys. Comm. **124**, 169 (2000)
12. http://www.cactuscode.org
13. K. Camarda, E. Seidel, Phys. Rev. D **59**, 064019 (1999)
14. P. Diener, Class. Quantum Grav. **20**, 4901,(2003)
15. O. Dreyer, B. Krishnan, D. Shoemaker, E. Schnetter, Phys. Rev. **D67**, 024018 (2002)
16. M.D. Duez, S.L. Shapiro, H.J. Yo, Phys. Rev. D **69**, 104016 (2004)
17. J.A. Font, T. Goodale, S. Iyer, M. Miller, L. Rezzolla, E. Seidel, N. Stergioulas, W. M. Suen, M. Tobias, Phys. Rev. D **65**, 084024 (2002)
18. I. Hawke, F. Löffler, A. Nerozzi, gr-qc/0501054.
19. K.D. Kokkotas, B.G. Schmidt, Living Rev. Relativity **2** (1999)
20. E.W. Leaver, Proc. R. Soc. Lond. A **402**, 285 (1985)
21. V. Moncrief, Annals of Physics **88**, 323 (1974)
22. E. Müller, M. Rampp, R. Buras, H.T. Janka, D.H. Shoemaker, Astrophys. J. **603**, 221 (2004)
23. L. Rezzolla, A. M. Abrahams, R. A. Matzner, M. Rupright, S. L. Shapiro Phys. Rev. D **59**, 064001 (1999)
24. M.E. Rupright, A.M. Abrahams, L. Rezzolla, Phys. Rev. D **58**, 044005 (1998)
25. E. Schnetter, S.H. Hawley, I. Hawke, Class. Quantum Grav. **21**, 1465 (2004)
26. M. Shibata, T.W. Baumgarte, S.L. Shapiro, Phys. Rev. D **61**, 044012 (2000)
27. M. Shibata, Phys. Rev. D **67**, 024033 (2003); Astrophys. J. **595**, 992 (2003)
28. R.F. Stark, T. Piran, Phys. Rev. Lett. **55**, 891 (1985); Proceedings of the workshop *Dynamical spacetimes and numerical relativity*, Philadelphia, CUP, Cambridge UK, pp. 40-73 (1986)
29. N. Stergioulas, J. L. Friedman, Astrophys. J. **444**, 306 (1995).
30. J. Thornburg,Class. Quantum Grav.,**21**, 743 (2004)
31. K. S. Thorne in *300 Years of Gravitation*, S. W. Hawking, W. Israel, eds., CUP, Cambridge, UK (1987)

THE ROLE OF DIFFERENTIAL ROTATION IN THE EVOLUTION OF THE r-MODE INSTABILITY

PAULO M. SÁ AND BRIGITTE TOMÉ

Departamento de Física and Centro Multidisciplinar de Astrofísica - CENTRA, Universidade do Algarve, Campus de Gambelas, 8005-139 Faro, Portugal
E-mail: pmsa@ualg.pt, btome@ualg.pt

We discuss the role of differential rotation in the evolution of the $l = 2$ r-mode instability of a newly born, hot, rapidly-rotating neutron star. It is shown that the amplitude of the r-mode saturates in a natural way at a value that depends on the amount of differential rotation at the time the instability becomes active. It is also shown that, independently of the saturation amplitude of the mode, the star spins down to a rotation rate that is comparable to the inferred initial rotation rates of the fastest pulsars associated with supernova remnants.

1. Introduction

The r-modes are non-radial pulsations modes of rotating stars that have the Coriolis force as their restoring force and a characteristic frequency comparable to the rotation speed of the star[1]. These modes are driven unstable by gravitational radiation in all rotating stars[2]. A deeper understanding of r-modes and their astrophysical relevance requires taking into account nonlinear effects in the evolution of the r-mode instability. One such a nonlinear effect, differential rotation induced by r-modes, has been studied by several authors. Rezzolla, Lamb and Shapiro[3] were the first to suggest that differential rotation drifts of kinematical nature could be induced by r-modes, deriving an approximate analytical expression for this differential rotation. Afterwards, the existence of such drifts was confirmed numerically both in general relativistic[4] and Newtonian hydrodynamics[5]. Recently, an exact r-mode solution, describing differential rotation of pure kinematic nature that produces large scale drifts along stellar latitudes, was found within the nonlinear theory up to second order in the mode's amplitude[6]. In this paper we discuss the role of differential rotation in the evolution of the $l = 2$ r-mode instability of a newly born, hot, rapidly-rotating neutron star.

2. The evolution model of Owen *et al.*

A few years ago, Owen *et al.*[7] proposed a simple model to study the evolution of the r-mode instability in newly born, hot, rapidly-rotating neutron stars. Within this model, it is assumed that the time evolution of the system (star and r-mode perturbation) is characterized by two parameters: the angular velocity of the star, $\Omega(t)$, and the amplitude of the mode, $\alpha(t)$. The total angular momentum of the star is then given by

$$J = I\Omega + J_c \tag{1}$$

where the momentum of inertia I of the equilibrium configuration is $I = \tilde{I}MR^2$, with $\tilde{I} = 0.261$, and the canonical angular momentum J_c of the r-mode perturbation is $J_c = -3/2\alpha^2\Omega\tilde{J}MR^2$, with $\tilde{J} = 1.635 \times 10^{-2}$. Here, only the $l = 2$ r-mode is being considered and it is assumed that the mass density ρ and the pressure p of the fluid are related by a polytropic equation of state $p = k\rho^2$, with k such that the mass and radius of the star take the values $M = 1.4M_\odot$ and $R = 12.53$ km, respectively.

Within this model it is also assumed that the total angular momentum of the star decreases due to the emission of gravitational radiation and that the physical energy of the r-mode perturbation (in the co-rotating frame) increases due to the emission of gravitational radiation and decreases due to the dissipative effect of viscosity. These assumptions lead then to a system of differential equations for the angular momentum of the star, Ω, and the amplitude of the r-mode, α:

$$\frac{d\Omega}{dt} = -\frac{2\Omega}{\tau_V}\frac{Q\alpha^2}{1+Q\alpha^2}, \tag{2}$$

$$\frac{d\alpha}{dt} = -\frac{\alpha}{\tau_{GR}} - \frac{\alpha}{\tau_V}\frac{1-Q\alpha^2}{1+Q\alpha^2}, \tag{3}$$

where $Q = 3\tilde{J}/(2\tilde{I}) = 0.094$, the gravitational-radiation and viscous timescales, τ_{GR} and τ_V, are given by[8]

$$\frac{1}{\tau_{GR}} = \frac{1}{\tilde{\tau}_{GR}}\left(\frac{\Omega^2}{\pi G\bar{\rho}}\right)^3, \tag{4}$$

$$\frac{1}{\tau_V} = \frac{1}{\tilde{\tau}_S}\left(\frac{10^9\text{K}}{T}\right)^2 + \frac{1}{\tilde{\tau}_B}\left(\frac{T}{10^9\text{K}}\right)^6\left(\frac{\Omega^2}{\pi G\bar{\rho}}\right). \tag{5}$$

In the above expressions, the fiducial timescales are $\tilde{\tau}_{GR} = -3.26\,\text{s}$, $\tilde{\tau}_S = 2.52 \times 10^8\,\text{s}$ and $\tilde{\tau}_B = 6.99 \times 10^8\,\text{s}$.

For a newly born, hot, rapidly-rotating neutron star there is an interval of relevant temperatures and angular velocities of the star for which the gravitational timescale is much smaller than the viscous timescale, $\tau_{\rm GR} \ll \tau_{\rm V}$. Therefore, for this interval of temperatures and angular velocities, we can neglect in the right-hand side of Eqs. (2) and (3) the terms proportional to τ_V^{-1} and obtain the solution $\Omega = \Omega_0$ and $\alpha = \alpha_0 \exp\{-(t-t_0)/\tau_{GR}\}$. If the initial angular velocity is chosen to be $\Omega_0 = \Omega_K$, where $\Omega_K = (2/3)\sqrt{\pi G \bar{\rho}}$ is the Keplerian angular velocity at which the star starts shedding mass at the equator, then $\tau_{GR} = -37.1\,{\rm s}$, implying that the perturbation grows exponentially from an initial amplitude $\alpha_0 = 10^{-6}$ to values of order unity in just about 500 s [7]. After this first stage of evolution of the r-mode instability, the amplitude α has to be forced, by hand, to take a certain saturation value $\alpha_{sat} \leqslant Q^{-1/2} = 3.26$; in the second stage of the evolution it is then assumed that $\alpha = \alpha_{sat}$ and Ω is determined from the equation

$$\frac{d\Omega}{dt} = \frac{2\Omega}{\tau_{GR}} \frac{\alpha_{sat}^2 Q}{1-\alpha_{sat}^2 Q}, \tag{6}$$

which yields the solution

$$\Omega(t) = \Omega_0 \left[\frac{0.030\alpha_{sat}^2}{1-Q\alpha_{sat}^2} \left(\frac{\Omega_0}{\Omega_K}\right)^6 \left(\frac{t-t_*}{\rm sec}\right) + 1 \right]^{-1/6}, \tag{7}$$

where t_* is the time at which occurs the transition from the first to the second stage of evolution. During the second stage of evolution, the star loses its angular momentum, spinning down to its final angular velocity. As can be seen from the above equations, the final angular velocity of the star depends critically on the saturation value of the mode's amplitude α_{sat}; for instance, after one year of evolution, for $\Omega_0 = \Omega_K$ and $\alpha_{sat} = 1$ one obtains $\Omega \simeq 0.1\Omega_K$, while for $\alpha_{sat} = 10^{-3}$ the angular velocity is $\Omega \simeq 0.9\Omega_K$. The fact that the growth of the mode's amplitude has to be stopped by hand at a certain saturation value introduces an element of arbitrariness into the solution, permitting, for instance, that agreement between the predicted final value of the angular velocity of the star and the value inferred from astronomical observations can always be achieved by fine-tuning the value of the saturation amplitude. As will be seen in the next section, this arbitrariness disappears when differential rotation is taken into account.

3. Evolution model with differential rotation

Differential rotation induced by r-modes *does* contribute to the physical angular momentum of the r-mode perturbation[6]. Therefore, a model for the evolution of the r-mode instability should include the effect of differential rotation. Here, we adopt the model of Owen *et al.*[7], described in the previous section, with the important difference[a] that the total angular momentum of the star is given by

$$J = I\Omega + \delta^{(2)}J, \tag{8}$$

where the physical angular momentum of the r-mode perturbation $\delta^{(2)}J$ is[6]

$$\delta^{(2)}J = \frac{1}{2}\alpha^2\Omega(4K+5)\tilde{J}MR^2. \tag{9}$$

In the previous expression, K is a constant fixed by the choice of initial data and gives the initial amount of differential rotation associated with the r-mode. The specific case $K = -2$, for which the physical angular momentum of the r-mode perturbation coincides with the canonical angular momentum, corresponds to the model of Owen *et al.*[7] described in great detail in the previous section. There is not, to our knowledge, any physical condition that forces K to take such a particular value $K = -2$. Therefore, we will consider in this paper arbitrary values of K in the interval $-5/4 \leqslant K \ll 10^{13}$ [b].

Following the procedure described in the previous section, we arrive at a system of two first-order coupled differential equations determining the time evolution of the amplitude of the r-mode $\alpha(t)$ and of the angular velocity of the star $\Omega(t)$:

$$\frac{d\Omega}{dt} = \frac{8}{3}(K+2)Q\frac{\Omega\alpha^2}{\tau_{GR}}, \tag{10}$$

$$\frac{d\alpha}{dt} = -\left[1+\frac{4}{3}(K+2)Q\alpha^2\right]\frac{\alpha}{\tau_{GR}}, \tag{11}$$

valid when the damping effect of viscosity can be neglected relatively to the driving effect of gravitational radiation.

---

[a]There is another difference, albeit less significant, between the model of Owen *et al.*[7] and the model we adopt here, namely, we deduce the evolution equations just from angular momentum considerations.

[b]The upper limit for K results from the fact that one wishes to impose the condition that $|\delta^{(2)}J(t_0)| \ll I\Omega_0$ (the initial amplitude $\alpha_0 \equiv \alpha(t_0)$ is considered throughout the paper to be 10^{-6}). The lower limit for K results from the fact that we do not want to saturate the amplitude of the mode by hand, a procedure needed for the case $K < -5/4$ in order to avoid that the total angular momentum of the star becomes negative.

This system of equations was solved analytically and discussed in great detail in Ref. [9]. In the initial stages of the evolution of the r-mode instability α increases exponentially[9],

$$\alpha(t) \simeq \alpha_0 \exp\left\{0.027 \left(\frac{\Omega_0}{\Omega_K}\right)^6 \left(\frac{t-t_0}{\text{sec}}\right)\right\}; \tag{12}$$

while for later times, α increases very slowly as

$$\alpha(t) \simeq 2.48 \left(\frac{\Omega_0}{\Omega_K}\right)^{3/5} \left(\frac{t-t_0}{\text{sec}}\right)^{1/10} \frac{1}{\sqrt{K+2}}. \tag{13}$$

The amplitude of the mode saturates in a natural way and a smooth transition between the regimes (12) and (13) occurs for $t - t_0 \simeq \text{few} \times 10^2$ seconds (see Fig. 1). This contrasts with the model of Owen *et al.*[7], in which, after the short initial period of exponential growth, the amplitude α has to be forced by hand to take a certain saturation value $\alpha_{sat} \leqslant Q^{-1/2}$. As can

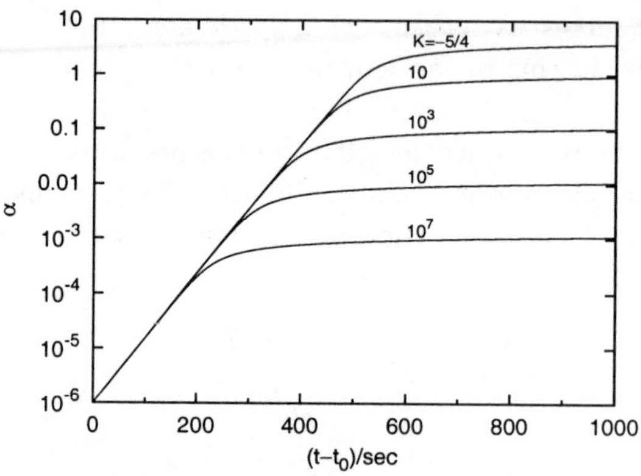

Figure 1. Time evolution of the amplitude of the r-mode for different values of K. The initial values of the amplitude of the mode and of the angular velocity of the star are, respectively, $\alpha_0 = 10^{-6}$ and $\Omega_0 = \Omega_K$.

be seen from Eq. (13), the saturation value of the amplitude of the mode depends crucially on the parameter K, namely, $\alpha_{sat} \propto (K+2)^{-1/2}$. If $K \simeq 0$, corresponding to a situation in which the initial amount of differential rotation is small, then the r-mode saturates at values of order unity. If, on the other hand, the initial differential rotation associated to r-modes is significant, then the saturation amplitude α_{sat} can be as small as $10^{-3} - 10^{-4}$.

Let us now turn our attention to the time evolution of the angular velocity of the star, Ω. In the initial stages of the evolution of the r-mode instability Ω decreases as[9]

$$\frac{\Omega(t)}{\Omega_0} \simeq 1 - \frac{4}{3}(K+2)Q\alpha_0^2 \exp\left\{0.054\left(\frac{\Omega_0}{\Omega_K}\right)^6\left(\frac{t-t_0}{\text{sec}}\right)\right\}; \qquad (14)$$

while for later times Ω decreases slowly as

$$\frac{\Omega(t)}{\Omega_0} \simeq 1.30\left(\frac{\Omega_0}{\Omega_K}\right)^{-6/5}\left(\frac{t-t_0}{\text{sec}}\right)^{-1/5}. \qquad (15)$$

The smooth transition between the regimes (14) and (15) occurs for $t - t_0 \simeq$ few $\times 10^2$ seconds (see Fig. 2). Remarkably, in the later phase of the

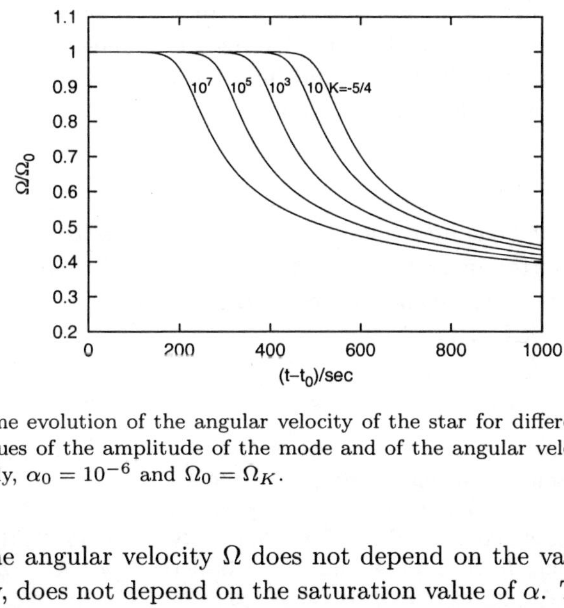

Figure 2. Time evolution of the angular velocity of the star for different values of K. The initial values of the amplitude of the mode and of the angular velocity of the star are, respectively, $\alpha_0 = 10^{-6}$ and $\Omega_0 = \Omega_K$.

evolution, the angular velocity Ω does not depend on the value of K and, consequently, does not depend on the saturation value of α. This contrasts with the results obtained in Ref. [7], where the value of Ω depends critically on the choice of α_{sat}.

After about one year of evolution, when the dissipative effect of viscosity becomes dominant and starts damping the mode, the angular velocity of the star becomes $\Omega_{\text{one year}} \simeq 0.042\Omega_K$ (for $\Omega_0 = \Omega_K$), in good agreement with the inferred initial angular velocity of the fastest pulsars associated with supernova remnants.

4. Conclusions

In this paper we have discussed the role of differential rotation in the evolution of the $l = 2$ r-mode instability of a newly born, hot, rapidly-rotating neutron star. We have shown that, a few hundred seconds after the mode instability sets in, the amplitude of the r-mode saturates in a natural way at values that depend on the initial amount of differential rotation associated to the r-mode. If the initial differential rotation of r-modes is small, then the r-mode saturates at values of order unity. On the other hand, if the initial differential rotation is significant, then the saturation amplitude can be as small as $10^{-3} - 10^{-4}$. These low values for the saturation amplitude of r-modes are of the same order of magnitude as the ones obtained in recent investigations on wind-up of magnetic fields[3] and on nonlinear mode-mode interaction[10]. We have also shown that the value of the angular velocity of the star becomes, after a short period of evolution of the r-mode instability, very insensitive to the saturation value of the mode's amplitude. After about one year of evolution the angular velocity is $0.042\Omega_K$ (for any α_{sat}), in good agreement with the inferred initial angular velocity of the fastest pulsars associated with supernova remnants.

Acknowledgments

We thank Óscar Dias and Luciano Rezzolla for helpful discussions. This work was supported in part by the *Fundação para a Ciência e a Tecnologia* (FCT), Portugal. BT acknowledges financial support from FCT through grant PRAXIS XXI/BD/21256/99.

References

1. J. Papaloizou and J. E. Pringle, *Mon. Not. R. Astron. Soc.* **182**, 423 (1978).
2. N. Andersson, *Astrophys. J.* **502**, 708 (1998).
3. L. Rezzolla, F. K. Lamb and S. L. Shapiro, *Astrophys. J.* **531**, L139 (2000).
4. N. Stergioulas and J. A. Font, *Phys. Rev. Lett.* **86**, 1148 (2001).
5. L. Lindblom, J. E. Tohline and M. Vallisneri, *Phys. Rev. Lett.* **86**, 1152 (2001).
6. P. M. Sá, *Phys. Rev. D.* **69**, 084001 (2004).
7. B. J. Owen, L. Lindblom, C. Cutler, B. F. Schutz, A. Vecchio and N. Andersson, *Phys. Rev. D.* **58**, 084020 (1998).
8. L. Lindblom, B. J. Owen and S. M. Morsink, *Phys. Rev. Lett.* **80**, 4843 (1998).
9. P. M. Sá and B. Tomé, *Phys. Rev. D* **71**, 044007 (2005).
10. P. Arras, E. E. Flanagan, S. M. Morsink, A. K. Schenk, S. A. Teukolsky and I. Wasserman, *Astrophys. J.* **591**, 1129 (2003).

ANALYTICAL r-MODE SOLUTION WITH GRAVITATIONAL RADIATION REACTION FORCE

ÓSCAR J. C. DIAS*

Perimeter Institute for Theoretical Physics, 31 Caroline St. N.,
Waterloo, Ontario N2L 2Y5, Canada
E-mail: odias@perimeterinstitute.ca

PAULO M. SÁ

Departamento de Física and Centro Multidisciplinar de Astrofísica – CENTRA,
Universidade do Algarve, Campus de Gambelas, 8005-139 Faro, Portugal
E-mail: pmsa@ualg.pt

We present and discuss the analytical r-mode solution to the linearized hydrodynamic equations of a slowly rotating, Newtonian, barotropic, non-magnetized, perfect-fluid star in which the gravitational radiation reaction force is present.

1. Introduction

The r-modes, which are pulsation modes of rotating stars that have the Coriolis force as their restoring force, are driven unstable by gravitational radiation (GR)[1]. In the frame co-rotating with the star, the energy of the unstable r-mode grows exponentially, $E = E_0 e^{-2t/\tau_{GR}}$, with the gravitational timescale $\tau_{\rm GR}$ given by[2,3]

$$\tau_{\rm GR} = -\left(\frac{2^{17}\pi}{5^2 \times 3^8}\,\frac{G}{c^7}\,\tilde{J}\Omega^6\right)^{-1}, \tag{1}$$

where G is the Newton's constant, c is the velocity of light, Ω is the angular velocity of the star and $\tilde{J} \equiv \int_0^R dr\hat{\rho}r^6$ with $\hat{\rho}$ and R being, respectively, the density and the surface's radius of the unperturbed star. For a wide range of relevant temperatures and angular velocities of newly born, hot, rapidly-rotating stars, bulk and shear viscosity do not suppress the exponential growth of the energy of the r-mode[2,3].

*also at Centro Multidisciplinar de Astrofísica – CENTRA.

In the above-mentioned investigations[2,3], the linearized hydrodynamics equations with the GR force were used to obtain an expression for the time evolution of the physical energy of the r-mode perturbation, dE/dt, from which the gravitational radiation and viscous timescales were determined. In this work, in which the main results of Ref. [4] are reported, we present an explicit expression for the r-mode velocity perturbations that solves the linearized hydrodynamics equations with the GR force. Our conclusions are that (i) these velocity perturbations are sinusoidal with the same frequency as the well-known GR force-free linear r-mode solution, (ii) the GR force drives the r-modes unstable with a growth timescale that agrees with the expression found in Refs. [2,3] and (iii) the amplitude of these velocity perturbations is corrected, relatively to the GR force-free case, by a term of order Ω^6.

2. Hydrodynamic equations with GR reaction force

The Newtonian hydrodynamic equations for a uniformly rotating, barotropic, non-magnetized, perfect-fluid star in the presence of the gravitational radiation (GR) reaction force are the Euler, continuity, and Poisson equations given, respectively, by

$$\partial_t \vec{v} + (\vec{v} \cdot \vec{\nabla})\vec{v} = -\rho^{-1}\vec{\nabla}P - \vec{\nabla}\Phi + \vec{F}^{\rm GR}, \tag{2}$$

$$\partial_t \rho + \vec{\nabla} \cdot (\rho \vec{v}) = 0, \tag{3}$$

$$\nabla^2 \Phi = 4\pi G \rho, \tag{4}$$

where ρ, P and $\vec{v}$ are, respectively, the density, the pressure and the velocity of the fluid, Φ is the Newtonian potential, and the GR reaction force,

$$\vec{F}^{\rm GR} = -\partial_t \vec{\beta} + \vec{v} \times (\vec{\nabla} \times \vec{\beta}), \tag{5}$$

is assumed to be given by the 3.5 post-Newtonian order expansion that includes the contribution of the current quadrupole moment, which is the main responsible for the GR instability that sets in[5,6]. In Eq. (5), $\vec{\beta}$ is the gravitational vector potencial whose components are given by

$$\beta_i \equiv \frac{16G}{45c^7} \epsilon_{ijk} x_j x_q S^{[5]}_{kq}, \tag{6}$$

where $S_{ij}(t)$ is the time-varying current quadrupole tensor,

$$S_{ij}(t) \equiv \int d^3x \epsilon_{kq(i} x_{j)} x_k \rho v_q, \tag{7}$$

ϵ_{ijk} is the Levi-Civita tensor, x_i is the Cartesian coordinate of the point at which the tensor is evaluated, and $S_{ij}^{[n]}(t)$ denotes the nth time derivative of S_{ij}.

3. Linearized hydrodynamics equations with GR reaction force

The hydrodynamic equations (2)–(4) can be linearized, yielding

$$\partial_t \delta^{(1)}\vec{v} + (\delta^{(1)}\vec{v} \cdot \vec{\nabla})\hat{\vec{v}} + (\hat{\vec{v}} \cdot \vec{\nabla})\delta^{(1)}\vec{v} = -\vec{\nabla}\delta^{(1)}U + \delta^{(1)}\vec{F}^{\rm GR}, \tag{8}$$

$$\partial_t \delta^{(1)}\rho + \hat{\vec{v}} \cdot \vec{\nabla}\delta^{(1)}\rho + \vec{\nabla} \cdot \left(\hat{\rho}\delta^{(1)}\vec{v}\right) = 0, \tag{9}$$

$$\nabla^2 \delta^{(1)}\Phi = 4\pi G \delta^{(1)}\rho, \tag{10}$$

where $\hat{\vec{v}} = \Omega r \sin\theta \vec{e}_\phi$ is the velocity of the unperturbed star, $\hat{\rho}$ its mass density, $\delta^{(1)}Q$ denotes the first-order Eulerian change in a quantity Q and we have defined $\delta^{(1)}U \equiv \delta^{(1)}P/\hat{\rho} + \delta^{(1)}\Phi$.

To compute the first-order Eulerian change in the GR force, $\delta^{(1)}\vec{F}^{\rm GR}$, we assume that the GR force-free r-mode velocity perturbations[a],

$$\begin{aligned} \delta^{(1)}v_r &= 0, \\ \delta^{(1)}v_\theta &= -\tfrac{i}{4}\sqrt{\tfrac{5}{\pi}}\tfrac{\alpha\Omega}{R} r^2 \sin\theta e^{i(2\phi+\omega t)}, \\ \delta^{(1)}v_\phi &= \tfrac{1}{4}\sqrt{\tfrac{5}{\pi}}\tfrac{\alpha\Omega}{R} r^2 \sin\theta\cos\theta e^{i(2\phi+\omega t)}, \end{aligned} \tag{11}$$

act as a source for the first-order Eulerian change in the current quadrupole tensor, $\delta^{(1)}S_{ij}$. In the above expression, α is the amplitude of the r-mode and we assume that

$$\omega = \omega_0 + i\varpi, \tag{12}$$

where the frequency of the r-mode, $\omega_0 \equiv \mathrm{Re}[\omega]$, and the small imaginary part that is related to the growth timescale of the instability of the mode, $\varpi = \mathrm{Im}[\omega] < 0$, are arbitrary parameters to be determined.

Under the above assumptions, $\delta^{(1)}S_{xx}$ is computed to be

$$\delta^{(1)}S_{xx} = -\alpha\Omega\sqrt{\frac{\pi}{5}}\frac{\tilde{J}}{R}e^{-\varpi t}e^{i\omega_0 t}, \tag{13}$$

where we have neglected the contribution coming from the terms $\hat{v}_i\delta^{(1)}\rho$ (of order $\alpha\Omega^3$) and retained only the dominant terms $\hat{\rho}\delta^{(1)}v_i$ (of order $\alpha\Omega$).

---

[a]In this article, we will be concerned exclusively with the $l = 2$ r-mode, which is the most unstable mode.

Similarly, it is straightforward to show that the first-order Eulerian change in the other components of the quadrupole tensor satisfy the relations

$$\begin{aligned} \delta^{(1)}S_{yy} &= -\delta^{(1)}S_{xx}, \\ \delta^{(1)}S_{xy} &= i\delta^{(1)}S_{xx}, \\ \delta^{(1)}S_{xz} &= \delta^{(1)}S_{yz} = \delta^{(1)}S_{zz} = 0. \end{aligned} \tag{14}$$

Using Eqs. (13) and (14), the first-order Eulerian change in the gravitational vector potential, $\delta^{(1)}\vec{\beta}$, is then computed to be

$$\begin{aligned} \delta^{(1)}\beta_r &= 0, \\ \delta^{(1)}\beta_\theta &= -\kappa\tfrac{\tilde{J}}{R}\alpha\Omega\omega_0^5 r^2 \sin\theta e^{-\varpi t} e^{i(2\phi+\omega_0 t)}, \\ \delta^{(1)}\beta_\phi &= -i\kappa\tfrac{\tilde{J}}{R}\alpha\Omega\omega_0^5 r^2 \sin\theta\cos\theta e^{-\varpi t} e^{i(2\phi+\omega_0 t)}, \end{aligned} \tag{15}$$

where the constant κ sets the strength of the GR reaction force and is defined as

$$\kappa \equiv \frac{16}{45}\sqrt{\frac{\pi}{5}}\frac{G}{c^7}. \tag{16}$$

Finally, taking into account that $\hat{S}_{ij} = 0$, implying that $\hat{\beta}_i = 0$, the first-order Eulerian change in the GR force, $\delta^{(1)}\vec{F}^{\rm GR}$, is computed to be

$$\begin{aligned} \delta^{(1)}F_r^{\rm GR} &= -3i\kappa\tfrac{\tilde{J}}{R}\alpha\Omega^2\omega_0^5 r^2 \sin^2\theta\cos\theta e^{-\varpi t} e^{i(2\phi+\omega_0 t)}, \\ \delta^{(1)}F_\theta^{\rm GR} &= i\kappa\tfrac{\tilde{J}}{R}\alpha\Omega\omega_0^5 \left(\omega_0 + 3\Omega\sin^2\theta\right) r^2 \sin\theta\, e^{-\varpi t} e^{i(2\phi+\omega_0 t)}, \\ \delta^{(1)}F_\phi^{\rm GR} &= -\kappa\tfrac{\tilde{J}}{R}\alpha\Omega\omega_0^6 r^2 \sin\theta\cos\theta e^{-\varpi t} e^{i(2\phi+\omega_0 t)}. \end{aligned} \tag{17}$$

4. The analytical r-mode solution with GR reaction force

The linearized hydrodynamic equations (8)–(10), with the first-order Eulerian change in the GR force, $\delta^{(1)}\vec{F}^{\rm GR}$, given by Eq. (17), admit the solution[4]

$$\begin{aligned} \delta^{(1)}v_r &= 0, \\ \delta^{(1)}v_\theta &= -\tfrac{i}{2}\alpha\Omega\left[\tfrac{1}{2}\sqrt{\tfrac{5}{\pi}}\tfrac{1}{R} + i\gamma\left(A-1\right)\Omega^5\right] r^2 \sin\theta\, e^{-\varpi t} e^{i(2\phi+\omega_0 t)}, \\ \delta^{(1)}v_\phi &= \tfrac{1}{2}\alpha\Omega\left[\tfrac{1}{2}\sqrt{\tfrac{5}{\pi}}\tfrac{1}{R} + i\gamma\left(A-1\right)\Omega^5\right] r^2 \sin\theta\cos\theta\, e^{-\varpi t} e^{i(2\phi+\omega_0 t)}, \end{aligned} \tag{18}$$

and

$$\delta^{(1)}U = \frac{1}{3}\alpha\Omega^2\left(\frac{1}{2}\sqrt{\frac{5}{\pi}}\frac{1}{R} + i\gamma A\Omega^5\right) r^3 \sin^2\theta\cos\theta\, e^{-\varpi t} e^{i(2\phi+\omega_0 t)}, \tag{19}$$

with ϖ and ω_0 given by

$$\varpi = -\frac{8}{9}\sqrt{\frac{\pi}{5}}\gamma R\Omega^6 \quad \text{and} \quad \omega_0 = -\frac{4\Omega}{3}. \tag{20}$$

The velocity perturbations given by Eq. (18) have a piece similar to the GR force-free solution, the difference being the factor $e^{-\varpi t}$ responsible for the exponential growth of the r-mode amplitude due to the presence of a GR reaction force, and another piece proportional to $\alpha\gamma(A-1)\Omega^6$, where A is a constant fixed by the choice of initial data and $\gamma \equiv 1024\kappa\tilde{J}/(81R)$.

The GR force-free limit is obtained when we set the parameter κ, defined in Eq. (16), equal to zero. In this limit, ϖ and γ also go to zero. Then, from Eqs. (18) and (19), we recover the GR force-free linear r-mode solution.

5. Discussion of the results and future directions

In this work we have presented the analytical r-mode solution to the linearized Newtonian hydrodynamic equations with the GR reaction force. The velocity perturbations $\delta^{(1)}\vec{v}$ are proportional to $e^{i(2\phi+\omega_0 t)}$, with $\omega_0 = -4\Omega/3$. Thus, they have the same sinusoidal behavior and the same frequency ω_0 as the GR force-free velocity perturbations given by Eq. (11). The amplitude of the velocity perturbations is proportional to $\exp\{-\varpi t\}$. Since $\varpi < 0$, the GR force induces then an exponential growth in the r-mode amplitude. The e-folding growth timescale $\tau_{\rm GR} = 1/\varpi$ agrees with the GR timescale (1) found in Refs. [2,3]. The velocity perturbations $\delta^{(1)}\vec{v}$ contain also a piece proportional to $\alpha\gamma(A-1)\Omega^6$, where A is an arbitrary constant fixed by the choice of initial data. If we choose this constant A to be of order unity, then this part of the solution could be neglected[4].

A quite interesting feature that has emerged from recent investigations on r-modes is the presence of differential rotation induced by the r-mode oscillation in a background star that is initially uniformly rotating. That differential rotation drifts of kinematical nature could be induced by r-mode oscillations of the stellar fluid was first suggested in Ref. [7]. The existence of these drifts was confirmed in numerical simulations of nonlinear r-modes carried out both in general relativistic hydrodynamics[8] and in Newtonian hydrodynamics[9]. Differential rotation was also reported in a model of a thin spherical shell of a rotating incompressible fluid[10]. Recently, an analytical solution, representing differential rotation of r-modes that produce large scale drifts of fluid elements along stellar latitudes, was found within the nonlinear Newtonian theory up to second order in the mode amplitude and in the absence of GR reaction[11]. This differential rotation plays a relevant role in the nonlinear evolution of the r-mode instability[12]. Two questions could be then naturally raised, namely, is differential rotation also induced by the gravitational radiation reaction and does this differential rotation play a relevant role in the nonlinear evolution of the r-mode instability? One of the aims of the investigation carried out in Ref. [4] is to initiate a programme that hopefully will allow to answer this question. The natural continuation of this investigation is then to try to find an analytical r-mode solution of the *nonlinear* hydrodynamic equations with the GR reaction force. This work is in progress[13].

Acknowledgments

It is a pleasure to thank Kostas Kokkotas and Luciano Rezzolla for interesting discussions. This work was supported in part by the *Fundação para a Ciência e a Tecnologia* (FCT), Portugal. OJCD acknowledges financial support from FCT through grant SFRH/BPD/2004.

References

1. N. Andersson, Astrophys. J. **502**, 708 (1998).
2. L. Lindblom, B. J. Owen and S. M. Morsink, Phys. Rev. Lett. **80**, 4843 (1998).
3. N. Andersson, K. D. Kokkotas and B. F. Schutz, Astrophys. J. **510**, 846 (1999).
4. O. J. C. Dias and P. M. Sá, Phys. Rev. D **72**, 024020 (2005).
5. L. Blanchet, Phys. Rev. D **55**, 714 (1997).
6. L. Rezzolla, M. Shibata, H. Asada, T. W. Baumgarte and S. L. Shapiro, Astrophys. J. **525**, 935 (1999).
7. L. Rezzolla, F. K. Lamb and S. L. Shapiro, Astrophys. J. Lett. **531**, L141 (2000).
8. N. Stergioulas and J. A. Font, Phys. Rev. Lett. **86**, 1148 (2001).
9. L. Lindblom, J. E. Tohline and M. Vallisneri, Phys. Rev. Lett. **86**, 1152 (2001).
10. Yu. Levin and G. Ushomirsky, Mon. Not. R. Astron. Soc. **322**, 515 (2001).
11. P. M. Sá, Phys. Rev. D **69**, 084001 (2004).
12. P. M. Sá and B. Tomé, Phys. Rev. D **71**, 044007 (2005).
13. O. J. C. Dias and P. M. Sá, in preparation (2005).

Space Radiation: Effects and Monitoring

PARTICLES FROM THE SUN

D. MAIA
CICGE, Faculdade de Ciências da Universidade do Porto, Observatório Astron'omico, Alameda do Monte da Virgem 4430-146 Vila Nova de Gaia, Portugal
E-mail: dmaia@fc.up.pt

Particle acceleration is ubiquitous in all forms of coronal activity. The question is where, when, and how are those particles accelerated? In the Sun this question can be answered by using spatially resolved observations of the electromagnetic emissions and in-situ measures of plasmas in the heliosphere. In order to relate the remote and the in-situ observations one needs to consider both the problem of the origin of the particles, and of the propagation of those particles. We present here the details of a kinematic approach to the particle propagation problem that allows us to infer the particle injection profile at the Sun from in-situ measures at any point in the heliosphere.

1. Introduction

Flares are the best observed energy input in the solar corona. The energy involved in a large flare is of the order of 10^{32} erg. A large fraction of that energy resides in the accelerated electrons and ions. In solar active regions processes not yet identified accelerate typically electrons to a few hundreds of keV and protons to tens of MeV, although some events show signatures of much higher energies. Accelerated particles interact with the coronal plasma and emit radiation in the a broad wavelength range that includes not only gamma and X-rays, but also microwaves and radio.

With a similar energy budget but very different in nature are Coronal mass ejections (CMEs). CMEs consist on a large volume of coronal magnetic fields and plasma (about 10^{15} g) that leaves the Sun with a velocity that can exceed 2000 $\mathrm{km\,s^{-1}}$. While in flares the energy is "spent" in heating the plasma and accelerating particles, in CMEs the energy is essentially in the bulk plasma motions. CMEs are relevant for particle acceleration because they drive shocks that can accelerate electrons and protons up to a certain energy. At 1AU that energy is about 1 MeV/nucleon, few keV/electron, close to the Sun is probably higher but it is not known.

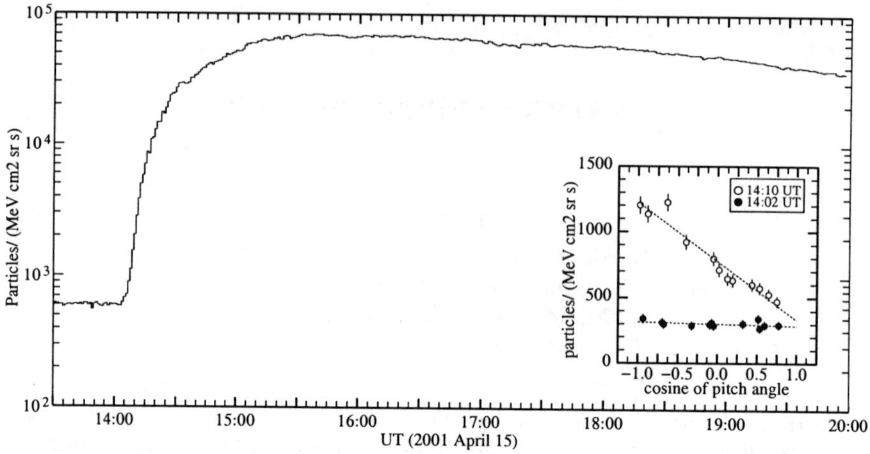

Figure 1. One-minute integration 178-290 KeV electron flux as seen by the EPAM experiment on the ACE spacecraft. The lower-left-corner box shows the flux as a function of cosine of pitch-angle for one minute integrations at times before the event onset (filled circle) and shortly after the onset (open circles).

Particles accelerated close to the Sun can gain access to open magnetic field lines and be detected by telescopes on spacecraft, originating the so called in situ particle events. Comparing in situ measures of charged particles with observation of electromagnetic emissions at the Sun is not straightforward because particles are affected by a series of propagation effects as they move throughout the heliosphere. We will be describe in some detail a kinetic treatment of particle propagation that allows one to determine, from in situ particle measurements, the particle injection function at the Sun. That injection function can then be used to establish which phenomena at the Sun is more likely to be at the origin of those particles.

2. Modelling propagation effects

One propagation effect to consider is the fact that the Sun rotates and as such magnetic field lines have a spiral nature. The length of that spiral depends on the Solar wind velocity: in the Earth vicinity it varies from about 1.1 to 1.4 AU. The magnetic field strength decreases with distance and conservation of the first adiabatic invariant will focuss the particles along the magnetic field direction as they propagate outwards. Small-scale irregularities in the magnetic field will the cause particles pitch angle to scatter. To determine the role of those effects on the particle profiles de-

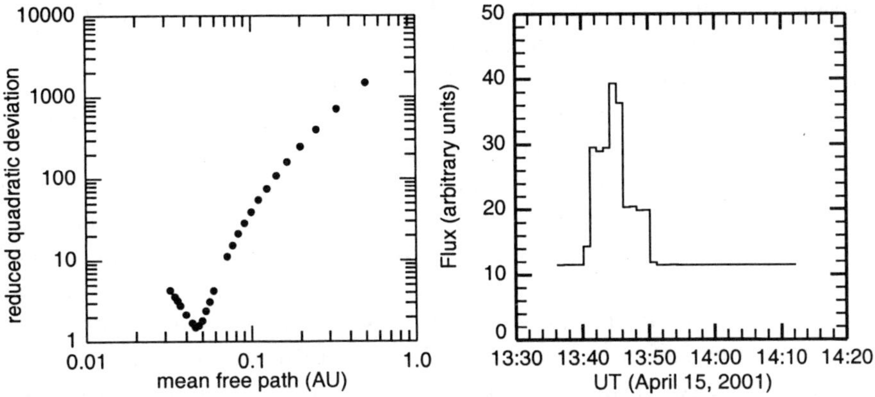

Figure 2. (left) Quadratic deviations in the anisotropy (model minus observations) versus mean free path. Note the pronounced minimum at about 0.04 AU. (right) The injection function corresponding to the best-fit mean free path. See text for details.

tected by particle telescopes onboard spacecraft we use a kinetic approach similar to that described in Ref. 2. We generate randomly a few million test particles, that are "released" simultaneously from near the Sun surface in a spiral magnetic field and "followed" for a few days, considering the effects of adiabatic focussing and pitch angle scattering. Different models are characterized by distinct values of the mean free path λ, and for each we compute the flux, and also the total anisotropy in pitch angle distribution, as a function of time and position.

To illustrate the results of the modelization we will infer the injection function for EPAM [1] electron observations on 2001 April 15. Figure 1 shows the electron flux measured by the EPAM instrument in the 178-290 KeV energy range. The 2001 April 15 event is clearly anisotropic, but the PADs in the few minutes following the rise are not comparable to the instrument resolution: the flux is seen rising at all pitch angles, including the channels in the anti-solar direction. This is and indication of considerable scatter in the immediate vicinity of the Earth. In order to determine the mean free path λ and injection function I we used a two step procedure. First, for each value of λ we get the injection function $I(\lambda)$ that reproduces the observed flux. No assumption is made on the injection function shape or duration. Then, for each pair λ and $I(\lambda)$ we compute the predicted average pitch and agle anisotropy and quantify its deviations from the observed anisotropy. This fitting procedure was fully tested for a wide range of artificial injection functions to be sure that is robust, efficient and unbiased. The results of this fitting procedure for the 2001 April 15 event are shown in Fig. 2. As we can see the average quadratic deviations have a rather pronounced minimum for a mean free path between 0.04 and 0.05 AU. The injection function corresponding to these rather small mean free paths is of relatively short duration, less than 10 mn and is shown in Fig. 2.

The fluxes of $\sim$1 GeV protons seen by ground based neutron monitors on 2001 April 15 were modelled by Ref. 3, taking in account propagation effects, including scatter. They find a release time of 13:42$\pm$1mn UT and obtain an injection profile at the Sun that agrees with the electron injection function obtained for EPAM data. The onset of this particle event was also determined by Ref. 4 by comparing the arrival times of ions and electrons of different energies, the so called $1/v$ method. The ions and electrons onset times fall roughly on the same line, with an estimated release time at the Sun of $13:44 \pm 1$mn UT, in remarkable agreement with ours and Ref. 3 model results.

3. Conclusion

The particles detected in situ by experiments onboard spacecraft are affected by propagation effects. We have accounted for those effects and developed a procedure to recover not only the release time but also the duration and shape of the injection function. The 2001 April 15 event was selected to illustrate the method because its characteristics imply a rather small mean free path at electron energies of a few hundred keV. We have obtained a good agreement with relase times obtained from our model with results from other authors, using other methods, namely the arrival of particles of different energies [4].

Acknowledgements

D. Maia was partially supported by Fundação para a Ciência e Tecnologia, through the program POCTI/FNU/43776/2001 and through grant SFRH/BPD/5521/2001.

References

1. R. Gould et al., *Sol. Phys.* **C34**, 1729 (1995).
2. R. Vainio, Kocharov *Astronomy and Astrophys.* **25**, L527 (2000).
3. V. Bieber, *Astrophys. J.* **A632**, 287 (2003).
4. A. Tylka, *Phys. Rev. Lett.* **86**, 4492 (2003).

SIMULATIONS OF SPACE RADIATION MONITORS

B. TOMÉ *

LIP, Av. Elias Garcia, 14-1
1000 Lisboa, Portugal
E-mail: bernardo@lip.pt

The radiation monitors to be included in the payloads of future space missions, must satisfy the scientific and safety requirements while meeting mass and power severe constrains. GEANT4 is a powerful tool to perform the development and the optimization of such detectors thanks to its capabilities for describing complex detector geometries and simulating the passage of particles through matter. The inclusion of an optical physics process category allows the full simulation of scintillation based detectors. A GEANT4 based simulation of a compact lightweight radiation monitor concept for future space missions is presented.

1. Introduction

Radiation monitors are becoming an essential component in space missions, providing crucial radiation environment information for the in-flight protection of the spacecraft and instruments onboard. In addition, the data acquired during the mission are a valuable input for space environment models; in particular data gathered in missions to other planets (e.g. particle fluxes and spectral distributions as a function of the distance to the Sun) are essential for models describing the injection and propagation of particles from the Sun. Several of the future space missions (e.g. LISA, Gaia, JWST, BepiColombo) are planned to carry radiation monitors. Given the limited resources on mass, power and accommodation on-board of a spacecraft, a new generation of compact and lightweight general purpose energetic particle detectors are being explored.

Monte Carlo simulation tools are essential to develop and optimize a given detector concept and to explore alternative detector concepts. The ability to predict the detector performance in realistic radiation environments is also crucial. This requires the implementation of software models

*This work was supported through FCT grant SFRH/BPD/11547/2002.

allowing to perform an end-to-end simulation, including the particle propagation from the source to the spacecraft location, followed by a detailed simulation of the detector response.

The GEANT4 toolkit is an ideal framework for simulating the particle interactions with the detector materials in complex geometries. A GEANT4 based simulation of a compact lightweight radiation monitor concept for future space missions is presented.

2. Space radiation environment

The high energy ionizing radiation[a] environment in the solar system consists of three main sources: the radiation belts, galactic cosmic rays and solar energetic particles.

The radiation belts are composed of electrons and protons trapped in the magnetosphere of the planets. In the radiation belts of the Earth (Van Allen belts) electrons have energies smaller than about 6 MeV and protons have energies up to 250 MeV (see Ref. 1). The Earth radiation belts are mostly relevant for low orbit missions.

The galactic cosmic rays (GCR) component consists of a continuous flux of electrons, protons and ions, arriving from beyond the solar system, which are originated mainly in the supernovae. The GCR flux is maximum at energies around 1 GeV/n, decreasing as a power law for higher energies. The lower energy particles are strongly affected by the 11-year solar cycle. Although the GCR flux is comparatively low, energetic heavy ions can locally produce significant energy depositions originating single-event upsets.

Solar energetic particles (SEP) events consist of a sudden and dramatic increase in the flux of particles from the sun, including electrons, protons and heavier ions. SEP events are associated with impulsive solar flares and Coronal Mass Ejections and occur with higher probability during the maximum of the solar cycle. Although the energies of SEP rarely exceed few hundred MeV they give rise to potential risks for space missions due to the high fluxes attained. On the other hand their study is relevant for developing and testing solar particle propagation models (Ref. 2).

[a] In the context of space radiation effects, particles are considered energetic if their energy is high enough to penetrate the outer skin of the spacecraft. This means electrons with energies above 100 keV and protons and ions with energies above 1 MeV.

3. The GEANT4 simulation toolkit

GEANT4 (Refs. 3, 4) is a Monte Carlo radiation transport simulation toolkit, with applications in areas as high energy physics, nuclear physics, astrophysics or medical physics research. It follows an Object–Oriented design which allows for the development of flexible simulation applications.

GEANT4 includes an extensive set of electromagnetic, hadronic and optics physics processes and tracking capabilities in 3D geometries of arbitrary complexity. The electromagnetic physics category cover the energy range from 250 eV to 10 TeV (up to 1000 PeV for muons) while hadronic physics models span over 15 orders of magnitude in energy, starting from neutron thermal energies. The optical physics process category allows the simulation of scintillation, Cherenkov or transition radiation based detectors. A distinct class of particles, the optical photons, is associated to this process category. The tracking of optical photons includes refraction and reflection at medium boundaries, Rayleigh scattering and bulk absorption. The optical properties of a medium, such as refractive index, absorption length, reflectivity coefficients, can be expressed as a function of the photon's wavelength. The characteristics of the interfaces between different media can be defined using the UNIFIED optical model (Ref. 5). Full characterization of scintillators include the emission spectra, light yield, the fast and slow scintillation components and associated decay time constants.

4. Simulation of a generic space radiation concept

The simulation of a simple space radiation monitor concept (Ref. 6) was implemented using the GEANT4 toolkit. The detector consists of a tracker made of two position sensitive silicon planes followed by a scintillating crystal surrounded by photodetectors (Fig. 1). An aluminum foil placed on the backside of the crystal is used to tune the energy range of the detector.

In the present simulation the silicon tracker planes were 0.065 mm and 0.5 mm thick, the crystal size was $3\times3\times3$ cm^3 and the aluminum absorber was 2 mm thick. The scintillation properties of the CsI(Tl) crystal were used. These included a light yield of 65000 photons per MeV of deposited energy and two scintillation components with decay time constants of 0.68 μs (64%) and 3.34 μs (36%). The scintillation light is readout by large area silicon PIN photodiodes, with a spectral sensitivity matching the CsI(Tl) emission spectrum. Since the silicon photodiodes are sensitive both to photons and charged particles, this makes them suitable to be used also as an anti-coincidence shield.

Figure 1. View of the simulated particle detector. From left to right are shown the two silicon planes, the scintillating crystal surrounded by photodetectors and the aluminum plane followed by a charged particle sensitive detector.

In the left plot of Fig. 2 is shown the fraction of the initial particle energy deposited in the crystal for protons of energies between 25 MeV and 125 MeV. Protons with energies of 25 MeV or 50 MeV are absorbed in the crystal. For 100 MeV protons the effect of straggling is visible.

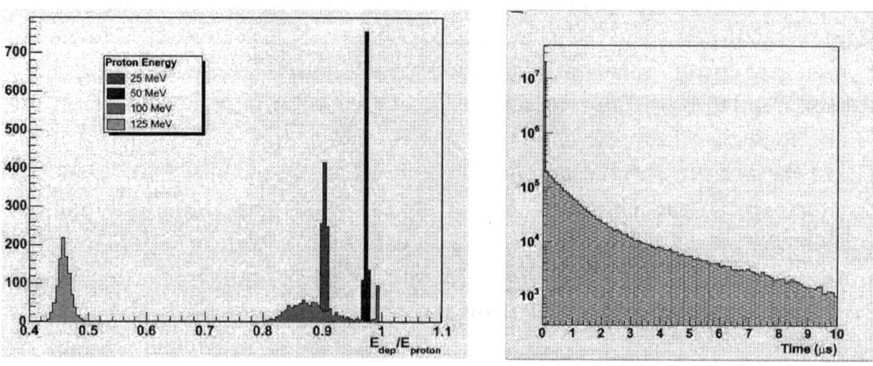

Figure 2. Left - Total energy deposited in the crystal by incident protons, expressed as the fraction of the initial particle energy. Right - Simulated time spectrum of the hits in the photodetectors.

The principle of the anti-coincidence operation mode is illustrated in the rigth plot of Fig. 2. It shows the time of the simulated hits in the photodiodes, produced by a 100 MeV proton that is not stopped inside the crystal. The fast signal at time zero is due to the direct ionization produced by the proton in a photodetector, while the slower signal comes from the scintillation light emitted by the crystal. Thus by a suitable signal processing it is possible to disentangle the direct ionization from the scintillation. This allows to identify the events where particles entered the detector from the sides or were not fully absorbed in the crystal.

Particle identification is performed by measuring the energy loss in the thin silicon trackers. Figure 3 shows the simulated energy loss in the first silicon plane, as a function of the initial particle energy, for electrons, protons and alphas.

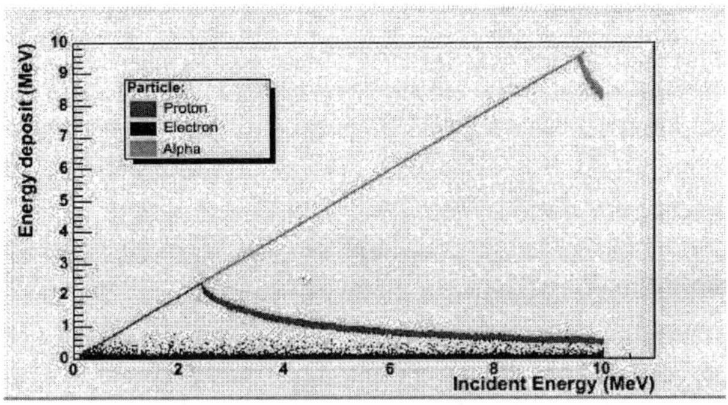

Figure 3. Energy deposited in the first silicon layer by electrons, protons and alphas, as a function of the initial particle energy.

References

1. E.R. Benton and E.V. Benton, "Space Radiation dosimetry in low-Earth orbit and beyond", *Nucl. Instrum. Methods* **B184**, 255 (2001).
2. See talk by D. Maia, these proceedings.
3. S. Agostinelli, J. Allison, K. Amako, J. Apostolakis, H. Araujo, P. Arce et al., "GEANT4 – a simulation toolkit," *Nucl. Instrum. Methods*, **A 506**, 250 (2003).
4. M. G. Pia, "The Geant4 Toolkit: simulation capabilities and application results," *Nucl. Phys. B – Proc. Suppl.*, **125**, 60 (2003).
5. A. Levin and C. Moisan, "A more physical approach to model the surface treatment of scintillation counters and its implementation in DETECT," in *Proc. IEEE 1996 NSS*, 2 (1996).
6. A. Owens et al. (ESA/ESTEC), private communication.

GEANT4 DETECTOR SIMULATIONS: RADIATION INTERACTION SIMULATIONS FOR THE HIGH-ENERGY ASTROPHYSICS EXPERIMENTS EUSO AND AMS*

PATRÍCIA GONÇALVES

LIP-Laboratório de Instrumentação e Física Experimental de Partículas,
Av.Elias Garcia nº14 - 1º, 1000-149 Lisboa

The system architecture of a GEANT4 based simulation framework and its application to EUSO/ULTRA and AMS/RICH performance studies are presented. ULTRA (Ultraviolet Light Transmission and Reflection in the Atmosphere) is an experimental support activity of EUSO (Extreme Universe Space Observatory), an experiment devoted to the study of extreme energy cosmic rays and neutrinos. Relevant aspects of the ULTRA simulation, namely the description of optical processes and the simulation of Fresnel lenses using parameterisation/replication techniques are described. The RICH (Ring Imaging CHerenkov detector) of the AMS (Alpha Magnetic Spectrometer) experiment, will incorporate a dual radiator, made of a low refractive index material, aerogel, and of sodium fluoride (NaF). A more realistic description of Cherenkov photon transmission through the aerogel surface, based on Atomic Force Microscopy images, was implemented in GEANT4.

1. The simulation framework

GEANT4 [1,2] is a toolkit for the simulation of particle transport and interaction with matter, featuring namely: the description of geometries of arbitrary complexity, standard electromagnetic physics processes (photons, electrons, positrons and muons). It describes *Optical physics* processes including the generation of photons by Scintillation, Cherenkov and transition radiation effects, Rayleigh scattering and media-boundary interactions (reflection, refraction). GEANT4 is based on an Object Oriented design, allowing the implementation of flexible simulation applications and of new/upgraded physics processes. The GEANT4 based part of the simulation framework entitled "GEANT4SpaceApplication" consisted of the description of different AMS and EUSO related detector geometries, of the interface with alternative sets of primary event generators and of the radiation transport processes. As for the integration of the readout electronics, signal digitisation and event reconstruction, it was done via the DIGITsim module [3] .The object oriented technology

* This work was supported by FCT under Grant SFRH/BPD/11500/2002.

for event data persistency and data analysis was handled through ROOT, LCG PI/AIDA and POOL.

2. The ULTRA experiment simulation

Within the EUSO experiment [4] there is an on-going program of critical design parameter studies, implying a set of dedicated experiments. The detection of the Cherenkov light associated with the Extensive Air Showers (EAS), measuring UV light diffusion coefficients of different types of media at the surface of the Earth, is the main goal of the ULTRA project (ULTRA - UV Light Transmission and Reflection in the Atmosphere) [5]. The ULTRA detector is a hybrid system consisting of a UV detection system, the UVscope, sensitive to the diffused Cherenkov light from EAS, in coincidence an array of scintillation detectors, the ETscope. A typical configuration of the ULTRA experiment is shown in Fig.1.

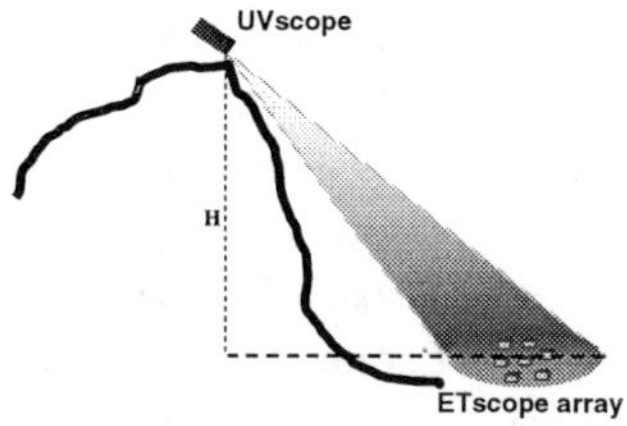

Figure 1. Typical configuration of the ULTRA experiment. The area covered by the ETscope array is seen in the UVscope field of view. The UVscope is located on top of a hill (represented by the thick line) at a height H.

2.1. *Simulation of the ULTRA ground detectors*

Each ULTRA ETscope station consists of a NUCLEAR NE 102A plastic scintillator, with dimensions 80x80cm^2 and 4cm thick, housed in a pyramidal stainless steel box. The inner walls of the box are coated with a white diffusing paint (similar to Oxide Titanium paint). The scintillation light is collected by two photomultipliers at approximately 30cm from the scintillator surface. The optical properties of the interface defined by the air and the white reflector covering the aluminum were specified within the UNIFIED model the "dielectricdielectric" TYPE and the "groundfrontpainted" FINISH were chosen. The scintillator emission spectrum, peaking at 423nm, and its light yield, about 10000 photons/MeV of deposited energy, were included in the material optical properties.

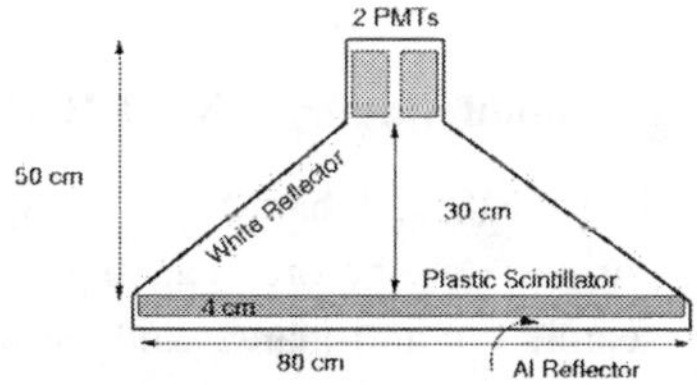

Figure 2. Schematic drawing of an ULTRA ETscope station.

2.2. *Simulation of the ULTRA UV telescope*

The Uvscope consists mainly of a Fresnel lens and one photomultiplier (PMT) located in the focal plane, enclosed in a cylindrical aluminum housing. The lens is 457 mm in diameter and is made of UV transmitting acrylic with 5.6 grooves per mm. A wide band filter (300 to 400 nm) is used to reduce the background. The external housing consists of a 1mm thick aluminum cylinder, with a length of 1030mm and a diameter of 518mm. To describe the geometry of the lens a parameterised replication of G4Cons volumes used. Each lens groove is described as a frustrum of a cone, with cross-section represented by a straight line with slope varying as a function of the radius. For this purpose the GEANT4 classes SpaceGEANT4FresnelLensParameterisation (inheriting from the G4VPVParameterisation class) and SpaceGEANT4FresnelLens were designed. A PMT with a 68mm diameter window was placed at a distance of 441.97mm from the lens, corresponding to the effective focal length at λ=400nm. The UNIFIED model was chosen to describe optical processes at material interfaces. For the "G4OpticalSurface" attributes of the Air/Aluminum interfaces, the "dielectricdielectric" TYPE and "groundfrontpainted" FINISH were used. All aluminum parts (cylinder walls and lens support) are assumed to have 5% reflectivity. Figure 3 represents the behavior of the UVscope when hit by a beam Ultraviolet photons.

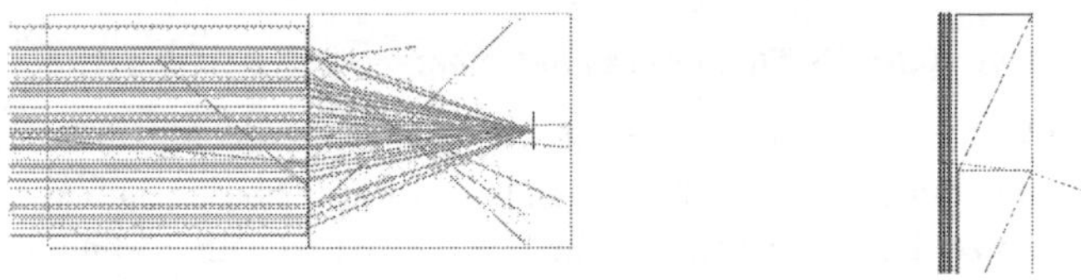

Figure 3. Left: ultraviolet photons focused by the Fresnel lens into the PMT. Right: the photons out of focus cross the lens surface in more than one point.

3. Simulating the AMS RICH radiator

The AMS (Alpha Magnetic Spectrometer) experiment aims at characterising cosmic rays before reaching Earth's atmosphere. Its main objectives are the search for anti-matter and dark matter and the study of the propagation and confinement of cosmic rays in the galaxy [6]. AMS had a precursor flight in June 1998 aboard the Discovery Space Shuttle, with around 100 million events recorded. The capabilities of the AMS spectrometer will be improved and extended through the inclusion both of new detectors and of a stronger magnetic field. Lip's collaboration in AMS is centred in the RICH (Ring Imaging CHerenkov) detector.

3.1. *RICH radiator simulation*

The AMS RICH will be built with a dual radiator, made of a low refractive index material, aerogel (n=1.03), and of sodium fluoride (NaF, n=1.33). It will provide an independent measurement of the velocity and charge of the cosmic rays. The RICH is a complex detector and performance assessment depends critically on the correct modeling of the light production, transmission and collection. A simplified design of the AMS RICH radiator setup, consisting of aerogel tiles supported by a plexiglas foil, was implemented in thc GEANT4 based simulation framework. A variable number of 11.3 x 11.3 x 11.3cm aerogel tiles, separated by a 0.1cm gap, were placed in a vacuum tank, of corresponding variable dimension, on top of a 0.1cm thick plexiglas foil. The gaps between the aerogel tiles can be alternatively left in vacuum, filled with plexiglas or with a material opaque to the Cherenkov photons. The implemented aerogel (SiO_2+vacuum) and plexiglas ($C_5H_8O_2$) properties, refractive index, absorption length, and clarity, in the case of aerogel, correspond to the AMS RICH radiator specifications. The setup implemented is shown in for a 3x3 aerogel tile array. This setup is being used to study more realistic models describing photon scattering in the aerogel surface, and to compare them with the results of the test beam of the AMS RICH prototype. In Fig. 4, shows the Cherenkov light produced by an 80GeV electron crossing the simulated radiator setup were also the effects of Rayleigh scattering and absorption can be observed.

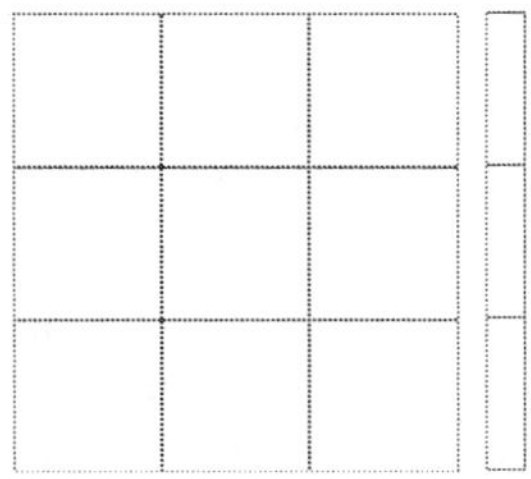

Figure 4. Implemented setup for the aerogel radiator

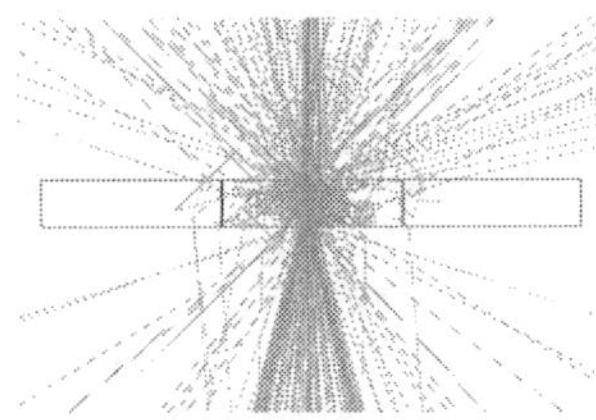

Figure 5. Cherenkov light produced by one 80GeV electron crossing the RICH radiator.

3.2. *AFM: Atomic Force Microscopy*

Previous test beam results revealed a disagreement between aerogel photon scattering in data and its description in the simulation. This effect, which cannot be described by the UNIFIED model in GEANT4, is thought to come from the interference of the photons with the aerogel surface defects. Although the UNIFIED model allows for a rather detailed description of photon reflection, this is not the case for transmitted photons, unfortunately the case of Cherenkov photons. Therefore, a more realistic description of the directional behavior of the transmitted Cherenkov photons was implemented with basis on Atomic Force Microscopy (AFM) scanning of the aerogel surface. In Fig. 6, a 2D view of the surface of an aerogel sample is shown, where surface defects can be observed. Two approaches were followed to obtain a realistic description within GEANT4 of the direction of Cherenkov photons after crossing this type of surface. In one approach, a new virtual class "G4VUserMapBoundaryModel" interfaces the surface maps to GEANT4. The other approach consisted of extending the UNIFIED model parameterisation by providing new methods to the GEANT4 class "G4OpBoundaryProcess".

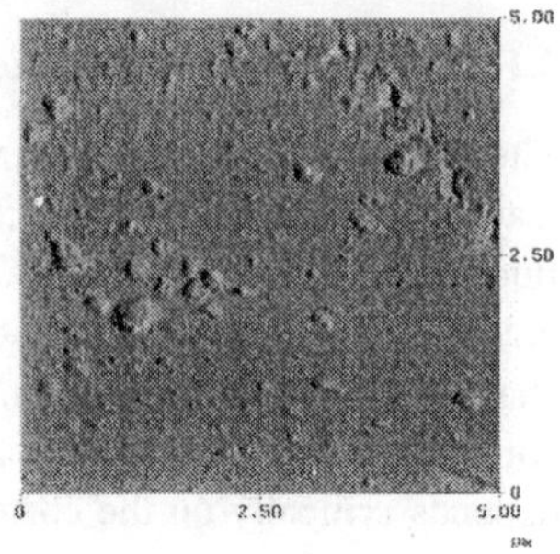

Figure 6. Atomic Force Microscopy image of the surface of an aerogel sample.

References

1. S. Agostinelli, J. Allison, K. Amako, J. Apostolakis, H. Araujo, P. Arce et al., "GEANT4 – a simulation toolkit," *Nucl. Instrum. Methods*, Vol. A506, 250 (2003).
2. M. G. Pia, "The Geant4 Toolkit: simulation capabilities and application results," *Nucl. Phys. B – Proc. Suppl.*, Vol. 125, 60 (2003).
3. M. C. Espírito-Santo, P. Gonçalves, M. Pimenta, P. Rodrigues, B. Tomé and A. Trindade, "GEANT4 Applications for Astroparticle Experiments", IEEE Transactions on Nuclear Science, Vol. 51, 1373 (2004).
4. L.Scarsi, "The Extreme Universe of Cosmic Rays: Observations from Space", Il Nuovo Cimento, Vol. C 24, 471 (2001).
5. O. Catalano, P. Vallania, D. Lebrun, P. Stassi, M. Pimenta, and C. Espirito Santo, "ULTRA technical report," Tech. Rep. EUSO–SEA–REP001, 2001.
6. S. P. Ahlen, V. M. Balebanov, R. Battiston, U. Becker, J. Burger, M. Capell et al, "An antimatter spectrometer in space," *Nucl. Instrum. Methods*, Vol. A 350, 351 (1994).

SOFTWARE FOR RADIOLOGICAL RISK ASSESSMENT IN SPACE MISSIONS

A. TRINDADE, P. RODRIGUES
LIP, Laboratório de Instrumentação e Física Experimental de Partículas,
Av. Elias Garcia 14-1,
1000–149 Lisboa, Portugal
E-mail: andreia@lip.pt

The radiation environment found in space missions present a potential risk to astronauts as well to electro–mechanics components of spacecrafts. In this paper we will present an overview of main sources and composition of the space radiation environment with relevance to astronauts dosimetry. Types and effects of space radiation in astronauts and radiation protection limits are discussed. A discussion on software tools for dose prediction is presented, with focus on how medical applications can help to improve space radiation risk assessment for mission planning.

1. Introduction

The radiation environment in space is characterized by a wide variety of primary particles covering an extended range of energies[1]. The three main sources of primary ionizing radiation are the galactic cosmic rays (GCR), trapped radiation in Earth's magnetosphere and solar particles events (SPE). GCR are mainly composed of protons (87%) and γ–rays, with a small fraction of electrons/positrons (2%) and heavy ions (1%). Trapped radiation mainly consists on electrons with energies up to 10 MeV and protons up to 100 MeV. Its presence is continous in time, with fluxes with variable intensity, function of Sun's solar activity. SPE radiation is composed of mostly low-energy protons and electrons, with fer percentage of higher-energy protons and heavy ions. In contrast to GCR, SPE are event driven, with occasional high fluxes over short periods, which coincide with increases in Sun's solar activity. The composition and intensity of these radiation sources can be modified by local planetary environments, distance of the location of exposure to the sun, existence of moons or local geology. Often, secondary ionizing radiation can be present, due to the interaction of the primary radiation with shielding materials (like the spacecraft hull).

Both primary and secondary particles can cause radiation damage in electronic components or biological damage when passing the body of an astronaut. The magnitude of the effects is however largely dependent on the type of space mission. GCR and SPE radiation are the main cause of concern for long-term mission duration, due to its constant presence (GCR) or high–fluxes (SPE). Severity of such event is largely enhanced by the fact that in long–term missions, immediate medical care would not be available. In order to provide an addition protection, shielded habitats have to be designed. Short–term missions in the Low–Earth Orbit are also associated with some risks, mainly caused by the trapped protons and electrons. Although the risk is reduced due to the shielding provided by spacecraft and habitat modules, extra-vehicular activities (EVA) present some concern due to the low attenuation provided by EVA spacesuits, against trapped electrons and protons.

2. Radiation Effects in Astronauts

Galactic cosmic heavy ions encountered in space are the more dangerous to human health due to their high linear energy transfer, which can break both strands of DNA. Besides this direct action against the cell DNA, an indirect action also takes place. This happens when after the irradiation, water molecules become ionized and form highly reactive free radicals. The free radicals can interact with strands of DNA causing them to break. In some conditions, mechanisms of DNA self–repair fail, leading to the appearance of cell mutations or cell death.

Depending on the exposure and type of the incident radiation, the effects on human health are classified in short and long term. Short term effects range from nausea and damage to the central nervous system and death (symptoms manifest within minutes to 30–60 days. Long term effects (symptoms do not manifest for months or even years) include the development of cataracts and cancer.

In order to correlate radiation exposure with molecular, biochemical and somatic effects several physical parameters were defined: absorbed dose, dose equivalent and effective dose. Absorbed dose is defined as absorbed energy per unit mass. In order to account for the different biological effectiveness of the incident radiation, the unit *dose equivalent* is used. Dose equivalent for a given tissue is given by the absorbed dose weighted by a quality factor Q. The normalization factor can range from 1 to X–rays, γ–

rays, β–rays to 20 for high–energy protons, α–particles, fission fragments and heavy nuclei. In order to account that for the same radiation type, different tissues and organs have their one radio–sensitivity value, the unit *effective dose* was defined. It is obtained by multiplying the individual dose equivalents by *risk weighting factors.* These factors reflect different organs radio–sensitivity to developing cancer and they represent the dose that the total body could uniformly receive that would give the same cancer risk as various organs getting different doses. In Table 1 the maximum dose equivalent admissible for U.S.A astronauts in short–term missions are shown. Data taken from Ref. 2.

Table 1. Ionising radiation exposure short-term limits (for U.S.A. crewmembers).

	Bone marrow (Gy–Eq.)	Eye (Gy–Eq.)	Skin (Gy–Eq)
Annual	50	2.0	3.0
30 days	0.25	1.0	1.0

Table 2 presents the effective dose limits for U.S.A astronauts. These limits provide the maximum effective dose that an astronaut with a given age at the time of exposure can receive without increase in more than three percent the risk of developing radiation–induced disease. Data taken from Ref. 2.

Table 2. U.S.A career effective dose (in Sievert) limits (based on three percent excess lifetime risk)

Age at exposure	Female (Sv)	Male (Sv)
25	0.4	2.0
35	0.6	1.0
45	0.9	1.5
55	1.7	3.0

3. From Medical to Space Applications

In order to guarantee that the current exposure limits are respected, dose predictions are required for mission planning. Such previsions need to taken into account effects such as the modulation effects in space-radiation environment (e.g. solar cycle), spacecraft shielding, spacecraft orbit and location.

Besides space dosimetry, the need to predict dose is a common requirement of several research fields such as external and internal radiotherapy, nuclear medicine imaging, radionuclide therapy and hadron therapy. Knowledge off the limitations presented by semi–analytical algorithms have favored the introduction of Monte Carlo techniques in dose calculations, since they allow a detailed treatment of the physics processes involved in the interaction of ionizing radiation with matter[3]. Geant4 is an object-oriented Monte Carlo simulation toolkit[4], developed by the RD44 (up to 1998) and Geant4 Collaboration, which provides a new approach for development of accurate dose calculations applications. Geometry modeling capability is also of paramount importance for realistic and efficient dose calculations, as required for astronauts space dosimetry. In this area, Geant4 provides a large set of solids of different complexity and allows an efficient repetitive structure representation. Such feature is vital for the implementation of realistic voxel–based anthropomorphic geometries which are used to define the setups for particle transport. In these phantoms, the human body is modeled as a large set of small elements with different chemical and density properties. Several applications, using this Geant4 feature were developed and they show how an important requirement of dose calculations in astronauts can be fulfilled. Example applications include the development of a Geant4-based framework for dose calculation in radiotherapy with external beams[5] and the Clear–PEM simulation framework for realistic scenarios for a Positron Emission Mammography detector optimization[6].

3.1. *Clear–PEM Positron Emission Mammography System*

The Clear-PEM detector is a positron emission mammography planar system designed to image both the breast and axillary lymph nodes. In order to optimize the final detector layout, realistic estimations of detector parameters, like detector efficiency, spatial resolution and background rate, must be obtained. A modular Geant4-based framework was developed, which includes the simulation of radioactive decay and photon tracking in defined phantoms, detector response, signal formation in the associated readout front-end electronics, and data acquisition by the DAQ/trigger system. Currently implemented phantoms in this module are a mathematical type torso-phantom, defined by simple mathematical volumes, and a voxelized NCAT phantom (developed by W.P Segars[7]) which models in detail a human body. The level of detail than can be included in the simulation can be observed from the following renderized images – Fig 1.

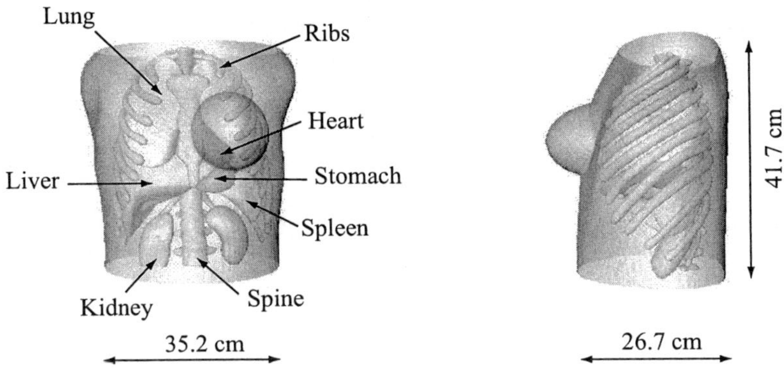

Figure 1. Volume rendering of the implemented NCAT torso phantom in Geant4 with $3.25 \times 3.25 \times 3.25$ mm^3 voxel resolution.

4. Conclusions

Dose predicament in astronauts is a complex task that requires both the correct description of space environment, simulation of shielding material provided by spacecraft, habitat and EVA suits as well as high resolution human body models within the spacecraft. Several developed applications for Monte Carlo dose calculations in Medicine have shown that this requirements can be fulfilled, making possible to improve the precision of dose calculation in space.

Acknowledgments

The authors would like to thank colleagues from the ClearPEM collaboration for their suggestions and contributions.

References

1. E.R. Benton and E.V. Benton, *Nucl. Instrum. Methods Phys.* **B184** 255 (2001).
2. Effects in Biological Material, *ECSS* ECSS–E–10–12 Draft 0.5.
3. P. Andreo, *Phys. Med. Biol.* **37** 861 (1991).
4. S. Agostinelli *et al*, *Nucl. Instrum. Methods Phys.* **A506** 250 (2003).

5. P. Rodrigues, A. Trindade, L. Peralta, C. Alves, A. Chaves and M.C. Lopes, *Appl. Rad. Isotope* **61** 1451 (2004).
6. P. Lecoq and J. Varela, *Nucl. Instrum. Methods Phys.* **A486** 1 (2002).
7. W.P. Segars, Development of a new dynamic NURBS–based cardiac–torso (NCAT) phantom, *PhD dissertation*, The University of North Carolina, USA 2000.

Neutrino Physics

RESULTS FROM K2K

S. ANDRINGA
IFAE/UAB

FOR THE K2K COLLABORATION

The K2K experiment was designed to confirm or disprove the atmospheric neutrino oscillations. Its experimental concept and detectors are described together with the full analysis of neutrino oscillations. The K2K data confirms neutrino oscillations, excluding the null oscillation hypothesis at 4σ, and favours maximal mixing with $\Delta m^2 \in [1.9, 3.6] \times 10^{-3}$ eV^2 at 90% Confidence Level.

1. The K2K Experiment

The KEK to Kamioka neutrino oscillation experiment, K2K, is the first neutrino long base line experiment. It was designed to confirm or disprove, in a controlled way, the neutrino oscillations of ν_μ to ν_τ, first seen by Super-Kamiokande[1] in athmospheric neutrino events.

A 98% pure muon neutrino beam with a known energy spectrum, peaked at 1.3 GeV, is produced at the KEK laboratory and detected at a distance of 250 km at the Super-Kamiokande detector. The fixed travelled length (L) and the reconstruction of the neutrino energy (E_ν) at both sites allows for a direct determination of the oscillation pattern of the ν_μ survival probability – which for two neutrinos oscillations is given by:

$$P(\nu_\mu \to \nu_\mu) = 1 - \sin^2(2\theta)\sin^2(1.27\Delta m^2 L/E_\nu)\,, \tag{1}$$

where Δm^2 is the mass difference of the two neutrino mass eigenstates involved in the atmospheric neutrino oscillations and θ is the mixing angle between the mass and the weak eigenstates.

With 10^{20} protons on target (POT) at the K2K beam line, similar contributions of statistical and systematic uncertainties are expected to be achieved in the oscillation parameters. Since Super-Kamiokande is used as the far detector and the K2K spectra is similar to that of atmospheric neutrinos, the large sample of atmospheric neutrino events can be used to control detection systematic uncertainties at the far detection site. On the other hand, a set of detectors at the production site collects very high statistics to reduce the systematic uncertainties on neutrino production, namely the beam intensity and energy spectra, and on neutrino interactions.

At the near site there are thus one water Čerenkov detector, to cancel detection uncertainties common to Super-Kamiokande; a set of Fine Grained detectors, to study the modelling of neutrino interactions and reduce the corresponding systematic uncertainties; and a set of beam monitors to verify, and control the neutrino spectra and flux and its extrapolation to the far site.

A detailed description of the K2K experiment is given in Ref. 2, and a schematic view of the near detector site is shown in Fig. 1.

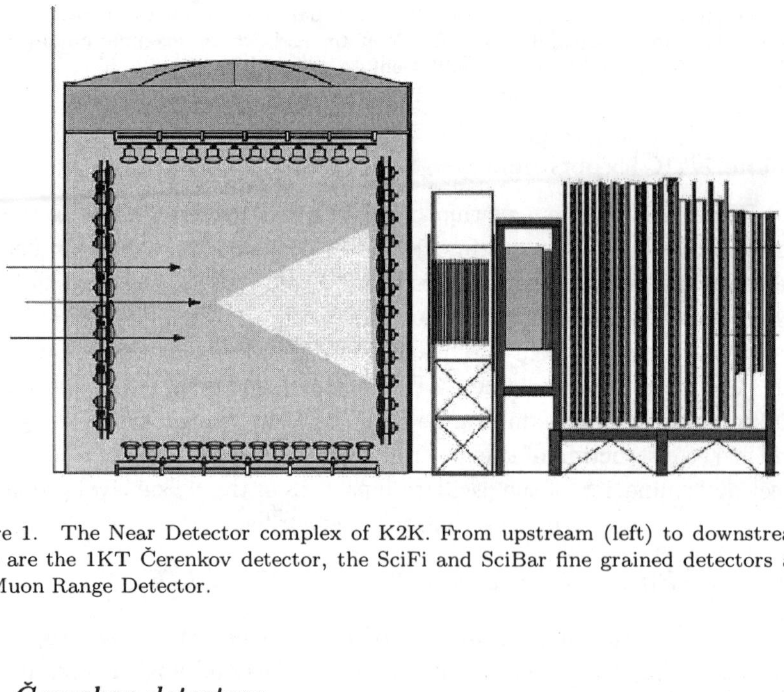

Figure 1. The Near Detector complex of K2K. From upstream (left) to downstream, there are the 1KT Čerenkov detector, the SciFi and SciBar fine grained detectors and the Muon Range Detector.

1.1. *Čerenkov detectors*

The Super-Kamiokande (SK)[1] is a 50 kt water Čerenkov detector with 40% of its area covered by PMTs; the outer part of the detector is used for vetoing cosmic-ray muons, leaving a fiducial volume of 22.5 kt. The K2K flux intensity is 10^6 neutrinos per spill. The charged current interactions (CC) producing a muon are detected with a 93% efficiency while neutral current interactions (NC) are detected with an efficiency of 68%.

At the near site, there is a water Čerenkov with a total volume of 1 kt, the 1KT detector, with an angular light coverage identical to that of SK. The fiducial volume is of 25 t but the flux intensity is 10^{11} neutrinos per spill. The efficiencies are of 87% (CC) and 55% (NC), lower than those of SK due to the presence of lower energy neutrinos (outside the SK acceptance) and an effective upper limit on the neutrino energy in CC events. This upper limit comes from the requirement that the event be totally contained in the fiducial volume, so that the muon momentum (P_μ) can be reconstructed - it corresponds to $P_\mu < 1.5$ GeV/c ($E_\nu <2$ GeV).

The same reconstruction and analysis methods are used for both Čerenkovs and the results at the 1KT give the prediction for the total number of events at SK. For energy reconstruction, events with only 1 Čerenkov ring are selected and this ring is required to be compatible with that produced by a muon. This results on a sample composed mostly of CC quasi-elastic (QE) events, for which E_ν can be reconstructed as:

$$E_\nu = \frac{(m_N - V)E_\mu - p_\mu^2/2. + m_N V - V^2}{(m_N - V) - E_\mu + p_\mu \cos\theta_\mu}, \tag{2}$$

where m_N is the free neutron mass, V is the nuclear binding energy, and the Fermi motion of the neutron is neglected. For 40% of the events, which are not QE, the reconstructed energy is biased to lower values.

1.2. *Fine grained detectors*

The fine grained detector complex is composed by two complementary scintillator trackers and a Muon Range Detector (MRD[a][3]) to contain the escaping muons so that their momentum (from 0.55 to 3.5 GeV/c) can be measured. In SciFi[4] the neutrinos interact predominantly in water, as at SK, while SciBar[5] is a fully active detector for neutrino interactions in carbon - allowing for better reconstruction of the interaction vertex.

The low threshold of the scintillators (for a proton, the momentum $P_p >$450-650 MeV/c) allows for the selection of large samples with a muon and a second track. Assuming QE kinematics, the direction of the second track can be compared with the one predicted from the measurement of the muon, $\Delta\theta_p$. The CC events are thus separated in three samples: 1 track, 2 track QE-like ($\Delta\theta_p <25°$) and 2 track nonQE-like.

[a]Due to its high density, the MRD is also used to monitor the stability of beam intensity, spectra and direction.

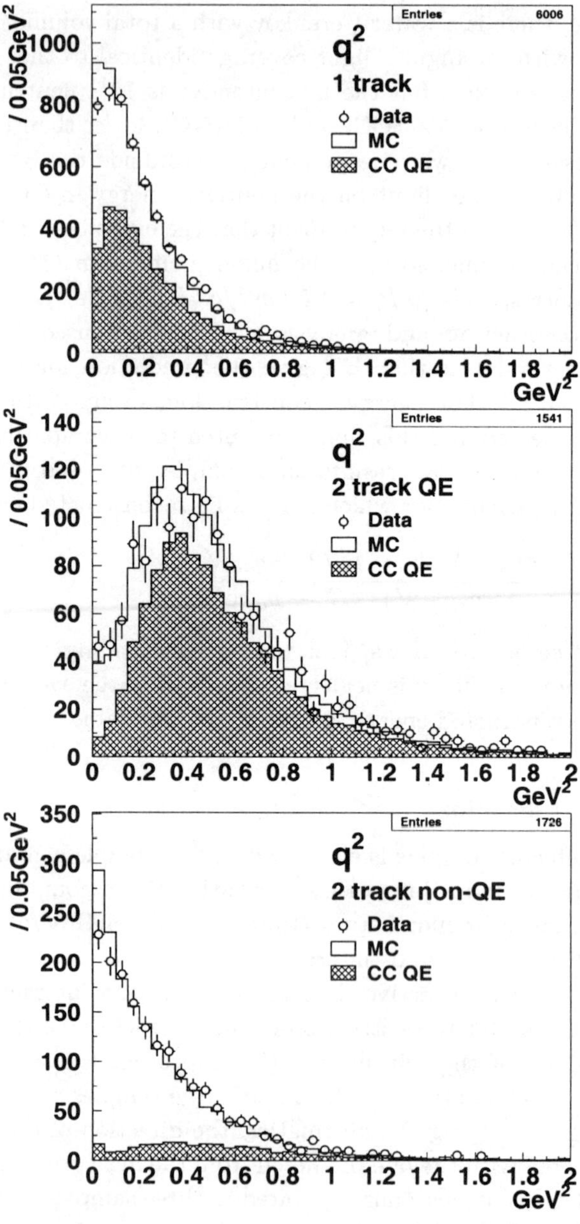

Figure 2. Distribution of q^2 (reconstructed assuming QE kinematics) in the 1 track, 2 track QE-like and 2 track nonQE-like samples selected by SciBar. The data (dots) are compared to the total expectation from the K2K simulation (open histograms). The QE fraction is shown by the shaded histograms.

Figure 2 shows the three samples selected in SciBar, with different QE purities. A clear disagreement is seen between the data and the expectations from simulation in the region of low transfered energy between the leptonic and hadronic currents (q^2), and for the samples dominated by non-QE interactions. This disagreement is seen in all K2K sub-detectors. Its origin is under study but it was shown that it can be either due to an excess in the prediction of coherent neutrino interaction with the nuclei (which is extrapolated down from higher energies) or to a lack of nuclear corrections in the production of Δ resonances (which are harder to calculate theoretically than those for the QE interaction).

1.3. *Simulation*

K2K uses two base simulations in its analyses: for the neutrino production at the K2K beam line and for the neutrino interaction in the K2K subdetector targets.

The K2K beam is produced by sending 12 GeV protons from the KEK Proton Synchrotron into an Aluminium target; two magnetic horns followed by a decay volume focus the produced π^+ which will decay to μ^+ and ν_μ. The μ^+ are later absorbed while the ν_μ follow the direction of SK. This is simulated as described in Ref.[6] and the main uncertainties come from the hadron production at the target.

The neutrino interactions with the nuclei of the different detectors are simulated with the NEUT package[7], also used by the Super Kamiokande collaboration. It includes the description of QE interactions[8], resonance production[9], deep inelastic scattering[10] and coherent interactions[11]. It uses the Fermi gas model to implement the nuclear corrections and includes final state interactions such as proton rescattering or pion absorption in the nuclei.

2. Neutrino Oscillation Analysis

The distribution of P_μ and θ_μ in the 1-track sample of 1KT, and the three samples of SciFi and SciBar are all used together in a fit to determine corrections to the K2K simulations. The fit is done first without the region of low θ_μ (i.e. the low q^2 region, where there is the discrepancy between data and expectations), to fix the experimental parameters – corrections to the K2K energy spectra and absolute energy scales of the near detectors – and then, including these events, to determine the ratio of nonQE to QE cross-sections. This last step is done under two different assumptions

– the absence of coherent interactions or the suppression of Δ resonance production by $q^2/0.1$ for $q^2 < 0.1$ – which give similar quality fits. The corrected description of the simulated spectra and interaction models is used to extract the predictions at SK[b].

At SK the events are selected following the atmospheric analysis[1]. The pulsed structure of the K2K beam allows for a very precise time selection, reducing the atmospheric neutrino background to a rate of 2 per mil.

Table 1. SK Event Selection, data and expectations in the case of no oscillation are shown separately for two K2K periods, the second corresponding to half light coverage of SK and addition of SciBar, with 47.9x10^{18} and 41.2x10^{18} POT, respectively.

	DATA (I+II)	MC no osc. (I+II)	
FC 22.5 kt	107 (55+52)	150.9 (79.1+71.8)	for Normalization
multi-ring	40 (22+18)	56.9 (30.2+26.7)	NC enhanced
1-ring	67 (33+34)	94.0 (48.9+45.1)	CC enhanced
e-like	10 (2+ 7)	8.6 (4.0+ 4.5)	NC π^0 dominated
μ-like	57 (30+27)	85.4 (44.9+40.5)	for Spectra Shape

The number of events seen and predicted in the absence of oscillations are shown in table 1, giving a clear indication of ν_μ disappearance. The compatibility with the null-oscillation hypothesis is 0.28% from the total number of events (107 events seen with 150.9 expected) and 0.11% considering only the shape of the energy spectra reconstructed in the sample of 1μ-ring events (dominated by QE), shown in Fig. 3.

2.1. $\nu_\mu \to \nu_\tau$ oscillation

The determination of the oscillation parameters is done by a likelihood fit using the two informations simultaneously. The likelihood includes three terms

$$\begin{aligned} L = & L_{normalization}(\sin^2(2\theta), \Delta m^2, systematics). \\ & L_{E_\nu shape}(\sin^2(2\theta), \Delta m^2, systematics). \\ & L_{systematics}(systematics)\,. \end{aligned} \quad (3)$$

[b]The suppression of Δ resonance production is used in what follows, but the oscillation results do not change if the suppression of coherent interactions is used instead.

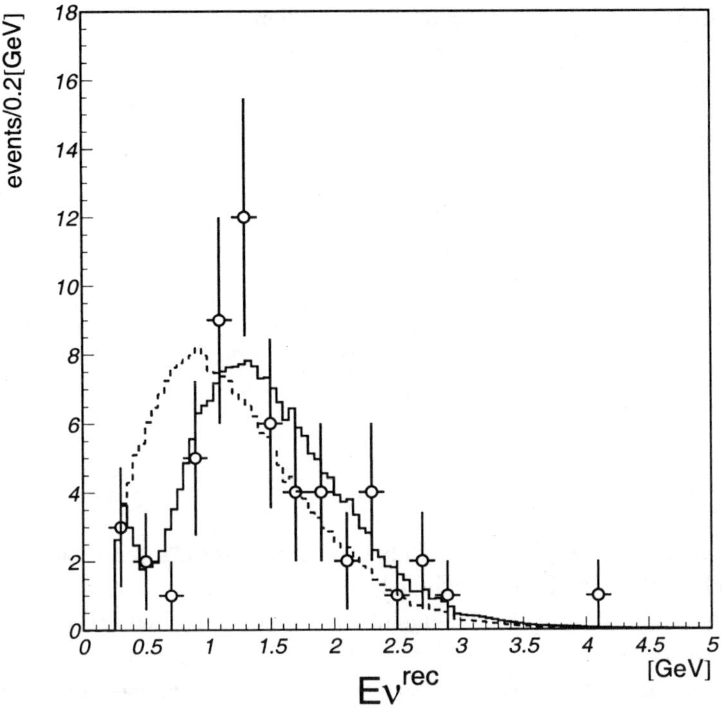

Figure 3. E_ν reconstructed according to Eq. 1, for the 1μ-ring sample selected at SK. Data (dots) are compared with the shapes expected in the hypothesis of null oscillation (dashed line) and the hypothesis of $\sin^2(2\theta) = 1.00$ and $\Delta m^2 = 2.79 \times 10^{-3}$ eV^2 (full line). In both cases the expectations are normalized to the number of events in data.

There are no constraints in the values of the neutrino oscillation parameters, while the systematic uncertainties include 5% from the extrapolation of the K2K flux and spectra to the far site, 5% from the overall normalization dominated by the fiducial volume determination in the Čerenkovs; 0.5% from the near detector flux and spectra determination and 0.5% from the nonQE to QE interaction cross-sections; and the experimental uncertainties on SK efficiency and energy scale, both of around 3%.

The best fit gives compatible numbers for the systematic parameters and values of $\sin^2(2\theta) = 1.51$ and $\Delta m^2 = 2.19 \times 10^{-3}$ eV^2; of $\sin^2(2\theta) = 1.00$ and $\Delta m^2 = 2.79 \times 10^{-3}$ eV^2 if only results in the physical region are considered. The probability that the first, non-physical, result is obtained as a statistical fluctuation if the true values are $\sin^2(2\theta) = 1.00$ and $\Delta m^2 = 2.79\times10^{-3}$ eV^2 is of 13%. This last set of parameters corresponds to a total number of events of 104.8 and an energy spectra shape, shown in Fig. 3, compatible in 36% with the one observed for 1μ-ring events.

Fits conducted using the two periods of K2K data separately or using the total number of events and the shape of energy spectra separately give results which are also compatible with the combined ones. At 90% confidence level these results are also compatible with the ones obtained for atmospheric neutrinos by the SuperKamiokande collaboration.

The allowed regions for Δm^2 and $\sin^2(2\theta)$ are shown in Fig. 4 and more details on the analysis can be found in Ref. [12]. The K2K data favour an oscillation with maximal mixing and $\Delta m^2 \in [1.9, 3.6] \times 10^{-3}$ eV^2; based on likelihood ratio they exclude the null oscillation hypothesis at 4σ.

3. Future Prospects

The final analysis of K2K will include 20% more data, after reaching the initial goal of 10^{20} POT, and reduced systematic errors. The main effect should come from a better knowledge of the flux obtained with the analysis of hadron production in a replica of the K2K target, done by the HARP[13] collaboration. In addition, K2K also conducts many analyses on neutrino interactions which will be useful to reduce the uncertainties in all the neutrino experiments with energies $\sim$1 GeV.

The long base line concept proved successful and will be followed soon by next generation experiments. There are two different ways to reach the oscillation maximum: with a high energy beam allowing for ν_τ production and reduction of nuclear effects uncertainties, but requiring a longer baseline and consequently lower statistics (as followed by MINOS[14] and, in the future, also by CNGS[15]), or with a low energy beam which enhances the CC QE interactions, increasing the sensitivity to the energy spectra, and allows for higher beam intensity – as will be followed by T2K[16]. T2K is expected to reach precisions of $\sin^2(2\theta) \sim 0.01$ and $\Delta m^2 \sim 10^{-4}$ eV^2, but also to be sensitive to the sub-dominant $\nu_\mu \rightarrow \nu_e$ oscillations.

4. Conclusions

The K2K collaboration confirms, in a controlled way, the neutrino oscillations first seen in atmospheric events, excluding the null oscillation hypothesis at 4σ. K2K data favours an oscillation with maximal mixing and $\Delta m^2 \in [1.9, 3.6] \times 10^{-3}$ eV^2, at 90% confidence level.

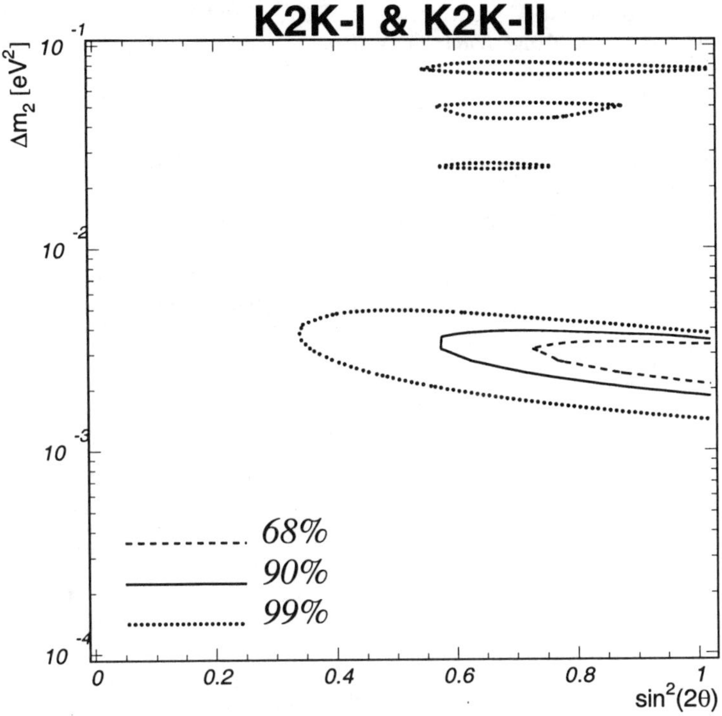

Figure 4. Regions of Δm^2 and $\sin^2(2\theta)$ allowed by the K2K data, at 68%, 90% and 99% confidence level.

K2K has reached its initial goal in terms of POT and will soon publish also results on ν_μ interactions at $E_\nu \sim 1$ GeV.

References

1. Super-Kamiokande Coll., Y. Ashie *et al*, *Phys. Rev. Lett.* **93**, 101801 (2004).
2. K2K Coll., S. H. Ahn *et al*, *Phys. Lett.* **B511**, 178 (2001).
3. K2K MRD Group, T. Ishi *et al*, *Nucl. Instrum. Meth.* **A482**, 244 (2002) [Erratum-ibid. **A488**, 673 (2002)].
4. K2K Coll., A. Suzuki *et al*, *Nucl. Instrum. Meth.* **A453**, 165 (2000).
5. K2K SciBar Group, K. Nitta *et al*, *Nucl. Instrum. Meth.* **A535**, 141 (2004).
6. K2K Coll., S. H. Ahn *et al*, *Phys. Rev. Lett.* **90**, 041801 (2003).
7. Y. Hayato, *Nucl. Phys. Proc. Suppl.* **112**, 171 (2002).
8. C. H. Lllewellyn Smith, *Phys. Rept.* **3**, 261 (1972).
9. D. Rein and L. M. Sehgal, *Annals Phys.* **133**, 79 (1981).

10. M. Gluck, E. Reya and A. Vogt, *Z. Phys.* **C67**, 433 (1995);
A. Bodek and U. K. Yang, *Nucl. Phys. Proc. Suppl.* **112**, 70 (2002).
11. D. Rein and L. M. Sehgal, *Nucl. Phys.***B223**, 29 (1983);
J. Marteau, J. Delorme and M. Ericson, *Nucl. Instrum. Meth.* **A451**, 76 (2000).
12. K2K Coll., E. Aliu *et al*, hep-ex/0411038, submitted to *Phys. Rev. Lett.*.
13. see http://harp.web.cern.ch/harp/, and CERN-SPSC/99-35 (1999).
14. see http://www-numi.fnal.gov/minwork/minoswk.html, and NuMI-L-337 (2000).
15. see http://proj-cngs.web.cern.ch/, and CERN AC/2000-03 (2000).
16. see http://neutrino.kek.jp/jhfnu/, and hep-ex/0106019 (2001).

SNO: SALT PHASE RESULTS AND NCD PHASE STATUS

J. MANEIRA ON BEHALF OF THE SNO COLLABORATION.
LIP-Lisboa
Av. Elias Garcia, 14, 1º ; 1000-149 Lisboa, Portugal
E-mail: maneira@lip.pt

The Sudbury Neutrino Observatory (SNO) is an underground heavy-water Cherenkov detector designed to measure neutral (NC) and charged (CC) current interactions of ^{8}B solar neutrinos on deuterium. This talk focused on the characterization of the detector response and on the Physics results from the second phase of SNO, in which 2 tonnes of NaCl were added to the heavy water, in order to enhance the sensitivity to NC interactions. This allowed for precision, energy-unconstrained measurements of the solar neutrino flux that confirmed solar model predictions and gave strong evidence for flavor change, excluding maximal flavor mixing at a level of 5σ. A status report was also given on the third phase of SNO, that aims to perform an event-by-event measurement of the NC rate.

1. Introduction

The discrepancy between measured[1,2,3] and predicted[4] solar neutrino fluxes, that lasted for over thirty years, was known as the solar neutrino problem. The solution was provided by the Sudbury Neutrino Observatory (SNO) results[7], that showed that the ν_e's produced in the Sun core undergo a flavor conversion on their way to the Earth. Neutrino oscillations[5,6] are an elegant mechanism for this flavor conversion that is strongly supported by other experiments as well.

SNO[9] is a heavy water Cherenkov detector located 2 km underground in a nickel mine near Sudbury, Canada. The active target of the detector consists of 1000 tonnes of heavy water(D_2O), contained in a 12 m diameter transparent acrylic vessel(AV) and observed by 9456 photomultiplier tubes (PMTs). The ^{8}B solar neutrinos are detected through the processes:

$$\begin{aligned} &\nu_e + d \rightarrow p + p + e^- && \text{(CC)}, \\ &\nu_x + d \rightarrow p + n + \nu_x && \text{(NC)}, \\ &\nu_x + e^- \rightarrow \nu_x + e^- && \text{(ES)}. \end{aligned}$$

While the charged current (CC) reaction is sensitive only to ν_e's, the neutral current reaction (NC) is equally sensitive to all active neutrino flavors ($x = e, \mu, \tau$), so its observation provides a model-independent measurement of the total flux of active ^{8}B solar neutrinos. The elastic scattering (ES) reaction is sensitive to all electron flavors as well, but with reduced sensitivity to ν_μ and ν_τ.

The NC neutron counting is carried out in different ways in each of the three phases of SNO: detection of a 6.25 MeV γ ray following capture on a deuteron in the first phase; detection of a 8.6 MeV γ ray cascade following capture on ^{35}Cl in the second phase, where 2 tonnes of salt were added to the D_2O. In the third phase, the neutrons will be detected by an array of 40 ^{3}He counters (*Neutral Current Detectors*, or NCDs), independently from Cherenkov events.

2. The Salt Phase of SNO

The salt data set presented here corresponds to 254.2 live days[8]. After removal of instrumental and non-Cherenkov like backgrounds, 3055 events were selected. The analysis used a kinetic energy threshold of $T_{eff} \geq 5.5$ MeV and a fiducial volume radius $R_{fit} \leq 550$ cm.

2.1. *Detector Response*

The times and positions of hit PMTs are used to calculate estimators for the vertex position, the event light isotropy and the particle direction and energy. The detector response was calibrated by using a source deployment system capable of reaching off-axis positions inside the acrylic vessel, as well as in vertical axes in the light water volume.

The energy estimator uses an energy normalization obtained with the ^{16}N γ source and an optical model calibrated with a laser diffusing source ("laserball") to estimate the event energy based on its number of hit PMTs (nhit), reconstructed position and direction. The laserball calibration measured an increase in optical attenuation in the D_2O volume throughout the salt phase, as shown in Figure 1(Left) and the ^{16}N calibration showed a steady drop in the detector response (about $-2\%\ yr^{-1}$). As shown in Figure 1(Right), the drift was predicted by Monte Carlo simulations using time-varying D_2O attenuation, so a correction was applied to the energy estimator. The uncertainty in the energy scale was measured with the ^{16}N calibration to be 1.1%.

A ^{252}Cf spontaneous fission source was used to calibrate the distribution of neutron detection efficiency in function of position in the detector.

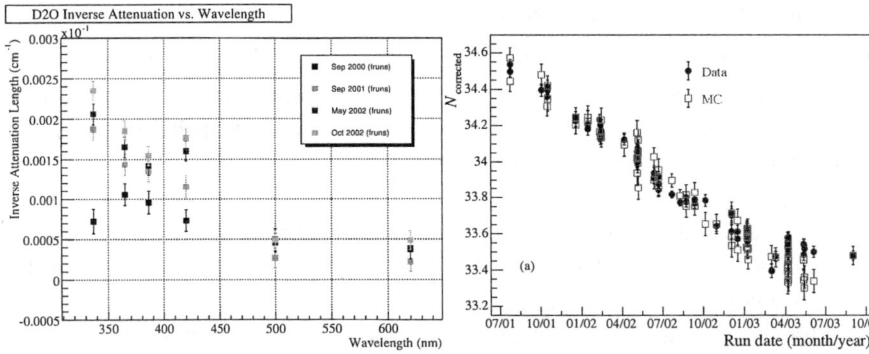

Figure 1. Optical and energy calibration. Left: Inverse attenuation length of D_2O in function of wavelength, for different calibration periods in the salt phase, as measured with the laserball source. Right: Number of hit PMTs (corrected to the center of the detector) in function of date, for data (dark points) and Monte Carlo simulations(light points) as measured with the ^{16}N source.

The volume-weighted efficiency within the analysis window is 0.399 ± 0.012, about a factor of three higher than in the pure D_2O phase.

The angular distributions of the PMT hit pattern of each event was expanded in terms of Legendre polynomials and the linear combination of coefficients β_{14} was chosen as the isotropy parameter. Single ring electron events are less isotropic than multiple ring γ cascades, so β_{14} can be used to separate these classes of events. The β_{14} distributions were calibrated by comparing Monte Carlo to ^{252}Cf and ^{16}N source data. The uncertainty in the mean value is 0.87%.

2.2. *Backgrounds*

The total photo-disintegration backgrounds were estimated to be 113 ± 26 neutrons. The largest contribution is from uranium and thorium chain activity in the D_2O and this was measured by *ex-situ* and *in-situ* methods throughout the data-taking period to be below the design goals.

Background neutrons produced at the acrylic vessel or in the H_2O can propagate into the fiducial volume, but the enhanced neutron capture efficiency in salt makes the external neutrons readily apparent, so an additional radial distribution function was included in the fits to extract the rate of this component. The contribution from low energy Cherenkov backgrounds was < 14.7 events (68% C.L.).

2.3. *Solar Neutrino Results*

The maximum likelihood method was applied to extract the fluxes of each neutrino interaction and the external neutron background rate from the salt data set. The likelihood function was constructed with probability density functions of volume-weighted vertex radius ($\rho = (R_{fit}/R_{AV})^3$), event angle with respect to the Sun ($cos\theta_{sun}$) and isotropy (β_{14}). As can be seen in Figure 2, the radial distribution allows the separation of backgrounds and β_{14} allows the separation of the CC and NC events, making possible an energy-unconstrained analysis.

The fitted number of events yield the following fluxes, in units of $10^6 cm^{-2}s^{-1}$:

$$\Phi_{CC} = 1.59^{+0.08}_{-0.07}(stat.)^{+0.06}_{-0.08}(syst.)$$
$$\Phi_{ES} = 2.21^{+0.31}_{-0.26}(stat.) \pm 0.10(syst.)$$
$$\Phi_{NC} = 5.21 \pm 0.27(stat.) \pm 0.38(syst.)$$

The systematic uncertainties are listed in Table II of reference [8].

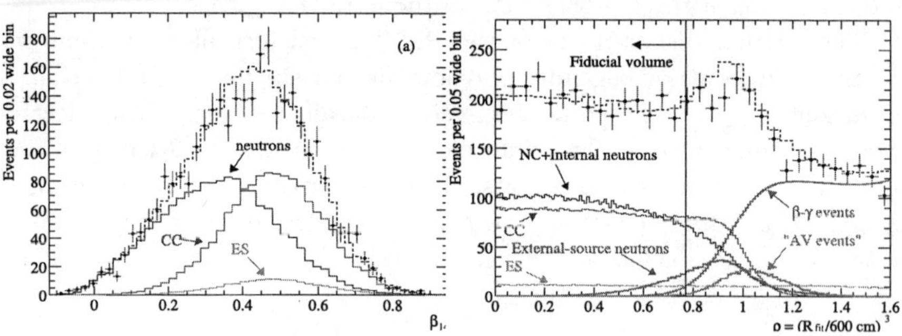

Figure 2. Distribution of (a) β_{14}, (b) volume-weighted radius for the selected events. Also shown are the Monte Carlo predictions for CC, ES, NC, internal and external neutrons scaled to the fit results. The dashed lines represent the summed components.

A combined χ^2 analysis of the shape-unconstrained fluxes from the salt phase, the day and night energy spectra from the pure D_2O phase and results from other solar neutrino experiments was carried in order to constrain the allowed regions of the neutrino mixing parameters Δm^2 and $tan^2\theta$. When the results from the reactor neutrino experiment KamLAND[10] are included, the best fit point is $\Delta m^2 = 7.1^{+1.2}_{-0.6}\times 10^{-5}$, $\theta = 32.5^{+2.4}_{-2.3}$, $f_B = 1.02$ (1σ errors). At 99% C.L., only the lower band of the LMA region is allowed, and maximal mixing is disfavored at a level of 5.4σ.

3. Status Report on the Third Phase of SNO

The main goal of the third phase of SNO is the event-by-event measurement of the neutral current rate, independently from Cherenkov detection. The neutral current detector (NCD) array consists in 40 sets ("strings") of proportional counters filled with (4 of them with ^{4}He). The counters are built of low background nickel(<10 ppt U, Th). The strings have a diameter of 5 cm and a length of 9-11 m, depending on the attachment position in a 1m×1m grid at the bottom of the AV. The capture of the neutron on ^{3}He yields a proton and a ^{3}H nucleus that can both be detected by ionization in the ^{3}He gas. The NCD electronics can acquire the full ionization waveform, that is used for the rejection of α/β radioactivity background.

As opposed to the first two phases of SNO, the NC and CC rate measurements in the NCD phase will be essentially independent. This will allow for a factor of 2 improvement in the precision of the CC/NC ratio and also improved θ_{12} constraints, in a global oscillation analysis.

After a one year period of NCD installation and commissioning, production data-taking for the third phase of SNO started in January 2005.

Acknowledgments

This work is supported by Canada: NSERC, Industry Canada, NRC, Northern Ontario Heritage Fund, Inco, AECL, Ontario Power Generation, HPCVL, CFI; USA: DOE; UK:PPARC; Portugal:FCT. We thank the SNO technical staff for their strong contributions.

References

1. B.T. Cleveland *et al.*, Astrophys. J. **496**, 505 (1998).
2. Y. Fukuda *et al.*, Phys.Rev. Lett. **77**, 1683 (1996); S. Fukuda *et al.*, Phys. Lett. **B539**, 179 (2002)
3. V. Gavrin, Nucl.Phys.Proc.Suppl. **138**, 87 (2005); W. Hampel *et al.*, Phys. Lett. **B447**, 127 (1999); T. Kirsten, Nucl.Phys.Proc.Suppl. **118**, 33 (2003)
4. J.N. Bahcall, A. Serenelli and S. Basu, Astrophys. J. **621**, L85 (2005).
5. V. Gribov and B. Pontecorvo, Phys. Lett. **B28**, 493 (1969)
6. L. Wolfenstein, Phys. Rev. **D17**, 2369-2374 (1978); S.P. Mikheyev, A.Y. Smirnov, Sov. Jour. Nucl. Phys. **42**, 913-917 (1985)
7. Q.R. Ahmad *et al.* (SNO Collaboration), Phys. Rev. Lett. **87**, 071301 (2001); Phys. Rev. Lett. **89**, 011301 (2002); Phys. Rev. Lett. **89**, 011302 (2002).
8. S.N. Ahmed *et al.* (SNO Collaboration), Phys. Rev. Lett. **92**, 181301 (2004); New paper on the full dataset submitted after the Workshop; nucl-ex/0502021
9. SNO Collaboration, Nucl. Instr. and Meth. **A449**, 172 (2000).
10. K. Eguchi *et al.*, Phys. Rev. Lett. **90**, 021902(2003)

THE ICARUS EXPERIMENT

S. NAVAS-CONCHA*

Dpto. de Física Teórica y del Cósmos & C.A.F.P.E., University of Granada,
Av. Severo Ochoa s/n, 18071 Granada, Spain
E-mail: navas@ugr.es

The ICARUS detector is a liquid argon time projection chamber. It provides three dimensional imaging and calorimetry of ionizing particles over a large volume, with high granularity. This multipurpose detector opens up unique opportunities to look for phenomena beyond the Standard Model through the study of atmospheric, solar and supernova neutrinos, nucleon decay searches and neutrinos from the CERN to Gran Sasso beam. The ICARUS technology has reached maturity with the construction and test (during summer 2001) of a 600 ton detector, demonstrating the feasibility of building large mass devices relevant for non-accelerator physics. The 600 ton detector was transported to the Gran Sasso laboratory (Italy) in December 2004. A summary of the most relevant results obtained during the test period as well as an overview of the general physics program of the ICARUS experiment are reported.

1. The ICARUS T600 detector

The ICARUS T600 liquid Argon (LAr) detector [1] consists of a large cryostat split in two identical, adjacent half-modules, each of $3.6 \times 3.9 \times 19.9$ m^3 internal dimensions. Each half-module is an independent unit housing an internal detector composed by two Time Projection Chambers (TPC), a field shaping system, monitors and probes, and by two arrays of photomultipliers. Externally the cryostat is surrounded by a set of thermal insulation layers. The TPC wire read-out electronics is located on the top side of the cryostat. The detector layout is completed by a cryogenic plant made of a liquid Nitrogen cooling circuit to maintain uniform the LAr temperature, and of a system of LAr purifiers.

A liquid Argon TPC allows to detect the ionization charge released at the passage of charged particles in the volume of LAr, for three dimensional image reconstruction and calorimetric measurement of ionizing events. The

*On behalf of the ICAURS Collaboration.

detector, equipped with an electronic read-out system, works as an "electronic bubble chamber" employing LAr as ionization medium. Unlike traditional bubble chambers, limited by a short window of sensitivity after expansion, the LAr TPC detector remains fully and continuously sensitive, self-triggerable and without read-out dead time.

A uniform electric field applied to the medium makes the ionization electrons drift onto the anode, following the electric field lines; thanks to the low transverse diffusion of the ionization charge, the electron images of ionizing tracks are preserved. Successive anode wire planes, biased at a different potential and oriented at different angles, make possible the three dimensional reconstruction of the track image. The wire pitch is 3 mm. The maximum drift path (distance between the cathode and the wire planes) is 1.5 m and the nominal drift field 500 V/cm (a 3 meter drift is foreseen for future ICARUS modules).

Each wire of the chamber is independently digitized every 400 ns. The electronics was designed to allow continuous read-out, digitization and independent waveform recording of signals from each wire of the TPC. Measurement of the time when the ionizing event occurred (so called "T_0 time" of the event), together with the electron drift velocity information, provides the absolute position of the tracks along the drift coordinate. The T_0 can be determined by detection of the prompt scintillation light produced by ionizing particles in LAr [2].

2. The T600 test run

A full above-ground test of the T600 experimental set-up was carried out in Pavia (Italy) during summer 2001. One T600 half-module was fully instrumented to allow a complete test in real experimental conditions. All technical aspects of the system, cryogenics, mechanics, LAr purification, read-out chambers, scintillation light detection, electronics, and DAQ were tested, and found to be satisfactorily in agreement with expectations.

During the test run, a very large amount of cosmic ray events was recorded [3] with different configurations of a dedicated trigger system: long, penetrating muons and spectacular, high multiplicity muon bundles, electromagnetic and hadronic showers and low energy events among other categories of events.

Like a bubble chamber, the ICARUS detector provides a measurement of the total ionization loss of a track with very high sampling. By extracting the physical information contained in the wires output signal, i.e. the energy deposited by the different particles and the point where such a deposition has occurred, it is possible to build a complete three dimensional spatial and calorimetric picture of the event. The measurement of the dE/dx and positions for a large number of points along a given track provides a way of estimating the particle momentum from range (for stopping particles) or multiple scattering, providing, in addition, a method for particle identification. As an example, Figure 1 shows a kaon decay candidate event acquired during the T600 test.

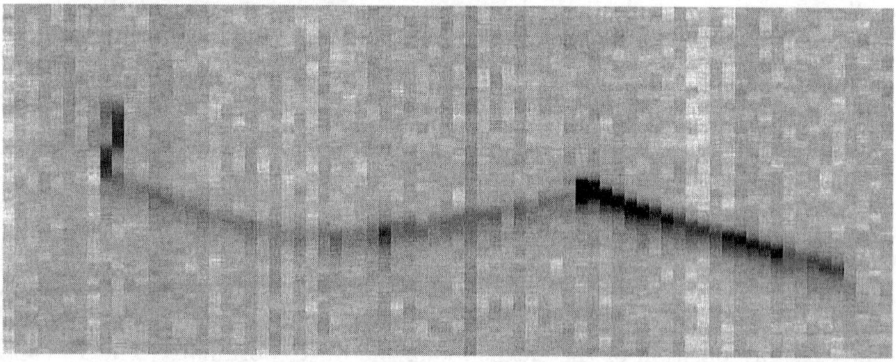

Figure 1. Charged Kaon decay candidate from the T600 test run. The Kaon, muon and electron are visible from right to left.

A fundamental requirement for the ICARUS technology is that electrons produced by ionizing particles might travel unperturbed in LAr from the point of production to the collecting wire planes. To this extend, the concentration of any kind of electro-negative impurity diluted in the liquid must be reduced at extraordinarily low levels. At the end of the run, the measurements provided by the *Purity Monitors* (dedicated devices immersed in the LAr) and from crossing muon tracks shown a maximum value of the drift electron lifetime of about 1.8 ms, equivalent to an electron mean free path greater than 280 cm, and was still increasing [4] (see Figure 2).

Another example that proves the ICARUS detector quality is given through the measurement of the muon decay energy spectrum from a sample of stopping muon events. The Michel ρ parameter was measured [5] by comparison of the experimental and Monte Carlo simulated μ spectra. The obtained value $\rho = 0.72 \pm 0.06 \pm 0.08$ is in agreement with the SM value $\rho = 0.75$.

The data coming from the test run provided also the possibility to per-

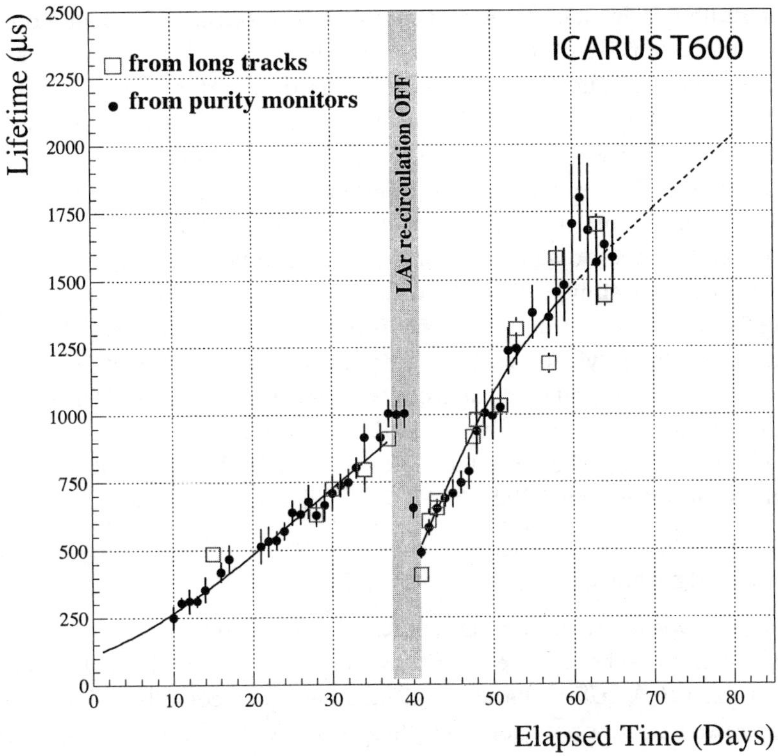

Figure 2. Evolution of the free electron lifetime during the T600 run.

form a precise measurement of the electron recombinacion in liquid Argon [6]. This parameter, crucial for the proper determination of the energy released by the particles in the medium, was measured by means of charged particle tracks. The values found for the normalization factor and the Birks law parameters are compatible with other liquid Argon TPC prototypes and other data published in the literature.

3. The ICARUS physics program

A liquid Argon time projection chamber (TPC), working as a electronic bubble chamber, continuously sensitive, self-triggering, with the ability to provide 3D imaging of any ionizing event, together with excellent calorimetric response, offers the possibility to perform complementary and simultaneous measurements of neutrinos, as those of the CNGS beam, those from cosmic ray events, and even those from the sun and from supernovae. The same class of detector can also be envisaged for high precision measurements at a neutrino factory [7,8] and can be used to perform background-free searches for nucleon decays. Hence an extremely rich and broad physics program, encompassing both accelerator and non-accelerator physics, will be addressed. These will answer fundamental questions about neutrino properties and about the possible physics of the nucleon decay.

The T600 detector was transported underground inside the Gran Sasso laboratory in December 2004. The first data is expected to be taken by spring 2005. Though it has a physics program on its own [9], the T600 construction was mainly motivated by technical issues. Since the concept of modularity was inserted starting from the very initial phases of the detector design, growing the detector mass to the several kton scale can be accomplished by "cloning" the actual T600 to the required number of units. The construction of the first 1200 ton module will start this year.

Acknowledgments

The author wishes to thank the conference organizers for preparing such an excellent workshop. The credit of the work presented on this report is due to the ICARUS Collaboration members that contributed to the detector construction, operation and physics analysis. The author gratefully acknowledges support from the Spanish Ministry of Education and Science and the *Ramon y Cajal* Programme.

References

1. S. Amerio *et al.* [ICARUS Collab.], *Nucl. Instrum. Meth.* **A527**, 329 (2004).
2. M. Antonello *et al.* [ICARUS Collab.], *Nucl. Instrum. Meth.* **A516**, 348 (2004).
3. F. Arneodo *et al.* [ICARUS Collab.], *Nucl. Instrum. Meth.* **A508**, 287 (2003) [Erratum-ibid. **A516**, 610 (2004)].
4. S. Amoruso *et al.* [ICARUS Collab.], *Nucl. Instrum. Meth.* **A516**, 68 (2004).
5. S. Amoruso *et al.* [ICARUS Collab.], *Eur. Phys. J.* **C33**, 233 (2004).
6. S. Amoruso *et al.* [ICARUS Collab.], *Nucl. Instrum. Meth.* **A523**, 275 (2004).
7. A. Bueno, M. Campanelli and A. Rubbia, *Nucl. Phys.* **B589**, 577 (2000).
8. A. Bueno, M. Campanelli, S. Navas-Concha and A. Rubbia, *Nucl. Phys.* **B631**, 239 (2002).
9. F. Arneodo *et al.* [ICARUS Collab.], `LNGS-P28/2001` (2001).

Cosmological Parameters Measurements

HIGH REDSHIFT SUPERNOVA SURVEYS

SÉBASTIEN FABBRO*
CENTRA - IST,
Avenida Rovisco Pais
1049-001 Lisbon, Portugal
E-mail: seb@ist.utl.pt

Few years ago, supernova observations have demonstrated their essential role in cosmology, by providing direct evidence for a dark energy component, responsible for the acceleration of the universe. We review how type Ia supernovae are used as distance indicators for cosmological parameter studies, and present the attempts of current surveys to turn pioneering distant supernova discoveries into daily routine, in order to provide a direct measurement of the dark energy equation of state.

1. Introduction

The quest for cosmological parameters is tightly linked to measuring distances. The greater the distances are, the more they give access to the geometry and the evolution of the universe. Since the first Hubble diagram in 1929, there has been several attempts to get good cosmological distance indicators. One type of astronomical source has been recently successful in converting redshifts to Megaparsecs with enough accuracy: Type Ia Supernovae (SNIa).

Two observations are needed to be able to use SNIa as cosmological distance indicators. From spectra we measure the redshift z and identify the supernova, and from multi-band photometry we reconstruct the light curve $F(t)$ of the SNIa.

Relating the two observations is done through the *luminosity distance* d_L derived directly from the inverse-squared relation $F(t) = L(t)/4\pi d_L^2(z;\boldsymbol{p})$, where $L(t)$ is the intrinsic luminosity and $F(t)$ the observed flux. The cosmological parameters $\boldsymbol{p}$ are only included in d_L, supernovae observations are therefore "cosmological-model-independent". In a standard concordance

*Work supported by project POCTI/CTE-AST/57664/2004 of the Fundação Ciênca e Tecnologia.

cosmological model, the luminosity distance relates with the reduced matter energy density : $\Omega_{\rm m}$, the reduced dark energy density $\Omega_{\rm x}$, and the ratio of pressure and density of the dark energy component $w(z) = p/c^2\rho(z)$. For a classic cosmological constant, we have simply $w(z) = -1$, but many other possibility[1] are speculated given the little observational constraints we have on $w(z)$. In practice, we need also to constrain $hL(t)$. This is done with nearby SNIa, and the same objects also allow corrections for real differences found in their light curves. Fortunately, these corrections are tractable: at a first order, the brighter objects are slower and bluer, and a semi-parameterized template can be constructed[2] to match pretty much all SNIa observations, and reduce scatter in the distance estimators down to 7%.

2. Requirements for a Supernova Survey

How well are we able to determine energy densities with the SNIa Hubble diagram? Apart from observational systematics and a SNIa intrinsic dispersion of 0.15mag, the precision on the energy densities and equation of state are limited. Energy densities and equation of state evolution parameters are

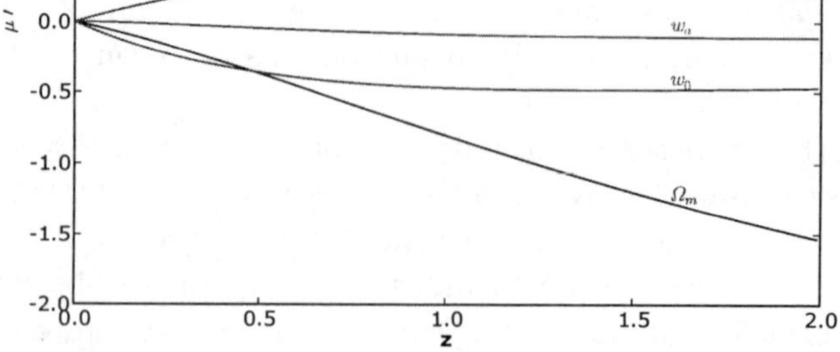

Figure 1. Distance modulus partial derivatives with respect to some cosmological parameters, in function of redshift.

strongly degenerated through any cosmological distance relations. Given a cosmological parameter set $\boldsymbol{p} = \{h, \Omega_m, \Omega_x, w_0, w_a\}$, where we chose a simple equation of state redshift evolution $w(z) = w_0 + w_a z/(1+z)$, we can compute a the Fisher matrix, which under reasonable assumptions, depends only on luminosity distance derivatives. In Figure 1, we show the derivatives of the distance modulus $\mu = 5\log(d_L(z;\boldsymbol{p})/10\text{pc})$ with respect to different cosmological parameters, with a fiducial set of cosmological parameters $\boldsymbol{p} = \{0.7, 0.3, 0.7, -0.8, -0.4\}$. The more similar the curve shapes are, the harder it is to break the degeneracy between the corresponding cosmological parameters. Supernovae observations (as well as any other observations so far) are therefore not self-sufficient to determine with decent accuracy the full set of cosmological parameters.

The situation is different for other parameterization of the equation of state or another set of fiducial values for the model. If the dark energy sector is only populated by a pure positive cosmological constant, the supernovae observations alone remain very strong to constrain cosmological parameters. As noted previously by several authors[3], the sensitive redshift region for a joint estimation of Ω_m and Ω_x is about $z \sim 0.7$. Even though distance measurements have their limits to trace a full history of cosmological parameters, supernova surveys are very effective to determine dark energy density and its evolution up to redshifts of 2. To allow constraints on the density parameters, we will then have to either combine other experiments, or assume priors on the parameters. Typical assumptions are a flat universe, justified e.g. by cosmic microwave background results[4], and a value of Ω_m which can be measured better than 10% with galaxy surveys[5].

To also achieve 10% on Ω_x and w_0, we need the good statistics of ongoing and future supernova surveys to beat the systematics resulting mostly from the current supernovae intrinsic dispersion. Obtaining a set of supernovae at various redshift is a dedicated task, on which two teams have concentrated their efforts for the last decade. SNIa only appear few times per millennium per galaxy and can be photo-metered for about 200 days in rest frame at low-z. Supernova surveys require both a wide field and a large mirror to cover a large volume of space, getting millions of galaxies. A discovery strategy has been set up through the nineties by the two collaborations, leading to the observational evidence of an accelerating universe driven by a repulsive component[6,7]. The use of 4-m class telescopes with the appearance of high quantum efficiency CCDs allowed such discoveries. Since a SNIa rises in about $\sim$ 20 days from undetectable light to maximum light, the strategy elaborated was to use a 3 weeks baseline between moonless reference images and discovery images of the same fields, and look for SNIa candidates in the subtracted images, using fine-tuned image processing software. Candidates were whereupon confirmed spectroscopically in 8m-10m class telescopes and followed-up photometrically in as many telescopes as allocated to build the light curves.

Today, this strategy does not hold up with the needs for greater precision and the pressure on telescope allocations. We will see in the next sections, current and future supernova projects elaborated to get much higher statistics.

3. Ground Surveys

The SuperNova Legacy Survey (SNLS, http://www.cfht.hawaii.edu/SNLS) is one of the Canada France Hawaii Telescope (CFHT) Legacy Survey that started in 2003. It uses the Megaprime imager, a wide-field camera (Megacam) mounted on the prime focus of the CFHT-3.6m telescope in Hawaii. The camera is a mosaic of 36 thinned 2K×4.5K CCDs covering one square degree. Four fields are observed in four bands (g_M, r_M, i_M, z_M), in average 18 nights centered on each new moon, until 2008. The observation strategy allows early discovery and well-sampled homogeneous follow-up photometry of candidates, since images are used by both online detection and follow-up photometry. To reach redshifts of $z \sim 1$, where the supernova reaches faint

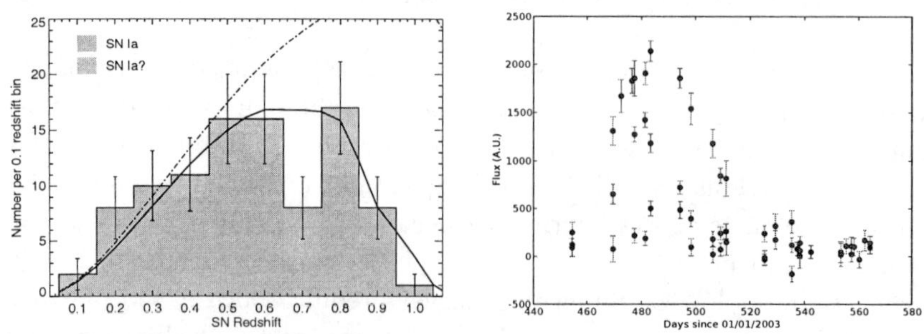

Figure 2. Left: SNIa redshifts acquired up to April 2004, with the SNLS. Right: a sample multi-color $griz$ SNIa light curve at $z = 0.91$ obtained with SNLS offline photometry.

magnitudes such as $i = 24.5$, 8-10 meter class telescopes are needed for spectroscopy. Spectroscopic follow-up has been acquired so far at the European Southern Observatory Very Large Telescope 8m, Gemini North and South 8m telescopes, as well as the Keck 10m telescopes. Accumulation of supernovae is shown in Figure 2, with a sample light-curve. If the SNLS continues on current tracks it already started, it could very well satisfy its objectives of cosmological constraints as shown in Figure 3.

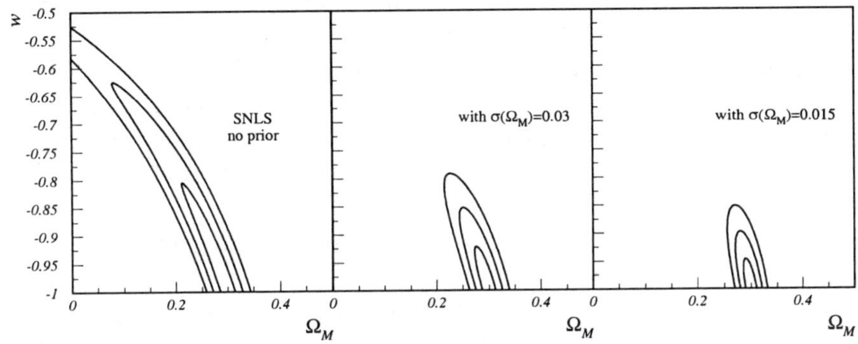

Figure 3. Expected $\Omega_m - w_0$ 39%, 86%, and 99% CL contours after 5 years of SNLS survey and 200 nearby SNIa. The fiducial cosmological model was for a cosmological constant and $\Omega_m = 0.3$, and the contours account for photometric errors and a 0.15 mag intrinsic dispersion.

The Equation of State: SupErNovae trace Cosmic Expansion (ESSENCE) project is a similar project in the Southern hemisphere, at the Cerro-Tololo 4-m telescope. It started in 2002 and aims to observe well 200 SNIa between in the redshift range $0.15 - 0.75$, in VRI. Supernovae at higher redshifts are expected to be followed with the Hubble Space Telescope (HST), and redder pass-bands photometric data acquired with other ground-based telescopes. More information on the project can be found on http://www.ctio.noao.edu/~wsne.

4. Space Surveys

For supernovae at redshift $z > 1$, needed to increase the leverage and beat the 5% measurements of cosmological parameters, space observations are required. Only space observations can give us good photometric precision and spectroscopy at higher redshift, where most of the SNIa flux is in infrared. There are two dependent program using currently the HST with the AFS camera: the Supernova Cosmology Project (SCP) has just completed a first search using HST-ACS (see http://supernova.lbl.gov) and the

PANS-GOODS survey, which has already published an analysis with 16 new supernovae followed with HST-ACS[8], and continue similar efforts. With the help of a supernova compilation of other surveys, the PANS-GOODS survey were able to get a first measurement of the epoch of deceleration.

The *SuperNova Acceleration Probe* (SNAP) is a project of a 2-m telescope in space, mounted with a wide field imager, an infrared imager and an integral field spectrometer. The observation strategy is similar to the SNLS, but covering 15 square degrees and reaching 2000 well followed SNIa/year, with an expected photometric precision of less than 2%.

SNAP data quality should be able to discriminate among dark energy models and alternative explanations to the acceleration of universe expansion. In principle, it would be able to detect time variation of the equation of state, together with prior information on Ω_m and low-redshift SNIa from the Nearby Supernova Factory. The SNAP design is still evolving. Depending on the availability of the funds, SNAP is expected to be launched in 2010 and its mission should last at least for three years. It is also designed to do together with supernovae, weak lensing observations, therefore nicely complementing the dark energy measurement.

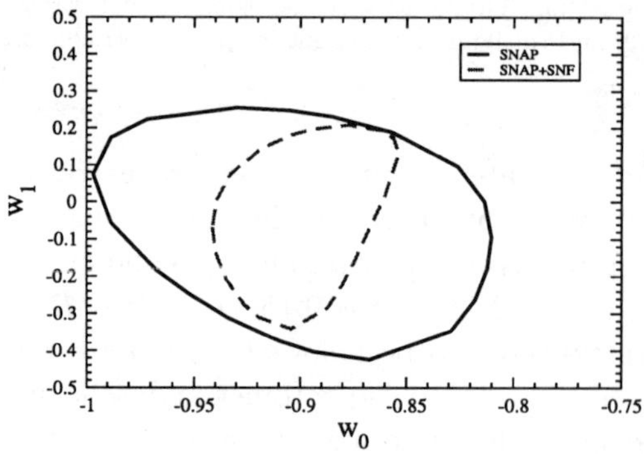

Figure 4. Expected confidence contours (68%) in the (w_0, w_1) plane for the SNAP experiment when the Nearby Supernova Factory SNIa are added (in red-dashed) and when they are not. A flat universe has been assumed with a Gaussian prior on Ω_m of $\sigma = 0.04$, and an equation of state evolution parameterized as $w(z) = w_0 + w_1 z$.

5. Conclusion

The golden age of supernova searches has started. We should be seeing wagons of high redshift supernovae data publications in the coming years, thanks to ground redshift large surveys as SNLS or ESSENCE. Space surveys can not yet be as productive with only HST, but each higher redshift SNIa brings more cosmological information. Future space surveys could bring even more. But to guaranty success in all these projects, we need a better supernova understanding. The Nearby Supernova Factory Project, described in this same volume[9], has been designed to satisfy this need.

References

1. Padmanabhan, T., *Physics Reports*, **380**, 235 (2003)
2. Guy, J. *et al.*, *astro-ph/0506583* (2005)
3. Astier, P., *Physics Letters B* **500**, 8 (2001)
4. Spergel, D. *et al.*, *Astrophysical Journal Supplement* **148**, 17 (2003)
5. Eisenstein, D. *et al.*, *astro-ph/0501171* (2005)
6. Perlmutter, S. *et al.*, *Astrophysical Journal* **517**, 565 (1999)
7. Riess, A. *et al.*, *Astronomical Journal* **116**, 1009 (1998)
8. Riess, A. *et al.*, *Astrophysical Journal* **607**, 665 (2004)
9. Antilogus, P. *this volume.*

SNFACTORY : NEARBY SUPERNOVA FACTORY

P. ANTILOGUS

The Nearby Supernova Factory: P. Antilogus, G. Garavini, S. Gilles, L-A. Guevara, D. Imbault, R. Pain, D. Vincent (LPNHE, Paris), G. Adam, R. Bacon, C. Bonnaud, L. Capoani, D. Dubet, F. Henault, B. Lantz, J-P. Lemonnier, A. Pecontal, E. Pecontal (CRAL, Lyon), N. Blanc, G. Boudoul, S. Bongard, A. Castera, Y. Copin, E. Gangler, G. Smadja (IPNL, Lyon), R. Kessler (KICP, Chicago), G. Aldering, B. C. Lee, S. Loken, P. Nugent, S. Perlmutter, J. Siegrist, R. Scalzo, R. C. Thomas, L. Wang (LBNL, Berkeley), C. Baltay, D. Rabinowitz, A. Bauer (Yale)

E-mail: P.Antilogus@in2p3.fr

The Nearby Supernova Factory (SNfactory) is an experiment designed to collect a large sample (300 or more) of low redshift ($0.03 \leq z \leq 0.08$) Type Ia supernovae, each followed up by 10-15 spectro-photometry (320-1000 nm) observations spaced at roughly 3 day intervals, and started $\sim$ 10 days before the maximum light. The goal of this experiment is to anchor the supernova Hubble diagram at low redshift and to provide a large sample of well studied SN Ia to determine those properties of SNe Ia affecting their use for cosmology. The status of the SNfactory experiment will be presented, including informations on the first spectro-photometric data collected by *SNIFS*, the dedicated Integral Field Unit (IFU) for SNe observation.

1. Introduction

The *SNfactory* project has been initiated by scientist at LBNL (Berkeley), KICP (Chicago), Yale in the United States; CRAL (Lyon), IPNL (Lyon) and LPNHE (Paris) in France. The *SNfactory* will concentrate on Type Ia supernovae (SNe Ia), the type which has been used to determine that the expansion of the universe is accelerating [1]. The *SNfactory* will lay the foundation for the new generation of experiments (SNLS ,ESSENCE) to measure the expansion history of the Universe. It will discover and obtain light curve of spectrophotometry quality for $\sim$ 300 Type Ia supernovae in the low-redshift end of the smooth Hubble flow.

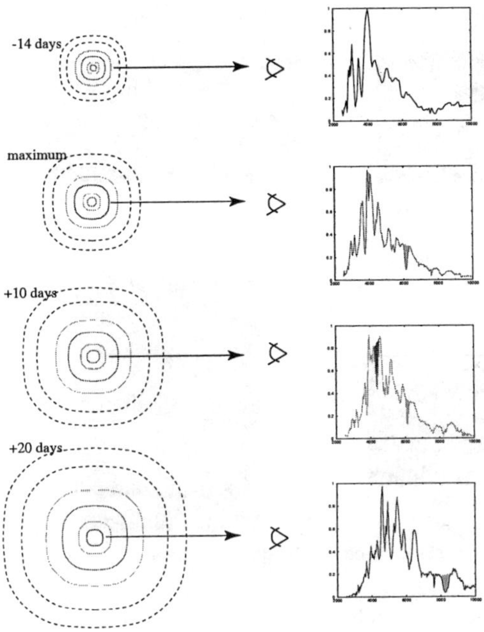

Figure 1. As the ashes of the SN Ia explosion expand with time, the light collected is sensitive to the composition (spectral features) and the speed (width of the spectral features) of deeper and deeper layers of the SN Ia. By observing such time series the *SNfactory* project will improve our understanding of the SN Ia explosion.

2. Science goals

The *SNfactory* science goals are (see Fig. 1 and Ref. 2 for details) :

- to anchor the low red-shift SN Ia hubble diagram : the minimization of the errors on the cosmological parameters by the high-redshift SNe experiments, such as SNLS or ESSENCE, can be improved by a factor ~ 2 using the unique large sample of SN Ia expected by *SNfactory*.
- to calibrate/measure technical quantities needed for the analysis of high red-shift SNe Ia (stretch-luminosity relation, K-corrections, intrinsic colors, etc.)
- to improve the quality and the understanding of the SNe Ia as distance indicator.

3. The *SNfactory* project

The supernova discovery for *SNfactory* is being carried out by the Palomar-QUEST project using the 1.2m aperture Samuel Oschin Schmidt Telescope

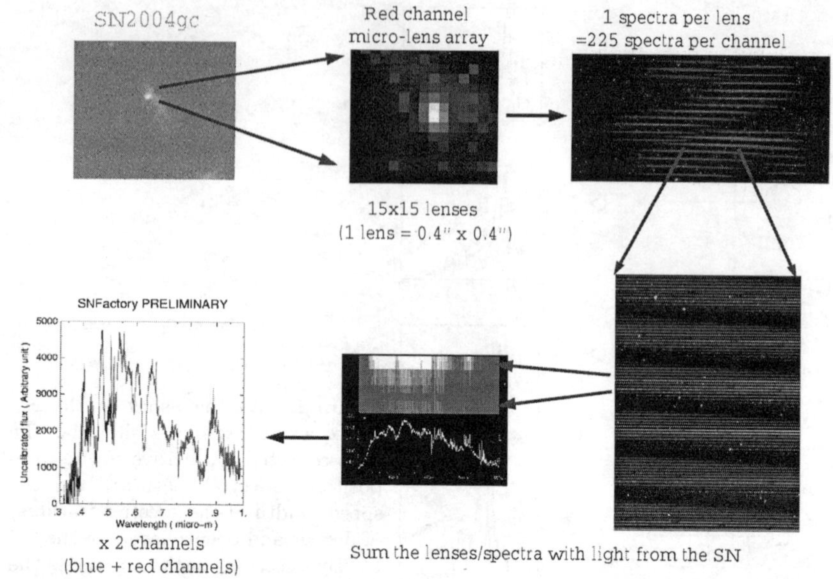

Figure 2. *SNIFS* principle: the SN and surrounding galaxy is sampled by two 15 × 15 micro-lens arrays (one in the blue , one in the red) covering a field of 6" × 6". A spectra is produced with the light collected by each lenses. From the data cube produced (= measured intensity for a given position in the field of view and a given wave length), a photometric spectra of the SN Ia is extracted. It should be noticed than an IFU allows spectro-photometric measurements. This is not the case of spectra obtained with a simple slit as the fraction of the spectered object masked by the slit is unknown. The SN Ia spectra obtained at different epochs, integrated in a given wave length domain, allows to produce the SN Ia light curves in any filter.

at the Palomar Observatory with the large area QUEST camera[a] built at Yale and Indiana Universities.

Traditionally supernovae have been followed with *BVRI* photometry, and spectra beyond the initial confirmation spectrum are rare. The *SNfactory* changes all that. Using a integral field unit [3] on a two-channel (blue 3200-5400Å& red 5000-10000Å) optical spectrograph with a resolution of respectively 4.6 Å and 6 Å, the *SNfactory*'s **S**uper**N**ova **I**ntegral **F**ield **S**spectrograph (*SNIFS*, build in France) allows spectroscopy of supernovae at all epochs. Because these spectra are spectrophotometric, *UBVRIZ* photometry can be synthesized from these spectra, without the uncertainties due to photometric color terms and K-corrections (see Fig. 2). *SNIFS* re-

[a]The QUEST camera covers $2.3^o \times 4.0^o$ with 112 CCD of 2.4k × 0.6k.

tains one advantage of the traditional approach, which allows surrounding field stars to be used for flux scaling when conditions are non-photometric, by also having an imager which integrates on the field immediately surrounding each supernova and having the exact same exposure as the integral field unit. The follow-up for the *SNfactory* is performed at the University of Hawaii's 2.2-m telescope, located on the summit of Mauna Kea. Mauna Kea has excellent atmospheric stability producing some of the best image quality attainable from the ground. The *SNfactory* has arranged for observation with *SNIFS* 20% of the time on the 2.2-m, allocated as the 2nd half of the night on every first and third night of a five-night cycle.

The *SNfactory* program plan to collect 300 nearby SN Ia ($0.03 < z < 0.08$) and to do a detailed study of the spectral evolution of each SN Ia: 10-15 spectra of spectro-photometric quality between 15 days before and 45 days after maximum.

4. First observations with *SNIFS*

The dedicated *SNIFS* IFU has been build in 2002–2003 and mounted on the UH 2.2m telecope in Hawaii in March 2004. *SNIFS* got its first light in April 2004, and was fully operational in July 2004. Since then, it is controlled remotely and has an automated target acquisition using directly the finding charts made with the discovery image. *SNfactory* science program started in August 2004 with a test run using supernova discovered by other observers. End of June 2005, the search at Palomar, which had been successfully tested in 2002 with a smaller camera, will provide a regular sample of Ia to follow with *SNIFS*.

Since August 2004, *SNIFS* has been collecting data 1/2 night 3 times a weak, each week. During this first test run (mid-August 2004 - end-December 2005) *SNIFS* has followed 5 SN :

- SN2004dt , SN Ia : 29 observations (4 Months, z=0.02)
- SN2004ef , SN Ia : 13 observations (2 Months, z=0.03)
- SDSS-SN191 , SN Ia : 3 observations (1 week, z=0.1)
- SN2004gk , SN Ic : 7 observations (1 Month, z=-0.0004)
- SN2004gs , SN Ia : 8 observations (4 weeks, z=0.027)

It has also screened 20 other SN, getting 1 spectra for each of them for identification (II,Ic/b,Ia).

The current *SNIFS* efficiency allows to follow 4-5 SN Ia each 1/2 night. An improvement up to factor 2 is foreseen, after implementing the hardware fixes and software changes than the test run has underlined.

Calibrating a given standard star by an other one collected the same photometric night, has shown photometric measurement at the 1% level between 320-850 nm. Work is underway to perform photometric calibration in non-photometric night using the photometric channel, and to improve the photometry above 850 nm, limited at the moment by fringing on the CCD.

5. Conclusion

The *SNIFS* IFU is fully operational and has started to take data. First spectro-photometry time series of SN Ia have been obtained.

During the summer 2005, the *SNIFS* instrument will start to follow SN Ia provided by the search pipeline based on the QUEST camera data at Palomar. At the fall 2005 *SNIFS* should follow 5 SN Ia in parallel, corresponding to one new SN Ia per week and a rate of 50 SN Ia time series per year.

References

1. S. Perlmutter *et al.*, Ap. J. **517**, 565 (1999).
2. G. Aldering *et al.*, Proceeding,SPIE **4836**, 61 (2002).
3. R. Bacon *et al*, A&AS **113**, 347 (1995).

A POLARIZED GALACTIC EMISSION MAPPING EXPERIMENT AT 5–10 GHZ

DOMINGOS BARBOSA[†]

*Centro de Física, Universidade do Porto, Rua do Campo Alegre 687
Porto, 4169-007, Portugal &
CENTRA, Instituto Superior Técnico, Av. Rovisco Pais
1049-001, Lisboa , Portugal*

RUI FONSECA

*Instituto de Telecomunicações, Campus Universitário de Santiago, 3810-193
Aveiro, Portugal &
CENTRA, Instituto Superior Técnico, Av. Rovisco Pais
1049-001, Lisboa , Portugal*

DINIS M. DOS SANTOS

*Dep. Electrónica e Telecomunicações, Universidade de Aveiro
3810-193, Aveiro , Portugal*

LUIS CUPIDO

*Centro de Fusão Nuclear, Instituto Superior Técnico, Av. Rovisco Pais
1049-001, Lisboa , Portugal*

ANA MOURÃO

*CENTRA, Instituto Superior Técnico, Av. Rovisco Pais
1049-001, Lisboa , Portugal*

GEORGE F. SMOOT

*Astrophysics Group, Lawrence Berkeley National Laboratory , 1 Cyclotron Road
94720 Berkeley, USA &
Physics Dept. , University of California, 366 Le Conte Hall,
94720 Berkeley, USA*

CAMILO TELLO

*Instituto Nacional de Astrofísica Espacial, Div. Astrofísica, Caixa Postal 515
12210-070 São José dos Campos SP, Brazil*

[†] Email: dsbarbos@fc.up.pt

We show in this paper the goals of the Galactic Emission Mapping (GEM) project at Portugal. This project aims to map the North Hemisphere low frequency polarized foreground contaminants of the Cosmic Microwave Background in the frequency range 5-10GHz. The resulting maps will be merged with a correspondent map of the southern Hemisphere obtained by the GEM-Brazil team to produce an almost all-sky template of the polarized synchrotron. This data will be important for proper polarized foreground extraction in the data processing of Planck Surveyor and other near-future Cosmic Microwave Background polarization B-mode probing experiments.

1. Towards CMB polarization measurements

Since 1992 [1], Cosmic Microwave Background (CMB) cosmology made a huge leap forward towards the era of high-precision measurements. The high-resolution map data returned from experiments like MAXIMA, Boomerang, DASI, NASA's WMAP [3–6] offer a direct glimpse into the physics at the surface of last scattering. These data provide strong constraints on cosmological parameters and theories of large scale structure formation and favors the inflationary paradigm. Recently, the predicted low level of CMB polarization (E-modes) was detected (DASI – 2002, WMAP – 2003) [7–9] and will constitute for the next decade the best probe of early Universe's physics. It will also offer insights on Large-Scale structure formation by measuring the effects of reionization produced by the first galaxies. Besides providing a consistency check on the inflationary scenario favored by observations, the great interest in polarization measurements comes with the recognition that CMB polarization carries the imprint of the primordial gravitational background (large scale B-mode).

The same inflationary mechanism responsible for the production of the small matter density fluctuations leading much latter to the Large Scale Structure, also produce a stochastic background of gravity waves (see Zaldarriaga [2] and references therein) . Yet, our view of the CMB is obscured by extragalactic and galactic foregrounds. Typically diffuse galactic emission is dominated by synchrotron radiation below 60GHz and by thermal dust emission above 60GHz [15]. Although good foreground templates exist for the main temperature anisotropies contaminants, namely galactic synchrotron emission and dust emission from galactic dust cirrus, little is known from templates about the polarized foregrounds.

2. The GEM project @ Portugal

Experiments like Planck Surveyor will require well estimated templates of polarized synchrotron (affecting the Planck Low Frequency Instrument 30–90 GHz) and polarized dust emission (affecting the Planck High Frequency Instrument 100–800 GHz). Existing polarization templates (for low frequencies - synchrotron) rely on a semi-empirical approach of extrapolating polarization substructure information from small sky area surveys with theoretical assumptions of depolarization of the synchrotron observed templates [10,13,14].

Furthermore, correlations of the WMAP first-year data with available foreground templates show another physical component appearing to be important – spinning dust emission which was not accounted for in the data processing analysis [17]. Thus, with foreground estimation as one of the main tasks currently being pursued by CMB teams, the Galactic Emission Mapper[1]

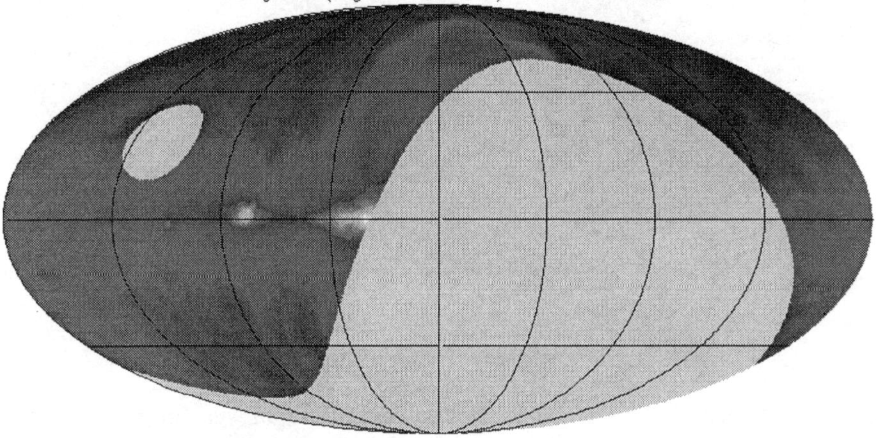

(GEM) – a portable 5.5-m dish originally aiming to determine the spatial distribution and absolute intensity in the radio and microwave spectrum of the radiation emitted by the Milky Way galaxy [11,12] – was upgraded towards polarized foreground cartography at 5-10GHz (C. Tello, in preparation) with the development of a new high-sensitivity pseudo-correlator receiver. It will map the polarized synchrotron radiation (polarized at a maximum of 60% level at 10 GHz) where this emission is dominant.

This dish is currently operated in Cachoeira Paulista, Brazil and will map I, Q, U Stokes parameters in the southern hemisphere sky down to a 1mK sensitivity at 5GHz. Recently, a new member was added to the GEM project (Barbosa *et al.*, in preparation): seeking complementarity to the very near-future GEM-Brazil maps, a new dish (9-m) equipped with a high-sensitivity polarizer will operate in central Portugal with similar scanning strategy (steady, slow azimuthal rotation at fixed elevation at 1rpm until the required sensitivity is achieved) aiming the coverage of the northern hemisphere at the same frequencies at a latitude of ~40° allowing a good coverage of high-galactic latitudes [17] (see Fig.1), target of future polarization B-mode probing experiments.

---

[1] USA site: http://aether.lbl.gov/www/projects/gem
Brazil site : http://www.cea.inpe.br/~cosmo/GEM/index_gem.htm

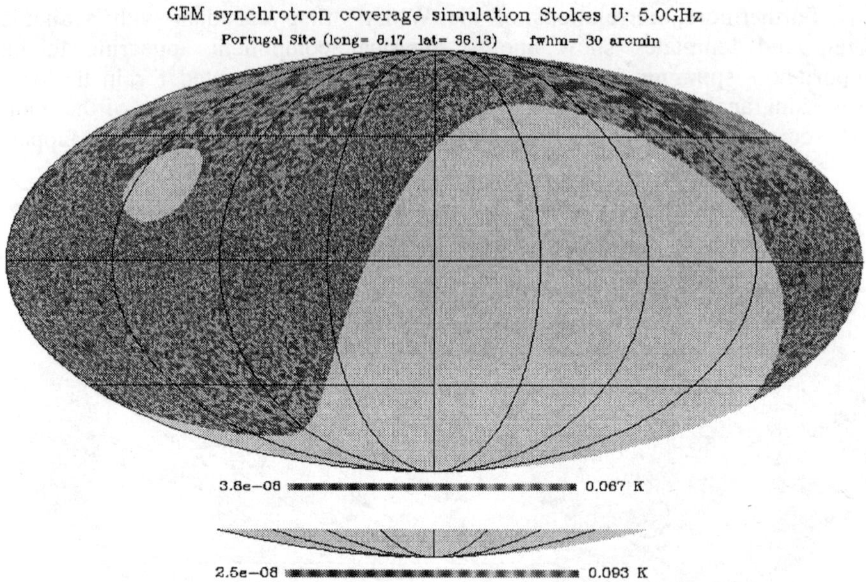

Figure 1. Maps of the I, Q, U Stokes parameters of synchrotron galactic radiation to be obtained at 5GHz in central Portugal with a resolution of 30 arcmin. The simulations used 30GHz templates [13] scaled down to 5GHz. The simulated scanning strategy was azimuthal rotation at 1rpm at a fixed elevation of 50°.

It would gather information on the I,Q,U Stokes parameters in the northern sky after 6 months of integration time over 40% of the sky with a resolution of 30 arcmin (the signal to detect is ~1 mK).

At the same time, continuous work on the receiver will prepare the upgrades towards 10GHz observations, where some signal from spinning dust is expected to appear mixed to synchrotron emission. At a later stage, by 2007, GEM-Brazil and GEM-Portugal maps will be merged to produce the biggest sky templates of the polarized synchrotron components (~80% sky). These templates will shed light on the synchrotron polarization fraction and substructure at scales down to 0.5°, covering a range of scales important for B-mode searching and allowing a better polarized foreground subtraction from large sky data sets such as the Planck Surveyor maps.

Acknowledgments

This work was supported by Fundação para a Ciência e Tecnologia – Portugal through projects POCTI/FNU/42263/2001 and POCI/CTE-AST/57209/2004. DB is supported by a rolling CFP grant. RF is supported by a BIC grant from IT – pólo Aveiro.

References

1. G. F. Smoot *et al.*, *Astrophy. J.* **369**, L1 (1992).
2. M. Zaldarriaga, *Carnegie Obs. S.*, **vol.2**, (2003), astro-ph/0305272.
3. P. de Bernardis, *Nature.* **404**, 955 (2001).
4. A.T. Lee *et al. Astroph. Jour.*, **561**, L1-L5 (2001).
5. N.H. Halverson *et al.*, *Astrophys.J.*. **568**, 38-45 (2002).
6. C.L. Bennet *et al.*, *Astroph. Journ. Sup.*, **148**, 1 (2003).
7. E.M. Leitch et al, *Nature.* **420**, 763 (2002).
8. J. Kovac *et al.*, *Nature.* **420**, 772 (2002).
9. A. Kogut *et al.*, *Astroph. Journ. Sup*, **148**, 161 (2003).
10. A. de Oliveira-Costa *et al.*, *New Astron. Rev.*, **47**, 1117 (2003) .
11. C. Tello *et al.*, *Astronomy & Astrophys.*, **145**, 495 (2000).
12. S. Torres *et al.*, *Astroph. Space Sci.*, **240**, 225 (1996).
13. G. Girardino *et al.*, *Astronomy & Astroph.*, **387**, 82, (2002).
14. C. Baccigalupi *et al.*, *Astronomy & Astrophys.* **372**, 8 (2001).
15. M. Tegmark *et al.*, *Astroph. J*, **530**, 133-165 (1999).
16. D. Finkbeiner, *Astrophys.J.,* **614**, 186-193 (2004).
17. R. Fonseca *et al.*, *New Astron. (submited.)*, astro-ph/0411477 (2004).

GALAXY CLUSTERS AS PROBES OF DARK ENERGY*

P. T. P. VIANA
Departamento de Matemática Aplicada,
Faculdade de Ciências da Universidade do Porto,
Rua do Campo Alegre, 687,
4169-007 Porto, Portugal

Centro de Astrofísica da Universidade do Porto
Rua das Estrelas,
4150-762 Porto, Portugal
E-mail: viana@astro.up.pt

Clusters of galaxies are the most massive virialized structures in the Universe. Given that the mass function of large-scale structures decreases exponentially at the high-mass end, galaxy clusters are a sensitive probe of its normalization and redshift evolution, and hence of the cosmological parameters that most influence it. It will be discussed to what extent the present amplitude of density perturbations, the matter density, and the density and equation of state of the dark energy, can be constrained using observational data on the present and past abundance of galaxy clusters. Current results will be reviewed, and the expected constraints from the *XMM-Newton Cluster Survey* (**XCS**) will be presented.

1. Introduction

Clusters of galaxies are one of the most important probes of the large scale structure and overall dynamical state of the Universe. Their present-day average statistical properties, and their evolution as a function of redshift, can be used to constrain the cosmological parameters that most influence the formation and evolution of large scale structures: the normalization of the power spectrum of density fluctuations, usually given as σ_8 - the dispersion of the density field on scales of $8\,h^{-1}$ Mpc (h is the present value of the Hubble parameter, H_0, in units of $100\ \mathrm{km\,s^{-1}\,Mpc^{-1}}$); the total matter density in units of the critical density, Ω_m; the energy density associated

*Work partially supported by grant POCTI/FNU/43753/2001 of Fundação para a Ciência e a Tecnologia.

with a possible cosmological constant, Ω_Λ. Recently, this last quantity as been often substituted by Ω_w, where w is a constant assumed to describe the equation of state, $w = p/\rho$, of a possible dark energy component ($w = -1$ in the case of a classical cosmological constant, Λ).

We will concentrate on the cluster number density: its present-day value and evolution with redshift. We start by briefly describing why and how it can be used to constrain cosmological parameters. Next, we review the estimation of σ_8 from data on the local cluster abundance and of Ω_m based on the redshift evolution of the cluster abundance. Finally, we discuss future prospects in this field, and show to what extent the *XMM-Newton Cluster Survey* (**XCS**) will be able to constrain cosmological parameters.

2. Cosmology with the cluster abundance

The number density of dark matter haloes as a function of mass M and redshift z, also known as the halo mass function, has long been considered an essential prediction of any credible large-scale structure formation model. The (comoving) halo mass function can be written as

$$n(M,\, z) = -F(\sigma)\frac{\rho_m}{M\sigma}\frac{d\sigma}{dM}\,, \tag{1}$$

with $M(R) = (4/3)\pi R^3\rho_0$, where $\rho_m = \Omega_m\rho_c$ is the present-day total matter density (ρ_c is the critical density), and

$$\sigma(R,z) = \sigma(R,0) * \frac{g[\Omega_m(z),\Omega_w(z),w]}{g[\Omega_m,\Omega_w,w]} * (1+z)^{-1}\,, \tag{2}$$

is the dispersion of the density field at some scale R at redshift z [g is a function that describes the growth of density perturbations in the linear regime, i.e. as long as $\sigma(R,z) < 1$]). The density parameters $\Omega_m(z)$ and $\Omega_w(z)$ depend only on Ω_m, Ω_w and w.

N-body simulations [1,2,3,4] have shown that if $F(\sigma)$ is written as

$$F(\sigma) = A\exp(-|\ln(1/\sigma) + B|^\epsilon) \tag{3}$$

then the parameters A, B and ϵ are independent of cosmology and redshift when halo masses are defined with respect to the background density.

From the previous expressions one finds that $n(M,\, z)$ depends essentially on σ_8, Ω_m, Ω_w and w (in decreasing order of importance). Given that $n(M,\, z)$ varies with M and z differently for distinct combinations of these parameters, in principle knowing how $n(M,\, z)$ changes can lead to the estimation of any of those parameters. Observationally, the most accessible interesting quantity is the number density of rich clusters of galaxies at the present epoch, providing an estimate for $n(M,\, z \sim 0)$ in the range of scales 10^{14} to $2 \times 10^{15}\, h^{-1}\, M_{\odot}$. Given that $n(M,\, z)$ depends most strongly on σ_8, traditionally the present-day abundance of galaxy clusters has been used to determine this parameter (as a function of the others). Data on the evolution with redshift of the cluster number density has in turn been used to try to break the degeneracy between σ_8 and the next most influential parameter, Ω_m, simultaneously estimating both (again as a function of the remaining parameters). This approach could be extended to the characterisation of the dark energy, but good enough data is not yet available.

3. Current results on σ_8 and Ω_m

The best method to find clusters of galaxies is through their X-ray emission, which is much less prone to projection effects than optical identification. The selection function of X-ray cluster catalogues is therefore usually well characterized, which is essential if the cluster abundance, and hence cosmological parameters, are to be determined with small uncertainty. Further, the X-ray temperature of a galaxy cluster is at present the most reliable estimator of its mass, thereby allowing the correct comparison between the observed cluster abundance and the halo abundance predicted by structure formation models.

Most determinations of σ_8 using the number density of rich clusters of galaxies at the present epoch have thus relied on X-ray selected galaxy cluster catalogues. Some recent results [5,6,7] are: $\sigma_8 = (0.7 \pm 0.06) \times (\Omega_m/0.35)^{-0.44}$; $\sigma_8 = (0.66 \pm 0.05) \times (\Omega_m/0.35)^{-0.25}$; $\sigma_8 = 0.60 + 0.17 \times (\Omega_m/0.35)^{-0.5} \pm 0.04$. All results assume $\Omega_w = 1 - \Omega_m$ with $w = -1$, and are at the 95% confidence level. We have concentrated just in the case of $\Omega_m = 0.35$ and found $\sigma_8 = 0.78$, being [0.72 - 1.08] the 90% confidence interval [8]. The discrepancy between the results shown is due essentially to distinct assumptions about the relation between the X-ray cluster observable, either luminosity or temperature, and the cluster mass. Note that weak lensing measurements of σ_8 show an even greater discrepancy [9–13].

For Ω_m to be be simultaneously estimated together with σ_8, a large catalogue of galaxy clusters with measured X-ray temperatures and covering a wide range of redshifts is needed. However, such a catalogue does not exist yet, with the closest being the *EMSS* [14,15,16,17], *RDCS* [18,19] and *Bright-SHARC* [20,21]. The estimates of Ω_m that have been obtained based on the

EMSS catalogue are the following: 0.55 ± 0.17 [22]; 0.45 ± 0.25 [23]; 0.95 ± 0.35 [24]; $0.4 - 0.8$ [25]; 0.25 ± 0.10 [26]; 0.85 (> 0.35 at 2σ) [27]; 0.44 ± 0.12 [28]; 0.2 to 0.5 [29]. The analysis of the *RDCS* catalogue has lead to $\Omega_m \simeq 0.25 \pm 0.15$ [30,31], while the analysis of the *Bright-SHARC* catalogue has lead to Ω_m within 0.85 to 1.00 [32]. All results have been obtained assuming $\Omega_w = 1 - \Omega_m$ and $w = -1$. The large dispersion in the results is to a great extent due to the limitations of the data, which exacerbates the sensitivity to differences in the various analysis.

4. The future is the XCS

There is a pressing need for a new galaxy cluster catalogue, of greater size, and in particular going to higher redshift, than existing ones, and with a well understood selection function. We have described [33] in considerable detail how such a catalogue may be constructed through serendipitous detections of galaxy clusters in archival data from the *XMM-Newton* satellite. By examining the many thousands of pointings which will be made, it will be possible to build a representative sample of randomly, and hence objectively, selected X-ray clusters. The *X-ray Cluster Survey* (**XCS**) will not only be an invaluable resource for cosmological studies, but will also have a variety of other applications [33].

Through a likelihood analysis we were able to estimate the uncertainty associated with the estimation of cosmological parameters using those clusters of galaxies in the **XCS** for which we will have their X-ray temperatures reliably estimated solely from the serendipitous data [34]. The likelihood calculation was performed by means of a Monte Carlo method [35], whereby 2000 realizations of the expected *XCS* catalogue were generated for an input fiducial cosmological model, with the addition of Poisson noise, and then for each realization the cosmological parameters σ_8, Ω_m and Ω_Λ, were allowed to vary so as to find their most probable values given each catalogue. The input fiducial model was the concordance cosmology with $\Omega_m = 0.3$ and $\Omega_\Lambda = 0.7$, for which we assumed $\sigma_8 = 0.9$. We found that the *XCS* will provide competitive estimates for the three most important cosmological parameters, σ_8, Ω_m and Ω_Λ, enabling their joint estimation to within respectively (at least) 5 per cent, 10 per cent and 15 per cent of their true values at the 95 per cent confidence. Introducing extra unknown parameters in the likelihood estimation, for example describing the equation of state of the dark energy component, would inevitably lead to some degradation in the accuracy with which all the other 3 cosmological parameters can be estimated from the **XCS**. An accurate characterisation of the equation of state of the dark energy, including its possible time dependence, using the redshift evolution of the cluster abundance will require a (near) full-sky galaxy cluster survey sensitive up to a redshift of unity. Although the **XCS** has this sensitivity, it only covers around 1% of the sky.

References

1. A. R. Jenkins et al., *MNRAS*, **321**, 372 (2001)
2. A. E. Evrard et al., *ApJ*, **573**, 7 (2002)
3. W. Hu and A. V. Kravtsov, *ApJ*, **584**, 702 (2003)
4. E. V. Linder and A. R. Jenkins, *MNRAS*, **346**, 573 (2003)
5. U. Seljak, *MNRAS*, **337**, 769 (2002)
6. S. W. Allen, R. S. Schmidt and A. C. Fabian, *MNRAS*, **328**, L37 (2001)
7. A. Voevodkin and A. Vikhlinin, *ApJ*, **601**, 610 (2004)
8. P.T.P. Viana et al., *MNRAS*, **346**, 319 (2003)
9. M. L. Brown et al., *MNRAS*, **341**, 100 (2003)
10. M. Jarvis et al., *AJ*, **125**, 1014 (2002)
11. C. Heymans et al., *MNRAS*, **347**, 895 (2004)
12. J. Rhodes et al., *ApJ*, **605**, 29 (2004)
13. L. Van Waerbeke, Y. Mellier and H. Hoekstra, *A&A*, **429**, 75 (2004)
14. I. M. Gioia et al., *ApJS*, **72**, 567 (1990)
15. J. P. Henry et al., *ApJ*, **386**, 408 (1992)
16. I. M. Gioia and G. A. Luppino, *ApJS*, **94**, 583 (1994)
17. A. D. Lewis et al., *ApJ*, **566**, 744 (2002)
18. P. Rosati et al., *ApJ*, **492**, L21 (1998)
19. B. P. Holden et al., *AJ*, **124**, 33 (2002)
20. A. K. Romer et al., *ApJS*, **126**, 209 (2000)
21. C. Adami et al., *ApJS*, **131**, 391 (2000)
22. J.P. Henry, *ApJ*, **489**, L1 (1997)
23. V. R. Eke et al., *MNRAS*, **298**, 1145 (1998)
24. D. E. Reichart et al., *ApJ*, **518**, 521 (1999)
25. P. T. P. Viana and A. R. Liddle, *MNRAS*, **303**, 535 (1999)
26. M. Donahue and G. M. Voit, *ApJ*, **523**, L137 (1999)
27. A. Blanchard et al., *A&A*, **362**, 809 (2000)
28. J. P. Henry, *ApJ*, **534**, 565 (2000)
29. J. P. Henry, *ApJ*, **609**, 603 (2004)
30. S. Borgani et al., *ApJ*, **517**, 40 (1999)
31. S. Borgani et al., *ApJ*, **561**, 13 (2001)
32. S. C. Vauclair et al., *A&A*, **412**, 37 (2003)
33. A. K. Romer et al., *ApJ*, **547**, 594 (2001)
34. A. R. Liddle et al., *MNRAS*, **325**, 875 (2001)
35. G. Holder, Z. Haiman and J. J. Mohr, *ApJ*, **560**, L111 (2001)

Black Hole Physics

ACOUSTIC BLACK HOLES

VITOR CARDOSO
McDonnell Center for the Space Sciences, Department of Physics, Washington University, St. Louis, Missouri 63130, USA
and
Centro de Física Computacional, Universidade de Coimbra, P-3004-516 Coimbra, Portugal
E-mail: vcardoso@wugrav.wustl.edu

We discuss some general aspects of acoustic black holes. We begin by describing the associated formalism with which acoustic black holes are established, then we show how to model arbitrary geometries by using a de Laval nozzle.

1. Introduction

The progress in understanding black holes has been immense, over these last forty years since their concept was born, and they now play a central role in modern physics. An important step to make black holes more accessible (from an experimental point of view) was given in 1981 by Unruh [1], who came up with the notion of analogue black holes. While not carrying information about Einstein's equations, the analogue black holes devised by Unruh do have a very important feature that defines black holes: the existence of an event horizon. The basic idea behind these analogue acoustic black holes is very simple: consider a fluid moving with a space-dependent velocity, for example water flowing throw a variable-section tube. Suppose the water flows in the direction where the tube gets narrower. Then the fluid velocity increases downstream, and there will be a point where the fluid velocity exceeds the local sound velocity, in a certain frame. At this point, in that frame, we get the equivalent of an apparent horizon for sound waves. In fact, no (sonic) information generated downstream of this point can ever reach upstream (for the velocity of any perturbation is always directed downstream, as a simple velocity addition shows). This is the acoustic analogue of a black hole, or a *dumb hole.* These objects are not true black holes, because the acoustic metric satisfies the equations of fluid dynamics and not Einstein's equations. One usually expresses this by saying that they are analogs *of* general relativity, because they provide an effective metric and so generate the basic kinematical background in which general relativity resides. They are not models *for* general relativity, because the metric is not dynamically dependent on something like Einstein's equations [2]. Following on Unruh's dumb hole proposal many different kinds of analogue black holes have been devised, based on condensed matter physics, slow light etcetera [2].

2. Effective acoustic geometry

Let us start with the equations of fluid dynamics, and try to arrange them in such a way that an effective metric stands out naturally. The fundamental equations of fluid dynamics are the equation of continuity

$$\partial_t + \nabla.(\rho \boldsymbol{v}) = 0\,, \tag{1}$$

and Euler's equation

$$\rho \frac{\boldsymbol{v}}{dt} \equiv \rho[\partial_t \boldsymbol{v} + (\boldsymbol{v}.\nabla)\boldsymbol{v}] = -\nabla p + \boldsymbol{F}\,, \tag{2}$$

where $\boldsymbol{F}$ are all the external forces acting on the fluid. Hereafter we make the following assumptions:(i) the external forces are all gradient-derived, or $\boldsymbol{F} = -\rho\nabla\Phi$; (ii) the fluid is locally irrotational, and introduce the velocity potential ψ, $\boldsymbol{v} = -\nabla\psi$; and (iii) the fluid is barotropic, i.e., the density ρ is a function of pressure p only. In this case, we can define

$$h(p) = \int_0^p \frac{dp'}{\rho(p')} \quad \text{or} \quad \nabla h = \frac{\nabla p}{\rho}\,. \tag{3}$$

Euler's equation can be written as

$$\partial_t \psi + h + \frac{1}{2}(\nabla\psi)^2 + \Phi = 0\,. \tag{4}$$

To study sound waves, we follow the usual procedure and linearize the continuity and Euler's equations around some background flow, by setting $\rho = \rho_0 + \epsilon\rho_1$, $p = p_0 + \epsilon p_1$, $\psi = \psi_0 + \epsilon\psi_1$, and discarding all terms of order ϵ^2 or higher. The continuity equation yields

$$\partial_t \rho_0 + \nabla.(\rho_0 \boldsymbol{v}_0) = 0\,, \quad \partial_t \rho_1 + \nabla.(\rho_1 \boldsymbol{v}_0 + \rho_0 \boldsymbol{v}_1) = 0\,. \tag{5}$$

Linearizing the enthalpy we get $h(p_0 + \epsilon p_1) \sim h(p_0) + \epsilon \frac{dh}{dp}_{p=p_0} = h_0 + \epsilon \frac{p_1}{\rho_0}$. Inserting this in Euler's equation one gets

$$-\partial_t \psi_0 + h_0 + \frac{1}{2}(\nabla\psi_0)^2 + \Phi = 0\,, \quad p_1 = \rho_0(\partial_t \psi_1 + \mathbf{v_0}.\nabla\psi_1)\,. \tag{6}$$

On the other hand, since the fluid is barotropic we have $\rho_1 = \frac{\partial \rho}{\partial p} p_1$, and using (6) this is the same as $\rho_1 = \frac{\partial \rho}{\partial p}\rho_0(\partial_t \psi_1 + \mathbf{v_0}.\nabla\psi_1)$. Finally, substituting this into (5) we get

$$0 = -\partial_t \left(\frac{\partial \rho}{\partial p}\rho_0(\partial_t\psi_1 + \mathbf{v_0}.\nabla\psi_1)\right) + \nabla.\left(\rho_0\nabla\psi_1 - \frac{\partial \rho}{\partial p}\rho_0\boldsymbol{v}_0(\partial_t\psi_1 + \mathbf{v_0}.\nabla\psi_1)\right) . \quad (7)$$

It can now easily be shown [2] that this equation can also be obtained from the usual curved space Klein-Gordon equation

$$\partial_\mu(\sqrt{-g}g^{\mu\nu}\partial_\nu\psi) = 0 \,, \quad (8)$$

with the effective metric $g^{\mu\nu}$ given by

$$g^{u\nu} \equiv \frac{1}{\rho_0 c}\begin{bmatrix} -1 & \vdots & -v_0^j \\ \cdots\cdots & . & \cdots\cdots\cdots \\ -v_0^i & \vdots & (c^2\delta_{ij} - v_0^i v_0^j) \end{bmatrix} . \quad (9)$$

Neither of the background quantities is assumed constant through the flow, and so they are in general dependent on the coordinates along the flow. Here we have used the definition of the local sound speed $c^{-2} = \frac{\partial \rho}{\partial p}$. We can see that the propagation of sound waves in a fluid is equivalent to the propagation of a scalar field in a generic curved spacetime described by (9). This means that *all* properties of wave propagation in curved space hold also for the propagation of sound waves. The power of this effective geometry should be clear: first, by changing the background flow, we change the effective acoustic metric. Second, since this geometry clearly has an apparent horizon at the point where $v_0 = c$, and since the existence of an apparent horizon implies Hawking radiation, then there should be Hawking radiation in this geometry, which takes the form of phonons [1]. The Hawking temperature can be computed to yield [2]:

$$kT_H = \frac{\hbar}{2\pi}\frac{\partial(c - v_\perp)}{\partial n} \,, \quad (10)$$

where $v_\perp$ is the component of the fluid velocity normal to the horizon, and $\boldsymbol{n}$ is the unit vector normal to the horizon. This can also be written as

$$T_H = 1.2 \times 10^{-9}\mathrm{K\,m}\left[\frac{c}{1000\,\mathrm{ms}^{-1}}\right]\left[\frac{1}{c}\frac{\partial(c - v_\perp)}{\partial n}\right] \quad (11)$$

This is, for all practical purposes a number too low to be detected. Despite the fact that one cannot observe Hawking radiation, one can still measure classical aspects of black holes. So we now turn to this, but first we explain how we can mimic several geometries by using a de Laval nozzle.

3. Shaping the nozzle

3.1. *The de Laval nozzle*

A de Laval nozzle is a device which can be used to accelerate a fluid up to supersonic velocities. They were first used in steam turbines, but they find many applications in rocket engines, nozzles in supersonic wind tunnels, etc. It consists of a converging pipe, where the fluid is accelerated, followed by a throat which is the narrowest part of the tube and where the flow undergoes a sonic transition, and finally a diverging pipe where the fluid continues to accelerate. It is sketched in Fig. 1.

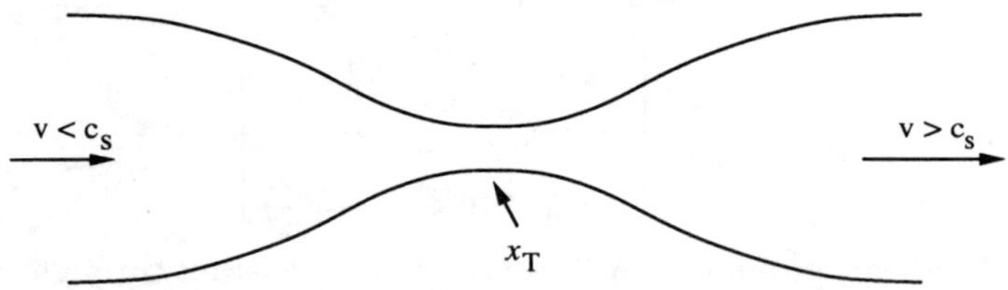

Figure 1. A sketch of a de Laval nozzle, used to make a smooth transition from subsonic to supersonic flow. The velocity of the fluid $v(x)$ equals the local velocity of sound $c(x)$ at the throat of the nozzle, $x = x_T$. The cross-section at this point is denoted by A_T.

Consider now a steady, isentropic flow through the nozzle, which has a varying cross-section $A(x)$, where x is the arc length along a streamline. Logarithmic differentiation of the continuity equation $\rho v A = \frac{dm}{dt} = \text{const}$ yields

$$\frac{1}{v}\frac{dv}{dx} + \frac{1}{A}\frac{dA}{dx} + \frac{1}{\rho}\frac{d\rho}{dx} = 0\,. \tag{12}$$

For isentropic flow $p = p(\rho)$ and $c^2 = \frac{dp}{d\rho}$. Using this in (12) we obtain

$$\frac{1}{v}\frac{dv}{dx} + \frac{1}{A}\frac{dA}{dx} + \frac{1}{c^2\rho}\frac{dp}{dx} = 0\,. \tag{13}$$

Using the component of Euler's equation along the streamline $\rho v \frac{dv}{dx} = -\frac{dp}{dx}$, and combining it with Eq. (13) we have finally

$$\frac{1}{v}(1 - \frac{v^2}{c^2})\frac{dv}{dx} = -\frac{1}{A}\frac{dA}{dx}\,. \tag{14}$$

According to (14), when the flow is subsonic ($v < c$), dv/dx and dA/dx have opposite signs. So narrowing the pipe will make the gas flow faster, which is what we expect from common experience. In fact, for very small v,

Eq. (14) can be written as $dv/v = -dA/A$ and thus vA is a constant, a well known result for incompressible fluids. The situation is opposite for $v > c$, when dv/dx and dA/dx have the same sign. This means that a region of increasing cross- section will accelerate the flow.

3.2. *Acoustic black holes by different nozzle configurations*

Let us now suppose that the dependence of c on x is small, i.e., that $\partial c/\partial x$ is not very large (as happens for example for perfect gases), and therefore that $c \sim$ const. Then, Eq. (14) can be solved yielding

$$\frac{1}{v} e^{v^2/(2c^2)-1/2} = A(x)/A(x_T)\,. \tag{15}$$

Let us choose the following generic form for A: $A(x) = \frac{1}{f(x)} e^{[A(x_T)f(x)]^2/2-1/2}$. For this to be a consistent solution we must have $dA/dx = 0$ at $x = x_T$, which results in the constraint $f(x_T) = \frac{1}{A_T}$. Eq. (15) is then trivially solved by $\frac{v}{c} = f(x)A(x_T)$ So we conclude that in order to mimic some metric, we have to be able to simulate the correct background flow.

4. An explicit example of an acoustic black hole and classical wave phenomena in its vicinities

Some classical aspects of wave propagation in acoustic black holes have been explored in [3,4]. Here I will summarize some of their results, focusing always on the $(2+1)$-dimensional rotating acoustic black hole [2], which I now describe. Consider a fluid having (background) density ρ. Assume the fluid to be locally irrotational (vorticity free), barotropic and inviscid. From the equation of continuity, the radial component of the fluid velocity satisfies $\rho v^r \sim 1/r$. Irrotationality implies that the tangential component of the velocity satisfies $v^\theta = B/r$. By conservation of angular momentum we have $\rho v^\theta \sim 1/r$, so that the background density of the fluid ρ is constant. We then take $v^r = A/r$, and A, B are constants. The acoustic metric describing the propagation of sound waves in this "draining bathtub" fluid flow is [2]:

$$ds^2 = -\left(c^2 - \frac{A^2+B^2}{r^2}\right)dt^2 + \frac{2A}{r}drdt - 2Bd\phi dt + dr^2 + r^2 d\phi^2\,. \tag{16}$$

The acoustic event horizon is located at $r_H = A/c$, and the ergosphere forms at $r_{ES} = (A^2+B^2)^{1/2}/c$.

4.1. *QNMs*

Black holes, like so many other objects, have characteristic oscillation or ringing modes, which are called quasinormal modes (QNMs) [5], the associated frequencies being termed QN frequencies, or ω_{QN}. The QN frequencies of the $(2+1)$-rotating acoustic black hole (16), described by the effective geometry just described were recently computed in [3,4], to which we refer the reader for further details. Considering a $m = 1$ mode, for which the lowest mode (this is the mode that controls the ringing phase) is approximately $\omega_{QN} \sim (0.4 - 0.33i)\frac{c}{r_H}$, with r_H the horizon radius. If one builds an acoustic black hole by making a $r_H = 1\,\mathrm{mm}$ hole in a tub with water, then this black hole should have a characteristic ringing frequency of $\omega \sim 4\times 10^5\,\mathrm{s}^{-1}$, and a typical damping timescale given by $\tau = \frac{1}{\mathrm{Im}[\omega]} \sim 3 \times 10^{-6}\,\mathrm{s}$.

4.2. *Late-time tails*

The evolution of a general set of initial data shows that shows that the signal from a black hole can roughly be divided in three parts: (i) the first part is the prompt response, at very early times, and the form depends strongly on the initial conditions.(ii) at intermediate times the signal is dominated by an exponentially decaying ringing phase, and its form depends entirely on the black hole characteristics, through its associated QNMs [5]. (iii) a late-time tail [6], usually a power law falloff of the field. This power law is also highly independent of the initial data. There is another case in which wave propagation develops tails: wave propagation in odd dimensional *flat* spacetimes (see Cardoso *et al.* in [6]). Analogue black holes also exhibit tails, shedding their hair in a power-law falloff manner. It was found in [3] that any perturbation Ψ in the vicinity of the $(2+1)$-dimensional analogue black hole (16) eventually decays as

$$\Psi \sim t^{-(2m+1)}\,. \tag{17}$$

On the other hand, this is precisely the tails that appear in any $(2+1)$-dimensional *flat* spacetime [6]. We thus have a consistent and elegant result.

4.3. *Superradiant amplification*

Rotating black holes can superradiate, in the sense that in a scattering experiment the scattered wave has a larger *amplitude* (the frequency is the same, this is *not* a Doppler effect as explained in Ref.7 and references therein) than the incident wave. Superradiance is a general phenomenon in physics. This was discovered by Zel'dovich [7], who pointed out that a cylinder made of absorbing material and rotating around its axis with frequency Ω can amplify modes of scalar or electromagnetic radiation of frequency ω, provided the condition

$$\omega < m\Omega \tag{18}$$

(where m is the azimuthal quantum number with respect to the axis of rotation) is satisfied. The explicit numerical calculation of reflection coefficients for the $(2+1)$-rotating acoustic black hole was done in Ref. 3, in the superradiant regime.

5. Conclusions

Analogue black holes have proven to be a very valuable tool for the investigation of problems related to Hawking radiation. It is also possible that will yield valuable information regarding *classical* phenomena involving black holes. We have shown here some aspects of classical phenomena involving acoustic black holes, that may prove useful for future experimental realization of these systems.

Acknowledgements

I would like to take this opportunity to thank Emanuele Berti, Óscar Dias, José Lemos, Mário Pimenta, Ana Sousa and Shijun Yoshida for many useful conversations and collaboration. I also acknowledge financial support from FCT through grant SFRH/BPD/2004.

References

1. W. G. Unruh, Phys. Rev. Lett. **46**, 1351 (1981).
2. M. Visser, Class. Quantum Grav. **15**, 1767 (1998).
3. E. Berti, V. Cardoso and J. P. S. Lemos, Phys. Rev. D **70**, 124006 (2004).
4. V. Cardoso, J. P. S. Lemos and S. Yoshida, Phys. Rev. D 70, 124032 (2004).
5. K. D. Kokkotas and B. G. Schmidt, Living Rev. Rel. **2**, 2 (1999); H.-P. Nollert, Class. Quantum Grav. **16**, R159 (1999).
6. R. H. Price, Phys. Rev. D**5**, 2419 (1972); E. S. C. Ching, P. T. Leung, W. M. Suen and K. Young, Phys. Rev. D**52**, 2118 (1995); V. Cardoso, S. Yoshida, O. J. C. Dias, J. P. S. Lemos, Phys. Rev. D **68**, 061503 (2003).
7. Ya. B. Zel'dovich, Pis'ma Zh. Eksp. Teor. Fiz. **14**, 270 (1971) [JETP Lett. **14**, 180 (1971)]; Zh. Eksp. Teor. Fiz **62**, 2076 (1972) [Sov. Phys. JETP **35**, 1085 (1972)].

SUPERRADIANT INSTABILITIES IN BLACK HOLE SYSTEMS*

ÓSCAR J. C. DIAS[†]

Perimeter Institute for Theoretical Physics, Canada.
(Also at CENTRA - Centro Multidisciplinar de Astrofísica)
E-mail: odias@perimeterinstitute.ca

VITOR CARDOSO, JOSÉ LEMOS, SHIJUN YOSHIDA[‡]

A wave impinging on a Kerr black hole can be amplified as it scatters off the hole if certain conditions are satisfied. This is known as superradiant scattering. By building a mirror around the black hole one can make the system unstable. This is the black hole bomb of Press and Teukolsky. We investigate in detail this process and compute the growing timescales and oscillation frequencies as a function of the mirror's location. It is found that in order for the system, black hole plus mirror, to become unstable there is a minimum distance at which the mirror must be located. We also show with an explicit example that such a bomb can be built, once an appropriate black hole is located. Now, a spacetime with a "mirror" naturally incorporated in it is anti-de Sitter (AdS) spacetime, since the AdS space behaves effectively as a box with a reflecting gravitational wall at its infinity. Thus, a small Kerr-AdS black hole is naturally unstable to superradiant scattering of a wave that is generated in its vicinity.

1. Superradiance. The black hole bomb

The first example of superradiant scattering , which would lead to the notion of superradiant scattering in black hole spacetimes, was given by Zel'dovich [1], by examining what happens when scalar waves impinge upon a rotating cylindrical absorbing object. Considering a wave of the form $e^{-i\omega t+im\phi}$ incident upon such a rotating object, Zel'dovich concluded that if the frequency ω of the incident wave satisfies $\omega < m\Omega$, where Ω is the angular velocity of the body, then the scattered wave is amplified. It was

*Based on Phys. Rev. D 70, 044039 (2004), [hep-th/0404096]; and Phys. Rev. D 70, 024002 (2004), [hep-th/0405006].

[†]I acknowledge financial support from FCT through grant SFRH/BPD/2004.

[‡]Respectively at: Washington University, USA; CENTRA, IST, Portugal; Waseda University, Japan.

also anticipated by Zel'dovich that by surrounding the rotating cylinder by a mirror one could make the system unstable.

A black hole is one of the most interesting rotating objects for superradiant phenomena. If the rotating object is a Kerr black hole, then superradiant scattering also occurs [2] for frequencies $\omega < m\Omega$, but where Ω is the angular velocity of the black hole. Feeding the amplified scattered wave again, one can extract as much energy as one likes from the black hole. Indeed, if one surrounds the black hole by a reflecting mirror, the wave will bounce back and forth, between the mirror and the black hole, amplifying itself each time. Then the total extracted energy should grow exponentially until finally the radiation pressure destroys the mirror. This is Press and Teukolsky's black hole bomb first proposed in Ref. [3], and studied in detail in Ref. [4]. Nature sometimes provides its own mirror: if one considers a massive scalar field (with mass μ) scattering off a Kerr black hole then, for $\omega < \mu$, the mass μ effectively works as a mirror [5]; the AdS infinity also provides a natural wall [6]; a wave propagating around rotating black branes or rotating black strings also finds itself trapped by a reflecting wall [7].

Here, following Ref. [3], we investigate in detail the black hole bomb, as proposed by Press and Teukolsky [3], by using a scalar field model. We study the oscillation frequencies and growing timescales as a function of the mirror's location, and as a function of the black hole rotation. We also discuss, following Ref. [6], the superradiant instability of the Kerr-AdS black hole.

2. Properties of the black hole bomb

A scalar wave evolving in the vicinities of a Kerr black hole (with mass M and angular momentum $J = aM$) is described by the curved spacetime Klein-Gordon equation $\nabla_\mu \nabla^\mu \Phi = 0$. This equation separates into an angular and radial wave equations under the ansatz $\Phi(t,r,\theta,\phi) = e^{-i\omega t + im\phi} S_l^m(\theta) R(r)$, where $S_l^m(\theta)$ are spheroidal angular functions, and the azimuthal number m takes on integer (positive or negative) values. For our purposes, it is sufficient to consider positive ω's. Near the event horizon, $r = r_+$, the scalar field behaves as $\Phi \sim e^{-i\omega t} e^{\pm i(\omega - m\Omega) r_*}$, where the tortoise r_* coordinate is defined implicitly by $dr_*/dr = (r^2+a^2)/(r^2+a^2-2Mr)$, and $\Omega = a/(2Mr_+)$ is the angular velocity of the event horizon. Requiring ingoing waves at the horizon, which is the physically acceptable solution, one must impose a negative group velocity $v_{\rm gr}$ for the wave packet. Since $v_{\rm gr} = \pm 1$ we must choose the minus sign in the equation that describes the scalar field near the horizon. However, notice that if

$$\omega < m\Omega\,, \tag{1}$$

the phase velocity $\frac{m\Omega-\omega}{\omega}$ will be positive. Thus, in this superradiance regime, waves appear as outgoing to an inertial observer at spatial infinity, and energy is in fact being extracted. Note that since we are working with positive ω, superradiance will occur only for positive m, i.e., for waves that are co-rotating with the black hole.

We consider a Kerr black hole bomb, a Kerr black hole surrounded by a spherical mirror (mirror placed at a constant Boyer-Lindquist radial r coordinate) with a radius r_0, so that the scalar field will be required to vanish at the mirror's location, i.e., $\Phi(r = r_0) = 0$. With these two boundary conditions, ingoing waves at the horizon and a vanishing field at the mirror, the problem is transformed into an eigenvalue equation for ω. The frequencies satisfying both boundary conditions will be called Boxed Quasi-Normal frequencies (BQN frequencies, ω_{BQN}) and the associated modes will accordingly be termed Boxed Quasi-Normal Modes (BQNM).

In the large wavelength limit, $1/\omega \gg M$, for which the wavelength of the scalar wave is much larger than the typical size of the black hole, one can compute the BQNMs analytically. The procedure is based on matched asymptotic expansions. The BQN frequencies of the scalar wave that are allowed by the presence of the mirror located at $r = r_0$ can then be shown to be, in this approximation,

$$\omega_{BQN} \simeq \frac{j_{l+1/2,n}}{r_0} + i\delta\,, \tag{2}$$

where $j_{l+1/2,n}$ are the zeros of the Bessel function $J_{l+1/2}$, and n is a natural number, labelling the mode overtone number. For example, the fundamental mode corresponds to $n = 1$. The imaginary component δ is given by

$$\delta \simeq -B\frac{|J_{-l-1/2}(j_{l+1/2,n})|}{|J'_{l+1/2}(j_{l+1/2,n})|}\frac{j_{l+1/2,n}/r_0 - m\Omega}{r_0^{2(l+1)}}\,, \tag{3}$$

where

$$B = \left(\frac{l!}{(2l-1)!!}\right)^2 \frac{r_+^2(M^2-a^2)^l}{(2l)!(2l+1)!}\,\frac{2^{2l+1}}{2l+1}\left(\prod_{k=1}^{l}(k^2+4\varpi^2)\right)\left[j_{l+1/2,n}\right]^{2l+1}\,,$$

and $4\varpi^2 = (j_{l+1/2,n} - m\Omega)^2 r_+^4/(M^2-a^2)$. The quantity $J'_{l+1/2}(j_{l+1/2,n})$ is the derivative of $J_{l+1/2}(x)$ evaluated at $x = j_{l+1/2,n}$. Note that for large overtone n, $j_{l+1/2,n} \sim (n + l/2)\pi$. Two highly important features of the black hole bomb can already be read from the equations above: first, since

$\delta \propto -(\text{Re}[\omega_{BQN}] - m\Omega)$. Therefore, $\delta > 0$ for $\text{Re}[\omega_{BQN}] < m\Omega$, and $\delta < 0$ for $\text{Re}[\omega_{BQN}] > m\Omega$. The scalar field Φ has the time dependence $e^{-i\omega t} = e^{-i\text{Re}(\omega)t}e^{\delta t}$ and thus, for $\text{Re}[\omega_{BQN}] < m\Omega$, the amplitude of the field grows exponentially and the BQNM becomes unstable, with a growth timescale given by $\tau = 1/\delta$. Second, $\text{Re}[\omega_{BQN}] \propto 1/r_0$, i.e., the wave frequency is proportional to the inverse of the mirror's radius. Thus, as one decreases the distance at which the mirror is located, the allowed wave frequency increases, and there will thus be a critical radius at which the BQN frequency no longer satisfies the superradiant condition (1). Since $\text{Re}[\omega_{BQN}] \sim 1/r_0$ this means there will be a critical radius r_0 below which the instability is no longer present. This is just a consequence of having no superradiance. Notice also that $\text{Re}[\omega_{BQN}]$ as given by (2) is equal to the normal mode frequencies of a spherical mirror in a flat spacetime. as expected.

The differential equation for $R(r)$ can be solved numerically to obtain the BQN frequencies. Some of our numerical results are shown in Figs. 1. Here we only show the data corresponding to the unstable BQNMs, which are the modes of interest for this discussion. ¿From Fig. 1.(a), where we show the imaginary part of the BQN frequency for the fundamental BQNM, one can already see two very important aspects of the black hole bomb: (i) the instability is weaker (i.e., the growing timescale $\frac{1}{\text{Im}[\omega_{BQN}]}$ is larger) for larger mirror radius, i.e., $\text{Im}[\omega_{BQN}]$ decreases as r_0 increases. This is also expected on physical grounds, as was remarked by Press and Teukolsky [3], if one views the process as one of successive amplifications and reflections on the mirror. (ii) As one decreases r_0 the instability gets stronger, as expected, but suddenly the BQNM is no longer unstable, i.e., the imaginary component of ω_{BQN} drops from its maximum value to zero, and the mode becomes stable. This feature was already noticed in the discussion on the analytical results, and it changes the way we face the whole process, although it can easily be explained. The numerical results show that the analytical treatment works quite well, even though it is a large wavelength approximation. Indeed, in Fig. 1.(b), we show $\text{Re}[\omega_{BQN}]$ as a function of r_0. One can see that $\text{Re}[\omega_{BQN}]$ behaves as $\frac{1}{r_0}$, which is consistent with equation (2). This means that it is indeed the mirror which selects the allowed vibrating frequencies. The reason behind this minimum distance is now easily explained. The oscillation frequency is given basically by the inverse of the mirror's radius. Thus, as one decreases the distance at which the mirror is located, the oscillation frequency increases, and there will therefore be a critical radius at which the frequency no longer satisfies

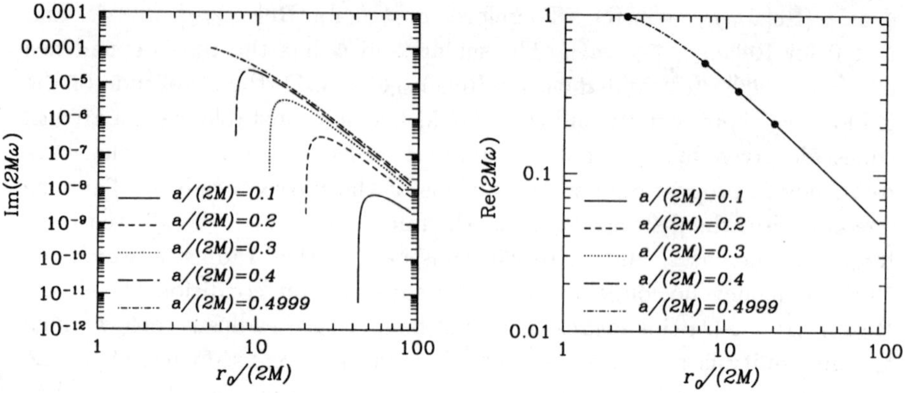

Figure 1. **(a)** The imaginary part of the fundamental (n=1) BQN frequency, as a function of the mirror's location, r_0. Here we show also the dependence on the rotation parameter. Note that at a certain critical value, this component of the BQN frequency suddenly decreases abruptly from its maximum value to zero. Therefore, the BQNM is no longer unstable. Tracking the mode to yet smaller distances shows that indeed it turns stable (i.e., the imaginary part of ω_{BQN} gets negative). The plot refers to a $l = m = 1$ wave. **(b)** The fundamental oscillation frequencies (real part of ω_{BQN}) as a function of the mirror's location and rotation parameter. The dots indicate at which r_0 the BQNM is no longer unstable [check with Fig. 1.(a)]. Note that there is no perceptive a-dependence (as matter of fact there is a very small a-dependence but too small to be noticeable). Thus, the oscillation frequency basically depends only on r_0, and for large r_0 goes as $1/r_0$, as predicted by the analytical formula (2). The plot refers to a $l = m = 1$ wave.

the superradiant condition (1). One can estimate the critical r_0 radius by equating the two sides, $r_{\rm crit} \sim \frac{j_{l+1/2,n}}{m\Omega}$. This estimate for the critical radius matches very well with our numerical data. In fact, to a great accuracy $r_{\rm crit}$ is given by the root of ${\rm Re}[\omega(r_{\rm crit})] - m\Omega = 0$. Note that according to this, it is not feasible to use this process to extract energy from a very slowly rotating black hole. In fact, for very small Ω, the critical radius is extremely large, thus making the task of building such a mirror an impossible one. In their description of the black hole bomb, Press and Teukolsky seem to assume that any frequency will be amplified and thus the system will become unstable no matter what frequency one deals with. For example, Press and Teukolsky consider a $M = M_\odot$ and, "for ease of construction" propose to build a $r_0 = 10^3$Km mirror. Our results show this bomb would explode if the rotation of the black hole satisfies $a > 6 \times 10^{-3}$, for a $m = 1$ mode, otherwise it is stable. We could try to go to higher m in order to remedy this, because of the $m\Omega$ factor. However, higher m implies higher l and our results, both numerical and analytical, show that the oscillating frequencies ($\text{Re}[\omega_{BQN}]$) scale with l. In fact, our numerical results show that $\text{Re}[\omega_{\text{BQN}}]$ behaves as $\text{Re}[\omega_{\text{BQN}}] \sim \pi/r_0(n + l/2)$. This behaviour is also predicted by the analytical study. Thus the results we have shown for $m = 1$ are basically the same for higher m's, apart from a scale factor.

Let us give an explicit example. One must compute the typical growing timescale, since if the growing timescale is too large, this could never be used as an energy source (because it would take too long to amplify the initial energy content). Take a black hole with mass $M = 1.4M_\odot$ and a rotation $a = 0.8M$. To better take advantage of the whole process, one should place the mirror at a position near the point of maximum growing rate (but larger than the point corresponding to the maximum growing rate), which in this case means $r_0 \sim 22M \sim 45.5\,\text{Km}$ [see Fig. 1.(a)]. This would give us a growing timescale of about $\tau \sim 0.8\,\text{s}$, i.e., every $0.8\,\text{s}$ the amplitude of the field gets approximately doubled. Notice that as energy is extracted from the black hole, its rotation a decreases, and therefore the critical radius r_{crit} increases [see Fig. 1.(a)]. One could worry that if the mirror is placed too close to r_{crit} the mode would soon become stable and therefore energy extraction would stop. Although this can happen, it is important to note that a very small decrease in a already gives us a huge amount of extracted energy. For the example above, the energy extraction stops when the rotation decreases to $a = 0.7M$. As a first order calculation one may assume the process to be adiabatic and thus one can set $dM \sim \Omega dJ$, where dM and dJ are the changes in mass and angular momentum of the black hole in this process. This then gives us a total amount of $\sim 0.01M$ of extracted energy before the bomb stops functioning, which is still a huge quantity.

3. Superradiant instability in AdS black holes

A spacetime with a "mirror" naturally incorporated in it is anti-de Sitter (AdS) spacetime, which has attracted a great deal of attention recently. As is well known, anti-de Sitter (AdS) space behaves effectively as a box, i.e., the AdS infinity works as a wall. Thus, one might expect that Kerr-AdS black holes can behave as the black hole bomb just described, and that they are therefore unstable. We have shown in [6] that for small Kerr-AdS black hole this instability is indeed present (small means that the size of the black hole horizon is much smaller than the cosmological length, $r_+ \ll \ell$), and in fact the properties of the instabilities present in the small Kerr-AdS black hole and in the black hole-mirror system are quite similar as will be discussed in the next section. As shown in Ref. [8], large Kerr-AdS black holes are stable.

4. Discussion of the results

An interesting, but easily understood feature born out in the work reported here [4] is that there is a minimum distance at which the mirror must be located in order for the Press-Teukolsky's black hole bomb [3] to work, i.e., for the system to become unstable. Basically this is because the mirror selects the frequencies that may be excited. For distances smaller than this, the system is stable and the perturbation dies off exponentially. This minimum distance increases as the rotation parameter decreases. We have shown by an explicit example that such a system may well be built, yielding a reliable source of energy.

The properties of the instabilities present in the small Kerr-AdS black hole and in the black hole-mirror system are quite similar. The AdS space behaves effectively as a box, i.e., the AdS wall with typical radius ℓ (the cosmological length) plays in the analogy the role of a mirror wall with radius $r_0 \equiv \ell$. In the Press-Teukolsky's black hole bomb the real part of the allowed frequency is proportional to the inverse of the mirror's radius [4], $\mathrm{Re}[\omega] \propto 1/r_0$, while in the small Kerr-AdS black hole case we have found that $\mathrm{Re}[\omega] \propto 1/\ell$ [6]. Moreover, in the Press-Teukolsky's system the growth timescale of the instability satisfies [4] $\delta^{-1} = 1/\mathrm{Im}[\omega] \propto r_0^{2(l+1)}$, while in the AdS black hole we have $\delta^{-1} \propto \ell^{2(l+1)}$ [6].

References

1. Ya. B. Zel'dovich, JETP Lett. **14**, 180 (1971); Sov. Phys. JETP **35**, 1085 (1972); J. D. Bekenstein and M. Schiffer, Phys. Rev. D **58**, 064014 (1998).
2. J. M. Bardeen, W. H. Press and S. A. Teukolsky, Astrophys. J. **178**, 347 (1972); A. A. Starobinsky, Sov. Phys. JETP **37**, 28 (1973).
3. W. H. Press and S. A. Teukolsky, Nature **238**, 211 (1972).
4. V. Cardoso, O. J. C. Dias, J. P. S. Lemos and S. Yoshida, Phys. Rev. D**70** 044039 (2004); `hep-th/0404096`.
5. T. Damour, N. Deruelle and R. Ruffini, Lett. Nuovo Cimento **15**, 257 (1976); S. Detweiler, Phys. Rev. D **22**, 2323 (1980).
6. V. Cardoso and O. J. C. Dias, Phys. Rev. D**70** 024002 (2004); `hep-th/0405006`.
7. V. Cardoso and J. P. S. Lemos, `hep-th/0412078`.
8. S. W. Hawking and H. S. Reall, Phys. Rev. D **61**, 024014 (1999).

MICROSCOPIC BLACK HOLE DETECTION IN UHECR: THE DOUBLE BANG SIGNATURE

M. PAULOS
LIP,
14–1, Av. Elias Garcia,
1000-149 Lisboa Portugal

According to recent conjectures on the existence of large extra dimensions in our universe, black holes could be produced during the interaction of Ultra High Energy Cosmic Rays with the atmosphere. However, and so far, the proposed signatures are based on statistical effects, not allowing identification on an event by event basis, and may lead to large uncertainties. In this talk, events with a double bang topology, where the production and instantaneous decay of a microscopic black hole (first bang) is followed, at a measurable distance, by the decay of an energetic tau lepton (second bang) are proposed as an almost background free signature. The characteristics of these events and the capability of large cosmic ray experiments to detect them are discussed.

1. Introduction

Recent attempts to solve the hierarchy problem rely on the existence of extra dimensions in our universe, thereby lowering the fundamental Planck scale down to TeV energies In such scenarios, gravity should get stronger at shorter distances, and therefore new phenomena should arise in TeV-scale experiments. One of the most interesting of these new phenomena is the production of black holes in collisions where the centre-of-mass energy is higher than 1 TeV. In this context, Ultra High Energy Cosmic Rays (UHECR) stand out naturally as the best candidates for black hole production, through the interaction with matter in the atmosphere

However, and so far, the proposed signatures of black hole production are based on statistical effects (rates and angular distributions), not allowing identification on an event by event basis, and may lead to large uncertainties In this talk events with a double bang topology are proposed as an almost background free signature, summarising the discussion done in Ref. 1, where a complete list of references can also be found.

2. The first bang: Production and decay of microscopic black holes

In the proposed scenario energetic neutrinos ($E_\nu \sim 10^6 - 10^{12}$ GeV) interact deeply in the atmosphere (cross-section $\sim 10^3 - 10^7$ pb) producing microscopic black holes with a mass of the order of the neutrino-parton center-of-mass energy ($\sqrt{s} \sim 1 - 10$ TeV). The rest lifetime of these black holes is so small ($\tau \sim 10^{-27}$ s) that an instantaneous thermal and democratic decay can be assumed. The average decay multiplicity ($< N >$) is a function of the parameters of the model (Planck mass M_D, black hole mass M_{BH}, number of extra dimension n) and typical values of the order of 5-20 are obtained in large regions of the parameter space. A large fraction of the decay products are hadrons ($\sim$ 75%) but there is a non negligible number of charged leptons ($\sim$ 10%).The energy spectra of such leptons in the black hole centre-of-mass reference frame peaks around M_{BH}/N.

The number of extra dimensions n is constrained by existing observational data to be $n > 2$ We shall take $n = 3$ and $n = 6$, but the conclusions apply equally well to higher dimensions. In this study, Planck mass values $M_D = 1,\ 2$ and 3 TeV were considered.

The production of tau leptons in black hole decays, in particular the tau energy spectrum, was simulated using the CHARYBDIS generator, while the black holes themselves were produced according to phase-space and without assuming an explicit $\nu N \rightarrow BH + X$ cross-section, as explained in Sec. 4.

3. The second bang: The decay of energetic tau leptons

A detectable second bang can be produced for tau leptons with a decay length large enough for the two bangs to be well separated, but small enough for a reasonable percentage of decays to occur within the field of view. This is of course determined by the tau energy.

Taus can decay leptonicaly ($\sim$ 34%) or hadronicaly ($\sim$ 66%). In both cases, a fraction of the energy escapes detection due to the presence of neutrinos. The average value of this energy, considering all decay modes, is of the order of 50%. Hadronic decays produce a large amount of visible energy, which will be seen as an extensive air shower. For leptonic decays, not only the fraction of energy not associated to neutrinos is lower, but also only decays into electrons originate extensive air showers, leading to observable fluorescence signals.

4. Observation prospects

The observation window for the double bang events is constrained by geometrical considerations (the two showers must be inside the field of view) and by the signal to noise ratio. EUSO [3] will be taken as a case study, without however making any detailed simulation. Double bang events were generated parameterising the shower development and the atmosphere response, in a model implemented in MATHEMATICA [7]. This approach was inspired in the method presented in the description of the SLAST simulation [8]. The GIL parameterisation [9] was used for the longitudinal shower development, while fluorescence yield follows Ref. 10. Attenuation in the atmosphere was included using tables produced with LOWTRAN [11]. In Fig. 1, the longitudinal fluorescence profile of a double bang event at the entrance of EUSO is shown as an example.

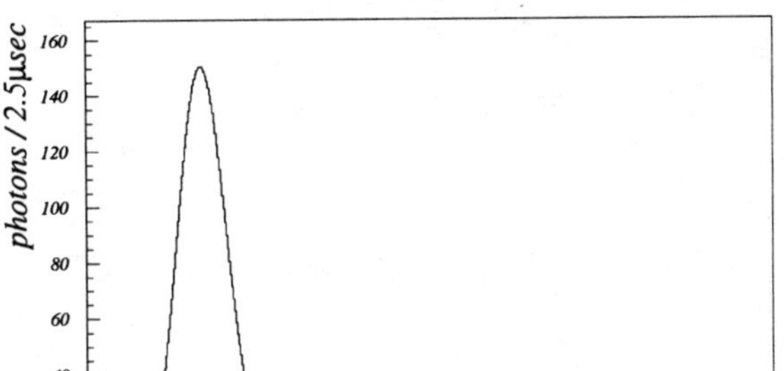

Figure 1. Longitudinal profile of double bang event, originating from an horizontal incoming neutrino with an energy of 10^{20} eV.

A CL of 99.7% (3σ) was chosen as the criterion of visibility of the second shower. The modified frequentist likelihood ratio method,which takes into account not only the total number of expected signal and background events but also the shapes of the distributions, was used. Signal events were obtained using the method described above. Threshold energies as low as 5×10^{18} eV (1×10^{18} eV) can be obtained for a photon detection efficiency of 0.1 (1.0) and a shower height of 10 km.

The fraction of the black hole events with a first bang within the EUSO field of view that also have an observable second shower is shown in Fig. 2(a), as a function of the primary neutrino energy, for (M_D=1 TeV, n=3, M_{BH}=5 TeV), and for detector efficiencies of 1.0 and 0.1.

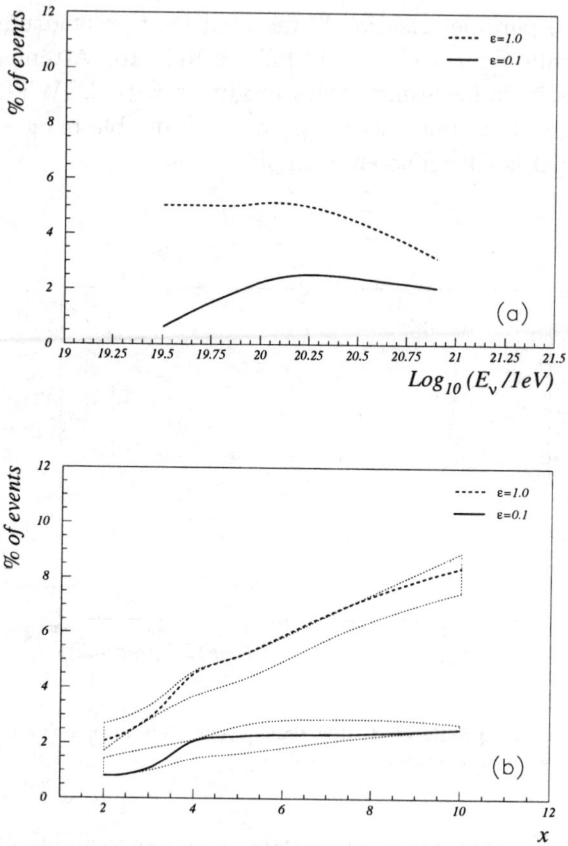

Figure 2. Fraction of the events with a first bang within the EUSO field of view that also have a visible second bang (a) as a function of the primary neutrino energy E_ν, for (M_D=1 TeV,$n=3$,M_{BH}=5 TeV), and for detection efficiencies ϵ =0.1 and ϵ =1, (b) as a function of $x = M_{BH}/M_D$, for E=10^{20} eV and detection efficiencies ϵ =0.1 and ϵ =1. The thick lines correspond to M_D=1 TeV,$n = 3$ and the dotted bands give the variation of the results when varying M_D between 1 and 2 TeV and n between 3 and 6.

These results take into account the fraction of events with taus in black hole decays, the tau energy spectrum and its decay length, the geometrical acceptance of EUSO and the visibility of the second shower. For a detector efficiency of 0.1 (1.), at $E_\nu = 10^{20}$ eV, of the order of 2% (5%) of the black hole induced events with a black hole decay visible in EUSO (first bang) are expected to have a visible second bang.

Figure 2(b) shows the expected fraction of double bang events as a function of $x = M_{BH}/M_D$, for $E_\nu = 10^{20}$ eV and two values of the detection efficiency, quantifying the dependence of the results on the values of the (M_D,n) parameters. The rate of black hole induced events depends strongly on the assumed cosmogenic and extragalactic neutrino fluxes. Values ranging from several tens to hundreds of events per year, for $M_D = 1$ TeV, have been predicted for the OWL telescope [6]. In the same reference the acceptance of EUSO was estimated to be 1/5 of OWL's.

Although the expected rate of double bang events due to black hole production in EUSO is rather small and would cover a relatively small region of the parameters space, the observation of just a few such events would be of the utmost importance, as almost no standard physics process has this signature.

References

1. Microscopic black hole detection in UHECR: the double bang signature", V. Cardoso et al., *Astroparticle Physics Journal* **22** (2005) 399-407, arXiv: hep-ph/0405056
2. Auger Collab., *The Pierre Auger Project Design Report*, FERMILAB-PUB-96-024, 252 (1996). http://www.auger.org.
3. O. Catalano, In Nuovo Cimento, 24-C, 2, 445 (2001); http://www.euso-mission.org.
4. J. F. Krizmanic et al., OWL/AirWatch Collab., Proc. of the 26th ICRC, Vol. 2, 388 (1999); http://owl.gsfc.nasa.gov
5. J. Tanaka, T. Yamamura, S. Asai and J. Kanzaki, *Study of black holes with the ATLAS detector at the LHC*, ATL-PHYS-2003-037, November 2003.
6. S. I. Dutta, M. H. Reno and I. Sarcevic, *On black hole detection with the OWL/Airwatch telescope*, hep-ph/0204218.
7. http://www.wolfram.com/products/mathematica/index.html
8. D. V. Naumov, *SLAST: Shower Initiated Light Attenuated to the Space Telescope*, LAPP-EXP-2004-02 (2002); D. Naumov, EUSO Report SDA-REP-015 (2003).
9. O. Catalano et al., Proc. of the 27th ICRC (2001).
10. F. Kakimoto et al., Nucl. Instrum. Meth. A372, 527 (1996).
11. F. X. Kneizys et al., *The MODTRAN 2/3 report and LOWTRAN 7 Model*, ed. L. W. Abreu and G. P. Anderson (1996).

GENERALIZED UNCERTAINTY PRINCIPLE AND HOLOGRAPHY

FABIO SCARDIGLI

CENTRA - Departamento de Fisica, Instituto Superior Tecnico, Av. Rovisco Pais 1, 1049-001 Lisboa, Portugal
E-mail: scardigli@fisica.ist.utl.pt

ROBERTO CASADIO

Dipartimento di Fisica, Università di Bologna and I.N.F.N., Sezione di Bologna, via Irnerio 46, 40126 Bologna, Italy
E-mail: casadio@bo.infn.it

We consider Uncertainty Principles which take into account the role of gravity and the possible existence of extra spatial dimensions. Explicit expressions for such Generalized Uncertainty Principles in $4+n$ dimensions are given and their holographic properties investigated. In particular, we show that the predicted number of degrees of freedom enclosed in a given spatial volume matches the holographic counting only for one of the available generalizations and without extra dimensions.

1. Introduction

During the last years many efforts have been devoted to clarifying the role played by the existence of extra spatial dimensions in the theory of gravity [1,2]. One of the most interesting predictions drawn from the theory is that there should be measurable deviations from the $1/r^2$ law of Newtonian gravity at short (and perhaps also at large) distances. Such new laws of gravity would imply modifications of those Generalized Uncertainty Principles (GUP's) (see Ref. 5) designed to account for gravitational effects in the measure of positions and energies.

On the other hand, the holographic principle is claimed to apply to all of the gravitational systems. The existence of GUP's satisfying the holography in four dimensions (one of the main examples is due to Ng and Van Dam [3]) led us to explore the holographic properties of the GUP's extended to the brane-world scenarios [4]. The results, at least for the examples we considered, are quite surprising. The expected holographic scaling indeed seems to hold only in four dimensions, and only for the Ng and van Dam's GUP. When extra spatial dimensions are admitted, the holography is destroyed. This fact allows two different interpretations: either the holographic principle is not universal and does not apply when extra dimensions are present; or, on the contrary, we take seriously the holographic claim in any number of dimensions, and our results are therefore evidence against the existence of extra dimensions. The four-dimensional Newton constant is denoted by $G_{\rm N}$ throughout the paper.

2. Ng and Van Dam GUP in four dimensions

An interesting GUP that satisfies the holographic principle in four dimensions has been proposed by Ng and van Dam [3], based on Wigner inequalities about distance measurements with clocks and light signals [6].

Suppose we wish to measure a distance l. Our measuring device is composed of a clock, a photon detector and a photon gun. A mirror is placed at the distance l which we want to measure and m is the mass of the system "clock + photon detector + photon gun". We call "detector" the whole system and let a be its size. Obviously, we suppose

$$a > r_{\rm g} \equiv \frac{2\,G_{\rm N}\,m}{c^2} = R_{\rm S}(m)\ , \tag{1}$$

which means that we are not using a black hole as a clock. Be Δx_1 the uncertainty in the position of the detector, then the uncertainty in the detector's velocity is

$$\Delta v = \frac{\hbar}{2\,m\,\Delta x_1}\ . \tag{2}$$

After the time $T = 2\,l/c$ taken by light to travel along the closed path detector–mirror–detector, the uncertainty in the detector's position (i.e. the uncertainty in the actual length of the segment l) has become

$$\Delta x_{\rm tot} = \Delta x_1 + T\,\Delta v = \Delta x_1 + \frac{\hbar T}{2\,m\,\Delta x_1}\ . \tag{3}$$

We can minimize $\Delta x_{\rm tot}$ by suitably choosing Δx_1, and we get

$$(\Delta x_{\rm tot})_{\rm min} = (\Delta x_1)_{\rm min} + \frac{\hbar T}{2\,m\,(\Delta x_1)_{\rm min}} = 2\left(\frac{\hbar T}{2\,m}\right)^{1/2}\ . \tag{4}$$

Since $T = 2\,l/c$, we have

$$(\Delta x_{\rm tot})_{\rm min} = 2\left(\frac{\hbar\,l}{m\,c}\right)^{1/2} \equiv \delta l_{\rm QM}\ . \tag{5}$$

This is a purely quantum mechanical result obtained for the first time by Wigner in 1957 [6]. From Eq. (5), it seems that we can reduce the error $(\Delta x_{\rm tot})_{\rm min}$ as much as we want by choosing m very large, since $(\Delta x_{\rm tot})_{\rm min} \to 0$ for $m \to \infty$. But, obviously, *here gravity enters the game.*

In fact, Ng and van Dam have also considered a further source of error, a gravitational error, besides the quantum mechanical one already addressed. Suppose the clock has spherical symmetry, with $a > r_{\rm g}$. Then the error due to curvature can be computed from the Schwarzschild metric surrounding the clock. The optical path from $r_0 > r_{\rm g}$ to a generic point $r > r_0$ is given by (see, for example, Ref. 7)

$$c\,\Delta t = \int_{r_0}^{r} \frac{d\rho}{1 - \frac{r_{\rm g}}{\rho}} = (r - r_0) + r_{\rm g}\,\log\frac{r - r_{\rm g}}{r_0 - r_{\rm g}}\,, \tag{6}$$

and differs from the "true" (spatial) length $(r - r_0)$. If we put $a = r_0$, $l = r$, the gravitational error on the measure of $(l - a)$ is thus

$$\delta l_{\rm C} = r_{\rm g}\,\log\frac{l - r_{\rm g}}{a - r_{\rm g}} \sim r_{\rm g}\,\log\frac{l}{a}\,, \tag{7}$$

where the last estimate holds for $l > a \gg r_{\rm g}$.

If we measure a distance $l \geq 2a$, then the error due to curvature is

$$\delta l_{\rm C} \geq r_{\rm g}\log 2 \simeq \frac{G_{\rm N} m}{c^2}. \tag{8}$$

Thus, according to Ng and van Dam the total error is

$$\delta l_{\rm tot} = \delta l_{\rm QM} + \delta l_{\rm C} = 2\left(\frac{\hbar\, l}{m\, c}\right)^{1/2} + \frac{G_{\rm N}\, m}{c^2}\,. \tag{9}$$

This error can be minimized again by choosing a suitable value for the mass of the clock, namely $m_{\rm min} = c(\hbar l)^{1/2}/G_{\rm N}$ and, inserting $m_{\rm min}$ in Eq. (9), we then have

$$(\delta l_{\rm tot})_{\rm min} = 3\,\left(\ell_{\rm p}^2\, l\right)^{1/3}\,. \tag{10}$$

The global uncertainty on l contains therefore a term proportional to $l^{1/3}$.

2.1. *Holographic properties*

We now see immediately the beauty of the Ng and van Dam GUP: it obeys the holographic scaling. In fact in a cube of size l the number of degrees of freedom is given by

$$n(V) = \left(\frac{l}{(\delta l_{\rm tot})_{\rm min}}\right)^3 = \left(\frac{l}{(\ell_{\rm p}^2\, l)^{1/3}}\right)^3 = \frac{l^2}{\ell_{\rm p}^2}\,, \tag{11}$$

as required by the holographic principle.

3. Models with n extra dimensions

We shall now generalize the procedure outlined in a previous section to a space-time with $4+n$ dimensions, where n is the number of space-like extra dimensions [4]. The link between the gravitational constant $G_{\rm N}$ in four dimensions and the one in $4+n$, henceforth denoted by $G_{(4+n)}$, of course depends on the model of space-time with extra dimensions that we consider. Models recently appeared in the literature mostly belong to two scenarios: (I) the Arkani-Hamed–Dimopoulos–Dvali (ADD) model [1], where the extra dimensions are compact and of size L; (II) the Randall–Sundrum (RS) model [2], where the extra dimensions have an infinite extension but are warped by a non-vanishing cosmological constant. A feature shared by (the original formulations of) both scenarios is that only gravity propagates along the n extra dimensions, while Standard Model fields are confined on a four-dimensional sub-manifold usually referred to as the *brane-world*.

In the ADD case the link between $G_{\rm N}$ and $G_{(4+n)}$ can be fixed by comparing the gravitational action in four dimensions with the one in $4+n$ dimensions. The space-time topology in such models is $\mathcal{M} = \mathcal{M}^4 \otimes \Re^n$, where $\mathcal{M}^4$ is the usual four-dimensional space-time and $\Re^n$ represents the extra dimensions of finite size L. From such comparison we obtain

$$G_{(4+n)} \sim G_{\rm N}\, L^n, \tag{12}$$

where we omit unimportant numerical factors.

The RS models are more complicated. It can be shown [2] that for $n=1$ extra dimension we have $G_{(4+n)} = \sigma^{-1}\, G_{\rm N}$, where σ is the brane tension with dimensions of $length^{-1}$ in suitable units. The gravitational force between two point-like masses m and M on the brane is obtained by perturbative calculations, not immediately applicable to a non-perturbative structure such as a black hole. Therefore we shall consider only the ADD scenario in this paper (see Ref. 4 for more details).

4. Ng and Van Dam GUP in $4+n$ dimensions

Ng and van Dam's derivation can be generalized to the case with n extra dimensions. The Wigner relation (5) for the quantum mechanical error is not modified by the presence of extra dimensions and we just need to estimate the error $\delta l_{\rm C}$ due to curvature.

We ought not to consider micro black holes created by the fluctuations ΔE in energy, as in Ref. 5, but we have rather to deal with (more or less) macroscopic clocks and distances. This implies that we have to distinguish four different cases: (1) $0 < L < r_g < a < l$; (2) $0 < r_{(4+n)} < L < a < l$; (3) $0 < r_{(4+n)} < a < L < l$; (4) $0 < r_{(4+n)} < a < l < L$; where $r_{(4+n)}$ is the Schwarzschild radius of the detector in $4+n$ dimensions, and of course $r_{\rm g} = r_{(4)}$. The curvature error will be estimated (as before) by computing the optical path from $a \equiv r_0$ to $l \equiv r$. Of course, we will use a metric which depends on the relative size of L with respect to a and l, that is the usual four-dimensional Schwarzschild metric in the region $r > L$, and the $4+n$ dimensional Schwarzschild solution in the region $r < L$ (where the extra dimensions play an actual role).

In cases (1) and (2) the length of the optical path from a to l can be obtained using just the four-dimensional Schwarzschild solution and the result is given by Eq. (10).

In cases (3) and (4) we instead have to use the Schwarzschild solution in $4+n$ dimensions [11],

$$ds^2 = -\left(1 - \frac{C}{r^{n+1}}\right) c^2 dt^2 + \left(1 - \frac{C}{r^{n+1}}\right)^{-1} dr^2 + r^2 d\Omega_{n+2}^2 , \qquad (13)$$

at least for part of the optical path. In the above,

$$C = \frac{16\,\pi\, G_{(4+n)}\, m}{(n+2)\, A_{n+2}\, c^2} , \qquad (14)$$

and A_{n+2} is the area of the unit $(n+2)$-sphere, that is

$$A_{n+2} = \frac{2\,\pi^{\frac{n+3}{2}}}{\Gamma\left(\frac{n+3}{2}\right)} . \qquad (15)$$

Besides, we note that, for $n = 0$,

$$C = \frac{2\, G_{\rm N}\, m}{c^2} = r_{\rm g} , \qquad (16)$$

that is, C coincides in four dimensions with the Schwarzschild radius of the detector. The $4+n$ dimensional Schwarzschild horizon is located where $(1 - C/r^{n+1}) = 0$, that is at

$$r = C^{1/(n+1)} \equiv r_{(4+n)} . \qquad (17)$$

Since measurements can be performed only on the brane, to the uncertainty Δx in position we can still associate an energy given by $\hbar\, c/(2\,\Delta x)$. The corresponding Schwarzschild radius is now given by Eq. (17) with $m = \Delta E/c^2$ and the critical length such that $\Delta x = r_{(4+n)}$ is the Planck length in $4+n$ dimensions,

$$\Delta x \simeq \left(\ell_{\rm p}^2\, L^n\right)^{\frac{1}{n+2}} \equiv \ell_{(4+n)} . \qquad (18)$$

In case (3) we obtain the length of the optical path from a to l by adding the optical path from a to L and that from L to l. We must use the solution in $4+n$ dimensions for the first part, and the four-dimensional solution for the second part of the path,

$$c\,\Delta t = \int_a^L \left(1 + \frac{C}{r^{n+1} - C}\right) dr + \int_L^l \left(1 + \frac{r_g}{r - r_g}\right) dr. \tag{19}$$

It is not difficult to show that from $r_{(4+n)} < L$ [which holds in cases (3) and (4)] we can infer

$$r_g < r_{(4+n)} < L\,. \tag{20}$$

We suppose $a^{n+1} \gg C = r_{(4+n)}^{n+1}$, that is $a \gg r_{(4+n)}$, so that we are not doing measures inside a black hole. Then $r_g < r_{(4+n)} \ll a < L < l$.

In case (4), the optical path from a to l can be obtained by using simply the Schwarzschild solution in $4+n$ dimensions. We get

$$c\,\Delta t = \int_a^l \left(1 + \frac{C}{r^{n+1} - C}\right) dr = (l - a) + C \int_a^l \frac{dr}{r^{n+1} - C}\,. \tag{21}$$

Also here we suppose, as before, that $a^{n+1} \gg C = r_{(4+n)}^{n+1}$, that is $a \gg r_{(4+n)}$ (i.e. our clock is not a black hole).

If the distance we are measuring is, at least, of the size of the clock ($l \geq 2\,a$), we can minimize δl_{tot} with respect to m in perfect analogy with the previous calculation and we obtain that the total minimum error (quantum mechanical + curvature) can be written in both cases in the form

$$(\delta l_{\text{tot}})_{\text{min}} = N(n) \left(\frac{\ell_{(4+n)}^{n+2}\, l}{a^n}\right)^{1/3}, \tag{22}$$

where $N(n)$ is an unimportant numerical factor (for detailed calculations see Ref. 4). Note that, for $n = 0$, Eq. (22) yields the four dimensional error given in Eq. (10).

4.1. *Holographic properties*

We finally examine the holographic properties of Eq. (22) for the GUP of Ng and van Dam type in $4+n$ dimensions.

Since we are just interested in the dependence of $n(V)$ on l and the basic constants, we can write

$$(\delta l_{\text{tot}})_{\text{min}} \sim \left(\frac{\ell_{(4+n)}^{n+2}\, l}{a^n}\right)^{1/3}. \tag{23}$$

We then have that the number of degrees of freedom in the volume of size l is

$$n(V) = \left(\frac{l}{(\delta l_{\rm tot})_{\rm min}}\right)^{3+n} = \left(\frac{l^2\, a^n}{\ell_{\rm p}^2\, L^n}\right)^{1+\frac{n}{3}} , \tag{24}$$

and the holographic counting holds in four-dimensions ($n = 0$) but is lost when $n > 0$. In fact we do not get something as

$$n(V) = \left(\frac{l}{\ell_{(4+n)}}\right)^{2+n} , \tag{25}$$

as we would expect in $4 + n$ dimensions. Even if we take the ideal case $a \sim \ell_{(4+n)}$ we get

$$n(V) = \left(\frac{l}{\ell_{(4+n)}}\right)^{2\left(1+\frac{n}{3}\right)} , \tag{26}$$

and the holographic principle does not hold for $n > 0$.

5. Concluding remarks

In the previous Sections, we have shown that the holographic principle seems to be satisfied only by uncertainty relations in the version of Ng and van Dam *and* for $n = 0$. That is, only in four dimensions we are able to formulate uncertainty principles which predict the same number of degrees of freedom per spatial volume as the holographic counting. This could be evidence for questioning the existence of extra dimensions. Moreover, such an argument based on holography could also be used to support the compactification of string theory down to four dimensions, given that there seems to be no firm argument which forces the low energy limit of string theory to be four-dimensional (except for the obvious observation of our world). In this respect, we should also say that the cases (3) and (4) of Section 4 do not have a good probability to be realized in nature since, if there are extra spatial dimensions, their size must be shorter than 10^{-1} mm [8]. Therefore, cases (1) and (2) of Section 4 are more likely to survive the test of future experiments.

References

1. N. Arkani-Hamed, S. Dimopoulos and G. Dvali, Phys. Lett. **B 429**, 263 (1998); Phys. Rev. D **59**, 0806004 (1999); I. Antoniadis, N. Arkani-Hamed, S. Dimopoulos and G. Dvali, Phys. Lett. **B 436**, 257 (1998).

2. L. Randall and R. Sundrum, Phys. Rev. Lett. **83**, 3370 (1999); Phys. Rev. Lett. **83**, 4690 (1999).
3. Y.J. Ng and H. van Dam, Mod. Phys. Lett. **A 9**, 335 (1994); Mod. Phys. Lett. **A 10**, 2801 (1995); Phys. Lett. **B 477**, 429 (2000); Y.J. Ng, Phys. Rev. Lett. **86**, 2946 (2001).
4. F. Scardigli and R. Casadio, Class. Quantum Grav. **20** (2003) 3915.
5. F. Scardigli, Phys. Lett. **B 452**, 39 (1999).
6. E.P. Wigner, Rev. Mod. Phys. **29**, 255 (1957); H. Salecker and E.P. Wigner, Phys. Rev. **109**, 571 (1958).
7. L.D. Landau and E.M. Lifshitz, *The classical theory of fields*, Pergamon Press, Oxford, 1975.
8. C.D. Hoyle et al., Phys. Rev. Lett. **86**, 1418 (2001).
9. S.B. Giddings, E. Katz and L. Randall, JHEP **0003**, 023 (2000); J. Garriga and T. Tanaka, Phys. Rev. Lett. **84**, 2778 (2000).
10. P.C. Argyres, S. Dimopoulos and J. March-Russell, Phys. Lett. **B 441**, 96 (1998).
11. R.C. Myers and M.J. Perry, Annals of Physics **172**, 304 (1986).

TESTING COVARIANT ENTROPY BOUNDS

SIJIE GAO AND JOSÉ P. S. LEMOS

Centro Multidisciplinar de Astrofísica - CENTRA
Departamento de Física, Instituto Superior Técnico
Av. Rovisco Pais 1, 1049-001 Lisboa, Portugal
E-mail: sijie@fisica.ist.utl.pt, lemos@fisica.ist.utl.pt

We give a brief review on Bousso's covariant entropy bound and its generalization. Bousso's bound is tested in the context of gravitational collapse. We also introduce a simple local condition for the generalized Bousso bound and apply it to a scalar field spacetime.

1. Covariant entropy bounds

The original entropy bound was proposed by Bekenstein in order to rescue the generalized second law in a gedanken experiment. Consider a matter system of energy E and entropy S. Let R be the linear dimension of the system. Then the Bekenstein bound takes the form [1]

$$S \leq 2\pi E R \,. \tag{1}$$

A related entropy bound conjecture is

$$S \leq \frac{A}{4} \,, \tag{2}$$

where A is the area surrounding the system. Both the bounds above are spatial entropy bounds and can be easily violated in cosmological spacetimes[2]. The covariant entropy bound, conjectured by Bousso, is the following [3]: Consider a connected two dimensional spatial surface B with area A. Let L be a hypersurface bounded by B and generated by orthogonal null geodesics with non-positive expansion. Then the total entropy, S_L, contained on L satisfies

$$S_L \leq \frac{A}{4} \,. \tag{3}$$

Evidences supporting this bound have been found in cosmological spacetimes and other matter systems [2–4]. In this conjecture, the null surface

L is required to be extended as far as possible unless a caustic is reached. This bound is tested in the context of gravitational collapse in section 2. A generalization of this bound allows L to be terminated at another spacelike 2-surface B' before a caustic is reached and the inequality (3) is replaced by [5]

$$S_L \leq \frac{1}{4}(A_B - A_{B'}) . \tag{4}$$

This is called the generalized covariant entropy bound. Some local conditions concerning this bound will be discuss in section 3.

2. Testing the Bousso bound in inhomogeneous gravitational collapse

In his original paper [3], Bousso discussed the bound in a process of gravitational collapse. In his model, a shell collapses into an existing Schwarzschild black hole. A new black hole with larger mass forms after the collapse. The apparent horizon L of the old black hole then moves in the interior of the new black hole and hits the singularity. Bousso showed that the entropy crossing L satisfies the covariant entropy bound. Deficiencies of his proof have been discussed in Ref. 6. This problem was re-examined with an exact solution of a shell filled with Tolman-Bondi dust[6]. Now we briefly review the arguments in Ref. 6.

In this solution, a spherically symmetric shell with gravitational mass M collapses into a black hole with mass M_0. Due to Birkhoff's theorem, the exterior of the shell is also Schwarzschild with mass $\tilde{M} = M_0 + M$. The shell consists of Tolman-Bondi dust. A marginally bound and self-similar solution for the shell is

$$ds^2 = -dT^2 + \frac{(3BR-T)^2}{R^{4/3}(6BR-6T)^{2/3}}dR^2 + \left(\frac{9}{2}R\right)^{2/3}(BR-T)^{4/3}d\Omega^2 \, , \tag{5}$$

where B is a constant and the comoving radial coordinate R is restricted by $R_1 < R < R_2$. By matching the shell with its Schwarzschild interior and exterior, one may find the simple relations $R_1 = M_0$ and $R_2 = \tilde{M}$.

The crucial step to test the entropy bound is to specify the entropy of the shell. We generalize the assumption in Ref. 3 where entropies for homogeneous cosmological models are defined. We assume that the entropy in a comoving volume does not change with time. In addition to this assumption, a boundary condition is needed to define the entropy. We then assume that the entropy density is equal to 1 at the hypersurface where the scalar curvature is equal to 1 in Planck units. By the two assumptions, the entropy of the shell takes the simple form

$$S = 8\pi(R_2 - R_1)\,. \tag{6}$$

The entropy bound is tested numerically by comparing S with the area of the original apparent horizon $A = 4\pi(2M_0)^2$. We find that the bound is satisfied for $M_0 \gtrsim 50$ and violated for $M \lesssim 50$. However, the apparent violation for the latter case can be interpreted as over-counting the entropy residing in the Planck regime. Therefore, there is no meaningful violation on the covariant entropy bound in classical relativity.

3. Local conditions for the generalized covariant entropy bound

Flanagan *et al.* [5] first proved Bousso's conjecture under two independent hypotheses concerning the local entropy of matter. In the spirit of Ref. 5, alternative local sufficient conditions have been proposed[7–8]. In Ref. 7, it was assumed that the matter entropy can be described in terms of a local entropy current s^a. Let k^a be the tangent vector of the light sheet L. The expansion of k^a is defined as $\theta = \nabla_a k^a$. Let $s = -k^a s_a$ be the flux of entropy that crosses L and $s' = k^a \nabla_a s$ be the rate of entropy flux on L. The stress-energy tensor of the matter is denoted by T_{ab}. Then the generalized Bousso bound can be derived from the following two conditions: (i)

$$s' \leq 2\pi T_{ab} k^a k^b\,. \tag{7}$$

(ii) On the initial 2-surface B of the light sheet,

$$s|_B \leq -\frac{1}{4}\theta|_B\,. \tag{8}$$

Inspired by these two conditions, we derived an even simpler local condition for the generalized entropy bound[9]. The following proposition was given in Ref. 9.

Proposition 3.1. *A necessary and sufficient condition for the generalized entropy bound to be satisfied for all light sheets in a region is that* $s \leq -\frac{1}{4}\theta$ *is satisfied everywhere in the region.*

This proposition applies to spacetimes where absolute entropy currents (entropy currents are independent of the light sheet) are well-defined. By checking $s = -\frac{1}{4}\theta$, we shall be able to identify the regions where the bound holds for all the light sheets lying inside. If the entropy density s does not vanish, proposition 3.1 indicates that there always exists a band region around the apparent horizon $\theta = 0$ such that the generalized bound is violated for light sheets in this region. It is worth mentioning that our proposition does not cover the case where a light sheet crosses the $s = -\frac{1}{4}\theta$ surface.

Proposition 3.1 is closely related to condition (ii) of Strominger and Thompson [7]. In Ref. 9, we showed that if both conditions in Ref. 7 are satisfied for a light sheet, then $s \leq -\frac{1}{4}\theta$ are satisfied everywhere on the light sheet. This implies that the role of condition (i) can be replaced by the single condition $s \leq -\frac{1}{4}\theta$. It can be shown that there exist light sheets on which $s \leq -\frac{1}{4}\theta$ is satisfied but condition (i) is nowhere satisfied. Therefore, condition (i) is not a necessary condition for the generalized entropy bound.

To demonstrate how Proposition 3.1 works, we consider a scalar field spacetime which was first discussed by Husain[4]. The spacetime is described by the metric

$$ds^2 = -t\, f(r)\, dt^2 + t\, f(r)^{-1}\, dr^2 + t\, r^2\, f(r)^{-\frac{1+\sqrt{3}/2}{\sqrt{3}/2}} \left(d\theta^2 + \sin^2\theta d\phi^2\right),$$

where

$$f(r) = \left(1 - \frac{2}{r}\right)^{-\frac{\sqrt{3}}{2}}, \tag{9}$$

and the corresponding scalar field in the spacetime is

$$\phi(t,r) = \frac{1}{4\sqrt{\pi}} \ln\left[t^{\sqrt{3}}\, f(r)^{1/\sqrt{3}}\right]. \tag{10}$$

The entropy flux is given by [4]

$$s^a = -\partial^a \phi\,. \tag{11}$$

Using this expression, we test the condition $s \leq -\frac{1}{4}\theta$ for future-directed ingoing light sheets. The result is shown in Fig.1. We see that there exists a band region between the apparent horizon and the $s = -\frac{1}{4}\theta$ surface where $s > -\frac{1}{4}\theta$. Proposition 3.1 then indicates that the generalized entropy bound is violated for any light sheet residing in this region. However, such kind of violation could be caused by the failure of hydrodynamic description of matter entropy[8]. To justify the violation, one needs to compare the proper length of the light sheet with the thermal wavelength of the scalar field. Detailed discussion can be found in Ref. 9.

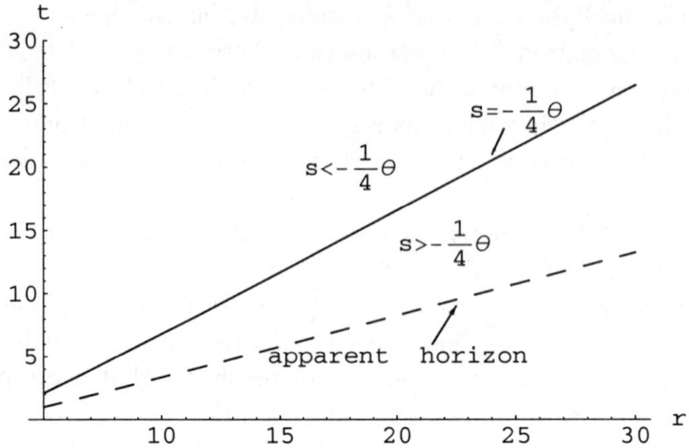

Figure 1. A spacetime diagram depicting the regions where $s < -\frac{1}{4}\theta$ is satisfied and violated. The generalized entropy bound is satisfied in the region $s < -\frac{1}{4}\theta$ and violated in the region $s > -\frac{1}{4}\theta$.

Acknowledgments

This work was partially funded by Fundação para a Ciência e Tecnologia (FCT) - Portugal through project POCTI/FNU/44648/2002. SG acknowledges financial support by the FCT grant SFRH/BPD/10078/2002 from FCT.

References

1. J. D. Bekenstein, *Phys. Rev. D* **9**, 3292 (1974).
2. R. Bousso, *Rev. Mod. Phys.* **74**, 825 (2004).
3. R. Bousso, *J. High Energy Phys.* **9907**, 004 (1999).
4. V. Husain, *Phys. Rev. D* **69**, 084002 (2004).
5. E. E. Flanagan, D. Marolf, and R. M. Wald, *Phys. Rev. D* **62**, 084035 (2000).
6. S. Gao and J. P. S. Lemos, *J. High Energy Phys.* **0404**, 017 (2004).
7. A. Strominger and D. Thompson, *Phys. Rev. D* **70**, 044007 (2004).
8. R. Bousso, E. E. Flanagan and D. Marolf, *Phys. Rev. D* **68**, 064001 (2003).
9. S. Gao and J. P. S. Lemos, *Phys. Rev. D*, accepted for publication, (2005).

Dark Matter and Dark Energy

DARK ENERGY - DARK MATTER UNIFICATION: GENERALIZED CHAPLYGIN GAS MODEL

ORFEU BERTOLAMI

Instituto Superior Técnico, Departamento de Física
Av. Rovisco Pais, 1049-001
Lisboa, Portugal
E-mail: orfeu@cosmos.ist.utl.pt

We review the main features of the generalized Chaplygin gas (GCG) proposal for unification of dark energy and dark matter and discuss how it admits an unique decomposition into dark energy and dark matter components once phantom-like dark energy is excluded. In the context of this approach we consider structure formation and show that unphysical oscillations or blow-up in the matter power spectrum are not present. Moreover, we demonstrate that the dominance of dark energy occurs about the time when energy density fluctuations start evolving away from the linear regime.

1. Introduction

The GCG model[1,2] is an interesting alternative to more conventional approaches for explaining the observed accelerated expansion of the Universe such as a cosmological constant[3] or quintessence[4]. It is worth remarking that quintessence is related to the idea that the cosmological term could evolve[5] and with attempts to tackle the cosmological constant problem.

In the GCG approach one considers an exotic equation of state to describe the background fluid:

$$p_{ch} = -\frac{A}{\rho_{ch}^{\alpha}} \quad , \tag{1}$$

where A and α are positive constants. The case $\alpha = 1$ corresponds to the Chaplygin gas. In most phenomenological studies the range $0 < \alpha \leq 1$ is considered. Within the framework of Friedmann-Robertson-Walker cosmology, this equation of state leads, after being inserted into the relativistic energy conservation equation, to an evolution of the energy density as[2]

$$\rho_{ch} = \left[A + \frac{B}{a^{3(1+\alpha)}} \right]^{\frac{1}{1+\alpha}} \quad , \tag{2}$$

where a is the scale-factor of the Universe and B a positive integration constant. From this result, one can understand a striking property of the GCG: at early times the energy density behaves as matter while at late times it behaves like a cosmological constant. This dual role is what essentially allows for the interpretation of the GCG model as an entangled mixture of dark matter and dark energy.

The GCG model has been successfully confronted with different classes of phenomenological tests: high precision Cosmic Microwave Background Radiation data[6], supernova data[7], and gravitational lensing[8]. More recently, it has been shown using the latest supernova data[9], that the GCG model is degenerate with a dark energy model with a phantom-like equation of state[10,11]. Furthermore, it can be shown that this does not require invoking the unphysical condition of violating the dominant energy condition and does not lead to the big rip singularity in future[10]. It is a feature of GCG model, that it can mimic a phantom-like equation of state, but without any kind of pathologies as asymptotically the GCG approaches to a well-behaved de-Sitter universe. Structure formation has been studied in Refs. [2, 12]. In Ref. [13], the results of the various phenomenological tests on the GCG model are summarized.

Despite these pleasing performance concerns about such an unified model were raised in the context of structure formation. Indeed, it has been pointed out that one should expect unphysical oscillations or even an exponential blow-up in the matter power spectrum at present[14]. This difficult arises from the behaviour of the sound velocity through the GCG. Although, at early times, the GCG behaves likc dark matter and its sound velocity is vanishingly small as one approaches the present, the GCG starts behaving like dark energy with a substantial negative pressure yielding a large sound velocity which, in turn, produces oscillations or blow-up in the power spectrum. In any unified approach this is inevitable unless the dark matter and the dark energy components of the fluid can be properly identified. These components are, of course, interacting as both are entangled within a single fluid. However, it can be shown that the GCG is a unique mixture of interacting dark matter and a cosmological constant-like dark energy, once one excludes the possibility of phantom-type dark energy[15]. It can be shown that due to the interaction between the components, there is a flow of energy from dark matter to dark energy. This energy transfer is vanishingly small until recent past, resulting in a negligible contribution at the time of gravitational collapse ($z_c \simeq 10$). This feature makes the model indistinguishable from a CDM dominated Universe till recent past. Subsequently, just before present ($z \simeq 2$), the interaction starts to grow yielding a large energy transfer from dark matter to dark energy, which leads to the dominance of the latter at present. Moreover, it is shown that the epoch of dark energy dominance occurs when dark matter perturbations start deviating from its linear behaviour and that the Newtonian equations for small scale perturbations for dark matter do not involve any k-dependent term. Thus, neither oscillations nor blow-up in the power spectrum do develop.

2. Decomposition of the GCG fluid

In Ref. [2], it is shown that the GCG can be described through a complex scalar field whose Lagrangian density can be written as a *generalized* Born-Infeld theory:

$$L_{GBI} = -A^{\frac{1}{1+\alpha}} \left[1 - (g^{\mu\nu}\theta_{,\mu}\theta_{,\nu})^{\frac{1+\alpha}{2\alpha}}\right]^{\frac{\alpha}{1+\alpha}} , \tag{3}$$

which reduces into the Born-Infeld Lagrangian density for $\alpha = 1$. The field θ corresponds to the phase of the complex scalar field[2].

Let us now consider the decomposition of the GCG into components. Introducing the redshift dependence and using Eqs. (1) and (2), the pressure is given by

$$p_{ch} = -\frac{A}{\left[A + B(1+z)^{3(1+\alpha)}\right]^{\frac{\alpha}{1+\alpha}}} \tag{4}$$

while the total energy density can be written as

$$\rho_{ch} = \left[A + B(1+z)^{3(1+\alpha)}\right]^{\frac{1}{1+\alpha}} , \tag{5}$$

where the present value of the scale-factor, a_0, has been set to 1.

We decompose the energy density into a pressure-less dark matter component, ρ_{dm}, and a dark energy component, ρ_X, with an equation of state $p_X = w_X \rho_X$; hence the equation of state parameter of the GCG can be written as

$$w = \frac{p_{ch}}{\rho_{ch}} = \frac{p_X}{\rho_{dm} + \rho_X} = \frac{w_X \rho_X}{\rho_{dm} + \rho_X} . \tag{6}$$

Therefore, from Eqs. (4), (5) and (6), one obtains for ρ_X

$$\rho_X = -\frac{\rho_{dm}}{1 + w_X \left[1 + \frac{B}{A}(1+z)^{3(1+\alpha)}\right]} . \tag{7}$$

¿From this equation one can see that requiring that $\rho_X \geq 0$ leads to the constraint $w_X \leq 0$ for early times ($z \gg 1$) and $w_X \leq -1$ for the future

($z = -1$). Thus, one can conclude that $w_X \leq -1$ for the entire history of the Universe. The case $w_X < -1$ corresponds to the so-called phantom-like dark energy, which violates the dominant-energy condition and leads to an ill defined sound velocity (see however Ref. [10]). Excluding this possibility, then the energy density can be uniquely split as

$$\rho = \rho_{dm} + \rho_\Lambda \tag{8}$$

where

$$\rho_{dm} = \frac{B(1+z)^{3(1+\alpha)}}{\left[A + B(1+z)^{3(1+\alpha)}\right]^{\frac{\alpha}{1+\alpha}}} , \tag{9}$$

and

$$\rho_\Lambda = -p_\Lambda = \frac{A}{\left[A + B(1+z)^{3(1+\alpha)}\right]^{\frac{\alpha}{1+\alpha}}} , \tag{10}$$

from which one finds the scaling behaviour of the energy densities

$$\frac{\rho_{dm}}{\rho_\Lambda} = \frac{B}{A}(1+z)^{3(1+\alpha)} . \tag{11}$$

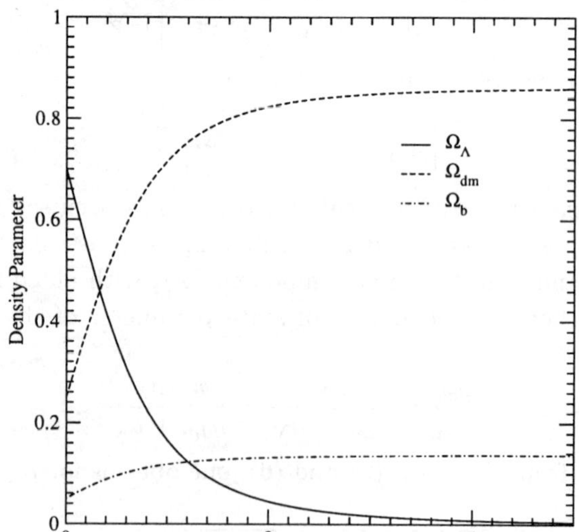

Figure 1. Density parameters Ω_{dm} and Ω_Λ and Ω_b as a function of redshift. It is assumed that $\Omega_{b0} = 0.05$, $\Omega_{dm0} = 0.25$, $\Omega_{\Lambda 0} = 0.7$ and $\alpha = 0.2$.

In what follows we express parameters A and B in terms of cosmological observables. From Eqs. (9) and (10), it implies that

$$\rho_{ch0} = \rho_{dm0} + \rho_{\Lambda 0} = (A+B)^{\frac{1}{1+\alpha}} \quad , \tag{12}$$

where ρ_{ch0}, ρ_{dm0} and $\rho_{\Lambda 0}$ are the present values of ρ_{ch}, ρ_m and ρ_Λ, respectively. Constants A and B can then be written as a function of ρ_{ch0}

$$A = \rho_{\Lambda 0}\ \rho_{ch0}^{\alpha}\ ; \quad B = \rho_{dm0}\ \rho_{ch0}^{\alpha} \quad . \tag{13}$$

It is also interesting to express A and B in terms of Ω_{dm0}, $\Omega_{\Lambda 0}$, the present values of the fractional energy densities $\Omega_{dm(\Lambda)} = \rho_{m(\Lambda)}/\rho_c$ where ρ_c is the critical energy density, $\rho_c = 3H^2/8\pi G$. From the Friedmann equation

$$3H^2 = 8\pi G \left[A + B(1+z)^{3(1+\alpha)}\right]^{\frac{1}{1+\alpha}} + 8\pi G \rho_{b0}(1+z)^3 \tag{14}$$

where ρ_{b0} is the baryon energy density at present, one obtains

$$A \simeq \Omega_{\Lambda 0}\ \rho_{c0}^{(1+\alpha)},\ B \simeq \Omega_{dm0}\ \rho_{c0}^{(1+\alpha)} \quad . \tag{15}$$

Therefore, with the present value of the Hubble parameter, H_0:

$$H^2 = H_0^2 \left[\left[\Omega_{\Lambda 0} + \Omega_{dm0}(1+z)^{3(1+\alpha)}\right]^{\frac{1}{1+\alpha}} + \Omega_{b0}(1+z)^3\right] \tag{16}$$

one can write the fractional energy densities Ω_{dm}, Ω_Λ and Ω_b as

$$\Omega_{dm} = \frac{\Omega_{dm0}(1+z)^{3(1+\alpha)}}{\left[\Omega_{\Lambda 0} + \Omega_{dm0}(1+z)^{3(1+\alpha)}\right]^{\alpha/(1+\alpha)} X} \tag{17}$$

$$\Omega_\Lambda = \frac{\Omega_{\Lambda 0}}{\left[\Omega_{\Lambda 0} + \Omega_{dm0}(1+z)^{3(1+\alpha)}\right]^{\alpha/(1+\alpha)} X} \tag{18}$$

$$\Omega_b = \frac{\Omega_{b0}(1+z)^3}{X} \tag{19}$$

where

$$X = \left[\Omega_{\Lambda 0} + \Omega_{m0}(1+z)^{3(1+\alpha)}\right]^{1/(1+\alpha)} + \Omega_{b0}(1+z)^3 \ . \tag{20}$$

Finally, as Ω_{dm0} and $\Omega_{\Lambda 0}$ are order one quantities, one can easily see that at the time of nucleosynthesis, Ω_Λ is negligibly small, and hence the model is not in conflict with known processes at nucleosynthesis.

It is important to realize that there is an explicit interaction between dark matter and dark energy. This can be understood from the energy conservation equation, which in terms of the components can be written as

$$\dot{\rho}_{dm} + 3H\rho_{dm} = -\dot{\rho}_{\Lambda} \ . \tag{21}$$

Thus, the evolution of dark energy and dark matter are coupled so that energy is exchanged between these components (see Refs. [16,17] for earlier work on the interaction between dark matter and dark energy). One can see from Figure 1, that until $z \simeq 2$, there is essentially no exchange of energy and the Λ term is vanishingly small. However, around $z \simeq 2$, the interaction starts to increase, resulting in a substantial growth of the dark energy term at the expense of the dark matter energy. Thus, by around $z \simeq 0.2$, dark energy starts dominating the energy content of Universe. Of course, these redshift values are α dependent and, in Figure 1, $\alpha = 0.2$ has been chosen. Nevertheless, the main conclusion is that in this unified model, the interaction between dark matter and dark energy is vanishing small for almost the entire history of the Universe making it indistinguishable from the CDM model. As can be clearly seen, the energy transfer has started in the recent past resulting in a significant energy transfer from dark matter to the Λ-like dark energy. In the next section we show that this energy transfer epoch is the one when dark matter perturbations start departing from its linear behaviour. But before that notice also that Eq. (21) expresses the energy conservation for the background fluid, which is reminiscent of earlier work on varying Λ cosmology[5,18,19] where the cosmological term decays into matter particles. In here, we have the opposite, as α is always positive, hence the energy transfer is from dark matter to dark energy. This is responsible for the late time dominance of the latter and ultimately to the observed accelerated expansion of the Universe.

3. Structure formation

Aiming to study structure formation, it is interesting to write the 0-0 component of Einstein's equation as

$$3H^2 = 8\pi G(\rho_{dm} + \rho_b) + \Lambda \ , \tag{22}$$

where Λ is given by

$$\Lambda = 8\pi G\rho_{\Lambda} \ . \tag{23}$$

We address now the issue of energy density perturbations. We first write the Newtonian equations for a pressure-less fluid with background density

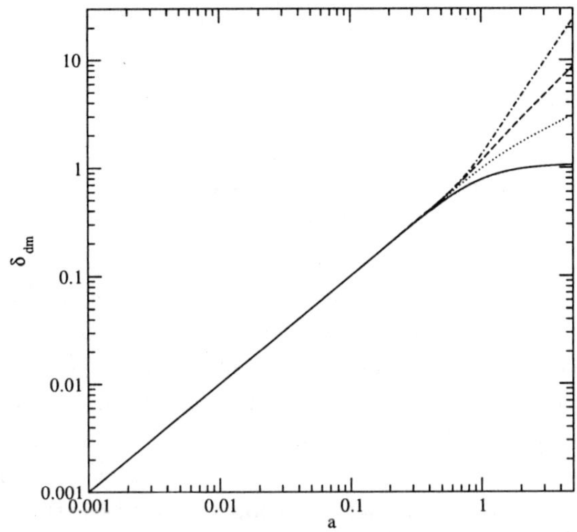

Figure 2. Density profile δ_{dm} as function of scale factor. The solid, dotted, dashed and dash-dot lines correspond to $\alpha = 0, 0.2, 0.4, 0.6$, respectively. It is assumed that $\Omega_{b0} = 0.05$, $\Omega_{dm0} = 0.25$ and $\Omega_{\Lambda 0} = 0.7$.

ρ_{dm} and density contrast δ_{dm}, with a source term due to the energy transfer from dark matter to dark energy. Assuming that both, the density contrast δ_{dm} and the peculiar velocity v are small, that is $\delta_{dm} << 1$ and $v << u$, where u is the velocity of a fluid element, one can write the Euler, the continuity and the Poisson's equations in the co-moving frame[19]:

$$\ddot{a}x + \frac{\partial v}{\partial t} + \frac{\dot{a}}{a}v = -\frac{\nabla\Phi}{a} \ , \tag{24}$$

$$\nabla \cdot v = -a\left[\frac{\partial \delta_{dm}}{\partial t} + \frac{\Psi \delta_{dm}}{\rho_{dm}}\right] \ , \tag{25}$$

$$\frac{1}{a^2}\nabla^2\Phi = 4\pi G \rho_{dm}(1 + \delta_{dm}) - \Lambda \ , \tag{26}$$

where Φ is the gravitational potential, and Ψ is the source term in the continuity equation due to the energy transfer between dark matter and the cosmological constant-type dark energy. The co-moving coordinate x is related to the proper coordinate r by $r = ax$. In here,

$$\Psi = -\frac{1}{8\pi G}\dot{\Lambda} \ . \tag{27}$$

One expects a perturbation also in the Λ term. However, it can be seen from the Euler equation, for a fluid with an equation state of the form

$p = w\rho,$

$$(w+1)\rho\left(\frac{\partial v}{\partial t} + v \cdot \nabla v\right) + w\nabla\rho + (w+1)\rho\nabla\Phi = 0 \tag{28}$$

so that, for $w = -1$, it follows that $\nabla\rho = 0$, from which implies that this cosmological constant like component is always homogeneous. We should mention that the Euler Eqs. (26) and (28) can have an extra term in the r.h.s. if the velocity of the created Λ-like particle has a different velocity from the decaying dark matter particle[19]. In this case, the Λ-like dark energy can have spatial variations which can be neglected for the Newtonian treatment. However, in our case, we are considering only the situation where both the decaying and created particles have the same velocity.

¿From the divergence of Eq. (24) and using Eqs. (25) and (26), one obtains the small scale linear perturbation equation for the dark matter in the Newtonian limit:

$$\frac{\partial^2 \delta_{dm}}{\partial t^2} + \left[2\frac{\dot{a}}{a} + \frac{\Psi}{\rho_{dm}}\right]\frac{\partial \delta_{dm}}{\partial t} - \left[4\pi G\rho_{dm} - 2\frac{\dot{a}}{a}\frac{\Psi}{\rho_{dm}} - \frac{\partial}{\partial t}\left[\frac{\Psi}{\rho_{dm}}\right]\right]\delta_{dm} = 0\ . \tag{29}$$

One sees that, if $\Psi = 0$, that is in the absence of energy transfer, one recovers the standard equation for the dark matter perturbation in the ΛCDM case. One can verify that this occurs for $\alpha = 0$. It can also be seen from the above equation that there is no scale dependent term to drive oscillations or to cause any blow up in the power spectrum.

We turn now to the evolution for the baryon perturbations in the Newtonian limit when the scales are inside the horizon. Given that our purpose is to consider the period after decoupling, the baryons are no longer coupled to photons and one can effectively consider baryons as a pressure-less fluid like the dark matter as there is no significant pressure due to Thompson scattering. We assume that there is no interaction between dark energy and baryons, which means that the Equivalence Principle is violated as, on its turn, dark matter and dark matter are strongly coupled. Given that it is parameter α that controls this interaction ($\alpha = 0$ means there is no interaction), it is a measure of the violation of the Equivalence Principle. One can also see from the behaviour of Ψ, that this violation also starts rather late in the history of the Universe. In the Newtonian limit, the evolution of the baryon perturbation after decoupling for scales well inside the horizon is similar to the one for dark matter however, as described earlier, the source term is absent as there is no energy transfer to or from baryons.

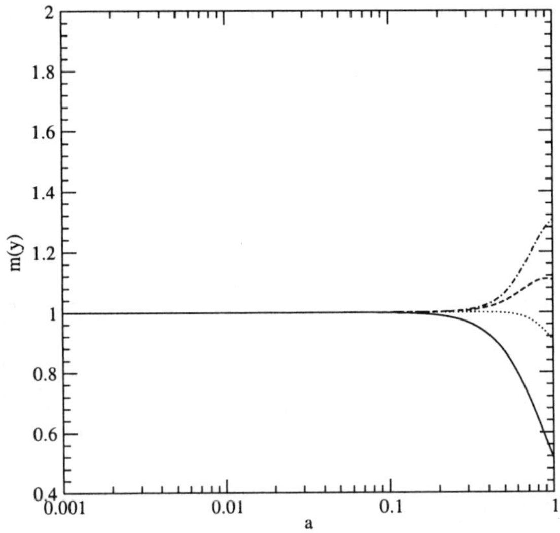

Figure 3. The growth factor $m(y)$ as a function of scale factor a. The solid, dotted, dashed and dash-dot lines correspond to $\alpha = 0, 0.2, 0.4, 0.6$, respectively. It is assumed that $\Omega_{b0} = 0.05$, $\Omega_{dm0} = 0.25$ and $\Omega_{\Lambda 0} = 0.7$.

Thus, the equation for the evolution of baryon perturbations is given by

$$\frac{\partial^2 \delta_b}{\partial t^2} + 2\frac{\dot{a}}{a}\frac{\partial \delta_b}{\partial t} - 4\pi G \rho_{dm} \delta_{dm} = 0 \quad , \tag{30}$$

where in the third term in the l.h.s., the contribution from baryons has been dropped as it is negligible compared to the one of dark matter.

It is convenient to define for each component the linear growth function $D(y)$,

$$\delta = D(y)\delta_0 \quad , \tag{31}$$

where $y = \log(a)$ and δ_0 is the initial density contrast (assuming a Gaussian distribution). It is also interesting to consider the so-called growth exponent $m(y) = D'(y)/D(y)$, where the prime denotes derivative with respect to the scale factor.

Asymptotically, given that dark matter drives the evolution of the baryon perturbations, then they grow with the same exponent $m(y)$. However, their amplitudes may differ and their ratio corresponds to the so-called bias parameter, $b \equiv \delta_b/\delta_{dm}$.

It is of course, phenomenologically interesting to study the behaviour of δ_{dm}, $m(y)$ and b as function of the scale factor a. While solving the

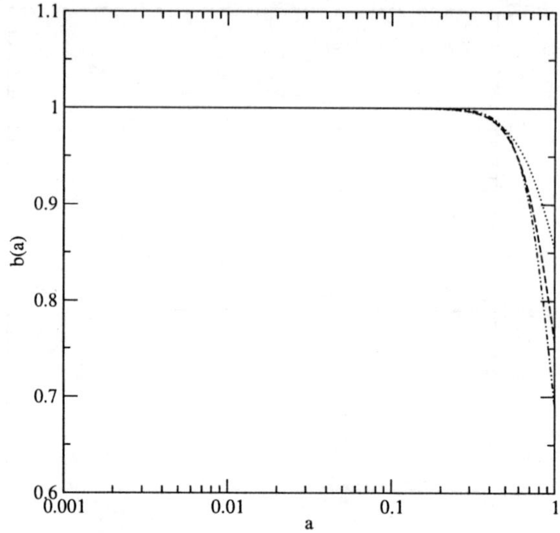

Figure 4. Bias parameter b as a function of the scale factor, a. The solid, dotted, dashed and dash-dot lines correspond to $\alpha = 0, 0.2, 0.4, 0.6$, respectively. It is assumed that $\Omega_{b0} = 0.05$, $\Omega_{dm0} = 0.25$ and $\Omega_{\Lambda 0} = 0.7$.

differential equations for the linear perturbation, the initial conditions are chosen so that at $a = 10^{-3}$, the standard linear solution $D \simeq a$ is reached. In Figure 2, it is shown the linear density perturbation for dark matter, δ_{dm}, as a function of α. One sees that, whereas for $\alpha = 0$ (the ΛCDM case), the perturbation stops growing at late times, for models with $\alpha > 0$ the perturbation starts departing from the linear behaviour around $z \simeq 0.25$, the very epoch when the Λ term starts dominating (cf. Figure 1). In view of this behaviour, it is tempting to conjecture that, in our unified model, the interaction between dark matter and Λ-like dark energy is related with structure formation, so that for a sufficiently high density contrast ($\delta_{dm} >> 1$), a significant energy transfer from dark matter to dark energy takes place. In any case, our proposal for GCG indicates that there is a connection between structure formation scenario and the dominance of dark energy, a link that ultimately results in the acceleration of the Universe expansion. This feature hints a possible way to understand why $\Omega_{dm} \simeq \Omega_\Lambda$ just at recent past, the so-called Cosmic Coincidence problem.

The behaviour of $m(y)$ is also interesting. One can infer from Figure 3 that from $z \simeq 5$ to the present, the growth factor is quite sensitive to the value of α. For $\alpha = 0.2$, $m(y)$ increases up to 40% at present in relation to

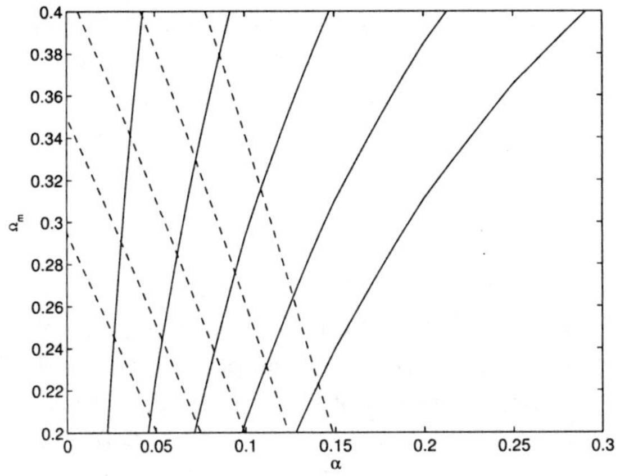

Figure 5. Contours for parameters b and m in the Ω_m-α plane. Solid lines refer to b whereas dashed lines refer to m. For b, contour values are 0.98, 0.96, ..., 0.9 from left to right. For m, contour values are 0.6, 0.65, ..., 0.8 from left to right.

the ΛCDM case.

Notice that $m(y)$ governs the growth of the velocity fluctuations in the linear perturbation theory as the velocity divergence evolves as $-Ham\delta_{dm}$; it follows then that large deviations of the growth factor with changing α are detectable via precision measurements of large scale structure and associated measurements of the redshift-space power spectrum anisotropy.

In what concerns the bias parameter, its behaviour is shown in Figure 4. From there one can see that it also changes sharply in the recent past as α increases. This bias extends to all scales consistent with the Newtonian limit, hence being distinguishable from the hydrodynamical or nonlinear bias which takes place only for collapsed objects. Therefore, from the observation of large scale clustering one can distinguish the non-vanishing α case from the $\alpha = 0$ (ΛCDM) case.

The growth factor and the bias parameter at $z \sim 0.15$ have been recently determined using the 2DF survey[20,21]. It is found for the redshift space distortion parameter, $\beta = 0.49 \pm 0.09$, and for the linear bias, $b = 1.04 \pm 0.14$. Notice that, as $\beta = m/b$, one can obtain $m = 0.51 \pm 0.11$. In Figure 5, it is shown contours for b and m in the Ω_m-α plane. ¿From the mentioned observational constraints on b and m, one can constrain α to a small but non-zero value ($\alpha \sim 0.1$). However, it is important to point out that our study refers to the properties of the baryons whereas the observations concern the fraction of baryons that collapsed to form bright galaxies; the relation between the two is still poorly known. As far as parameter β is concerned, one should bear in mind that this constraint is obtained in the context of the standard ΛCDM model in order to convert redshift to distance. Thus, a full analysis in the context of the GCG model is still to be performed.

Furthermore, as can be seen from Figure 2, there is no suppression of δ_{dm} at late times for any positive value of α, and hence one should not expect the corresponding suppression in the power spectrum normalization, σ_8, for the total matter distribution. This was one major problem in the previous GCG model approach which, as pointed out in Ref. [14], cannot be solved even after the inclusion of baryons. In the approach developed in Ref. [15] and described here, one can overcome this difficulty.

Another interesting cosmological test for our model is the study of the M/L ratio for clusters of galaxies. The most recent average value[22], $\Omega_m = 0.17 \pm 0.05$, has been extracted from the observation of 21 clusters with $z \sim 1$. The fact that nearby cluster data seem to prefer smaller values for Ω_m than the one obtained from WMAP data, can be interpreted as a signal in favour of a decaying dark matter model such as the GCG.

4. Conclusions

In this contribution, we have presented a setup where the GCG has been decomposed in two interacting components. The first one behaves as dark matter since it is pressure-less. The second one has an equation of state, $p_X = \omega_X \rho_X$. It has been shown that $\omega_X \leq -1$. Thus, once phantom-like behaviour is excluded the decomposition is unique. Apparently the model does not look different from the interacting quintessence models where one has two different interacting fluids; however, an interesting feature of our proposal is that it can be described through a single fluid equation. Hence, as far the background cosmology is concerned, we have an unified GCG fluid behaving as dark matter in the past and as a dark energy in the present. Nevertheless, when studying structure formation in this model one should consider it as an interacting mixture of two fluids to achieve a proper description. In any unified model, one expects an entangled mixture of interacting dark matter and dark energy. In the case of the GCG, we can uniquely identify the components of this mixture and the interaction. Moreover, we find that one does not need anything besides an evolving cosmological term to describe dark energy. This is consistent with recent studies that show that a combination of WMAP data and observations of high redshift supernovae can be described via a cosmological constant-like dark energy[23]. One can also consider the GCG as a decaying dark matter model where the decay product is a cosmological constant.

Obviously it remains to be seen how one can obtain such a decaying dark matter model from a fundamental theory. Given the fact that the GCG equation of state arises from a generalized Born-Infeld action, it is possible that D-brane physics can shed some light into this issue (see eg. Ref. [24]).

Furthermore, we have demonstrated that in the context of our setup, the so-called dark energy dominance is related with the time when matter fluctuations become large ($\delta_{dm} > 1$), a possibility has actually been previously conjectured[25]. Moreover, we have shown that in what concerns structure formation, the linear regime ($\delta_{dm} \sim a$) is valid till fairly close to the present, meaning that at the time structure formation begins, $z_c \simeq 10$, the influence of the dark energy component was negligible and that clustering occurs very much like in the CDM model. We have shown that the growth factor as well as the bias parameter have a noticeable dependence on the α parameter. We have implemented a model which exhibits a violation of the Equivalence Principle, as dark energy and baryons are not directly coupled. This may turn out to be an important observational signature of our approach.

Acknowledgments

It is a pleasure to thank Maria Bento, Anjan Sen, Somasri Sen and Pedro Silva for sharing the fun on the research of the GCG properties.

References

1. A. Kamenshchik, U. Moschella and V. Pasquier, *Phys. Lett.* **511** (2001) 265.
2. M.C. Bento, O. Bertolami and A.A. Sen, *Phys. Rev.* **D66** (2002) 043507.
3. See e.g. M.C. Bento and O. Bertolami, *Gen. Relat. and Gravitation* **31** (1999) 1461; M.C. Bento, O. Bertolami and P.T. Silva, *Phys. Lett.* **B498** (2001) 62.
4. B. Ratra and P.J.E. Peebles, *Phys. Rev.* **D37** (1988) 3406; *Ap. J. Lett.* **325** (1988) 117; C. Wetterich, *Nucl. Phys.* **B302** (1988) 668; R.R. Caldwell, R. Dave and P.J. Steinhardt, *Phys. Rev. Lett.* **80** (1998) 1582; P.G. Ferreira and M. Joyce, *Phys. Rev.* **D58** (1998) 023503; I. Zlatev, L. Wang and P.J. Steinhardt, *Phys. Rev. Lett.* **82** (1999) 986; P. Binétruy, *Phys. Rev.* **D60** (1999) 063502; J.E. Kim, *JHEP* 9905 (1999) 022; J.P. Uzan, *Phys. Rev.* **D59** (1999) 123510; T. Chiba, *Phys. Rev.* **D60** (1999) 083508; L. Amendola, *Phys. Rev.* **D60** (1999) 043501; O. Bertolami and P.J. Martins, *Phys. Rev.* **D61** (2000) 064007; *Class. Quantum Gravity* **18** (2001) 593; A.A. Sen, S. Sen

and S. Sethi, *Phys. Rev.* **D63** (2001) 107501; A.A. Sen and S. Sen, *Mod. Phys. Lett.* **A16** (2001) 1303; A. Albrecht and C. Skordis, *Phys. Rev. Lett.* 84 (2000) 2076; Y. Fujii, *Phys. Rev.* **D61** (2000) 023504; M.C. Bento, O. Bertolami and N.C. Santos, *Phys. Rev.* **D65** (2002) 067301.

5. M. Bronstein, *Phys. Zeit. Sowejt Union* **3** (1933) 73; O. Bertolami, *Il Nuovo Cimento* **93B** (1986) 36; *Fortschr. Physik* **34** (1986) 829; M.Ozer and M.O Taha, *Nucl. Phys.* **B287** (1987) 776.
6. M.C. Bento, O. Bertolami and A.A. Sen, *Phys. Lett.* **B575** (2003) 172; *Phys. Rev.* **D67** (2003) 063003; *Gen. Relat. and Gravitation* **35** (2003) 2063; D. Caturan and F. Finelli, *Phys. Rev.* **D68** (2003) 103501; Amendola, F. Finelli, C. Burigana and D. Caturan, *JCAP* **0307** (2003) 005.
7. J.C. Fabris, S.B.V. Gonçalves and P.E. de Souza, astro-ph/0207430; A. Dev, J.S. Alcaniz and D. Jain, *Phys. Rev.* **D67** (2003) 023515; V. Gorini, A. Kamenshchik and U. Moschella, *Phys. Rev.* **D67** (2003) 063509; M. Makler, S.Q. de Oliveira and I. Waga, *Phys. Lett.* **B555** (2003) 1; J.S. Alcaniz, D. Jain and A. Dev, *Phys. Rev.* **D67** (2003) 043514.
8. P.T. Silva and O. Bertolami, *Ap. J.* **599** (2003) 829; A. Dev, D. Jain and J.S. Alcaniz, *Astron. Astrophys.* **417** (2004) 847.
9. J.L. Tonry *et al. Ap. J.* **594** (2003) 1; B.J. Barris *et al.*, *Ap. J.* **602** (2004) 571; A.G. Riess *et al.* [Supernova Search Team Collaboration], *Ap. J.* **607** (2004) 665.
10. O. Bertolami, A.A. Sen, S. Sen and P.T. Silva, *Mont. Not. R. Astr. Soc.* **353** (2004) 329.
11. M.C. Bento, O. Bertolami, A.A. Sen, and N.C. Santos, *Phys. Rev.* **D71** (2005) 063501.
12. N. Bilić, G.B. Tupper and R.D. Viollier, *Phys. Lett.* **B535** (2002) 17; J.C. Fabris, S.B.V. Gonçalves and P.E. de Souza, *Gen. Relat. and Gravitation* **34** (2002) 53.
13. O. Bertolami, astro-ph/0403310.
14. H. Sandvik, M. Tegmark, M. Zaldarriaga and I. Waga, *Phys. Rev.* **D69** (2004) 123524.
15. M.C. Bento, O. Bertolami and A.A. Sen, *Phys. Rev.* **D70** (2004) 083519.
16. L. Amendola, *Phys. Rev.* **D62** (2000) 043511; D. Tocchini-Valentini and L Amendola, *Phys. Rev.* **D65** (2002) 063508.
17. L.P. Chimento, A.S. Jakubi, D. Pavón and W. Zimdahl, *Phys. Rev.* **D67** (2003) 083513.
18. K. Freese, F.C. Adams, J.A. Frieman and E. Mottola, *Nucl. Phys.* **B287** (1987) 797.
19. R.C. Arcuri and I. Waga, *Phys. Rev.* **D50** (1994) 2928.
20. E. Hawkins *et.al.*, *Mont. Not. R. Astr. Soc.* **346** (2003) 78.
21. L. Verde *et.al.*, *Mont. Not. R. Astr. Soc.* **335** (2002) 432.
22. N. Bahcall and J.M. Comerford, *Ap. J. Lett.* **565** (2002) L5.
23. H.K. Jassal, J.S. Bagla and T. Padmanabhan, *Mont. Not. R. Astr. Soc. Lett.* **356** (2005) L11.
24. R. Neves and C. Vaz, *Phys. Lett.* **B568** (2003) 153.
25. U. Alam, V. Sahni, T.D. Saini and A.A. Starobinsky, *Mont. Not. R. Astr. Soc.* **354** (2004) 275.

COSMOLOGY AND SPACETIME SYMMETRIES

RALF LEHNERT

Department of Physics and Astronomy
Vanderbilt University, Nashville, Tennessee, 37235
E-mail: ralf.lehnert@vanderbilt.edu

Cosmological models often contain scalar fields, which can acquire global nonzero expectation values that change with the comoving time. Among possible consequences of these scalar-field backgrounds, an accelerated cosmological expansion, varying couplings, and spacetime-symmetry violations have recently received considerable attention. This talk studies the interplay of these three key signatures of cosmologically varying scalars within a supergravity framework.

1. Introduction

Sizable efforts are currently directed towards the search for a more fundamental theory that must resolve a variety of theoretical issues left unaddressed in present-day physics. Many approaches along these lines require novel scalars as a key ingredient. Moreover, scalar fields are often invoked in cosmological models[1,2,3] to explain certain phenomenological questions, such as the observed late-time accelerated expansion of the Universe, the horizon and the flatness problem, or the claimed variation of the fine-structure parameter. The search for observational consequences of such scalars can therefore yield valuable insight into new physics.

Scalar fields can acquire global expectation values that vary with the expansion of the Universe. Measurable effects of variations in the scalar background are typically feeble because of the cosmological time scales involved. This suggests ultrahigh-precision studies as a promising tool for experimental investigations. The remaining task is to identify suitable tests.

Varying scalars determine a spacetime-dependent background that violates translation invariance. This effect could be measurable because symmetries are typically amenable to high-precision tests. Moreover, translation-invariance breakdown is typically associated with Lorentz violation[4,5] since translations, rotations, and boosts are intertwined in the Poincaré group. This perhaps less appreciated result opens a door for alternative experimental investigations of cosmologically varying scalars and is the primary focus of this talk.

Lorentz and CPT breakdown has also been suggested in other contexts as a candidate Planck-scale signature including strings,[6] spacetime foam,[7,8] nontrivial spacetime topology,[9] loop gravity,[10] and noncommutative geometry.[11] The emergent low-energy effects are described by the Standard-Model Extension (SME),[12] which has provided the basis for studies of Lorentz and CPT breaking with mesons,[13,14,15,16] baryons,[17,18,19] electrons,[20,21,22] photons,[23] muons,[24] neutrinos,[12,25] and the Higgs.[26]

2. Connection between translation and Lorentz symmetry

Consider the angular-momentum tensor $J^{\mu\nu} = \int d^3x \, \left(\theta^{0\mu}x^\nu - \theta^{0\nu}x^\mu\right)$, which generates boosts and rotations. Its construction involves the energy–momentum tensor $\theta^{\mu\nu}$, which is no longer conserved when spacetime-translation symmetry is broken. Consequently, $J^{\mu\nu}$ will typically depend on time, so that the usual time-independent boost and rotation generators cease to exist. As a result, Lorentz and CPT invariance is no longer assured.

Lorentz violation through varying scalars can also be understood in a more intuitive way. A varying scalar is always associated with a nonzero 4-gradient. This background gradient determines a preferred direction on scales comparable to the variation: Consider, for instance, a particle species that is coupled to such a gradient. The propagation features of these particles might now depend upon whether the motion is perpendicular or parallel to the background gradient. This implies physically inequivalent directions, and thus the violation of rotation invariance. Since rotations are contained in the Lorentz group, Lorentz symmetry must be broken.

We finally establish the effect at the Lagrangian level. Consider a model with a varying coupling $\xi(x)$ and two scalars ϕ and Φ. Suppose that the Lagrangian contains a term $\xi(x)\,\partial^\mu\phi\,\partial_\mu\Phi$. We next integrate the action by parts with respect to the partial derivative acting on ϕ. This produces an equivalent Lagrangian $\mathcal{L}' \supset -K^\mu\phi\,\partial_\mu\Phi$. Here $K^\mu \equiv \partial^\mu\xi$ is a nondynamical background 4-vector violating Lorentz invariance. Note that for cosmological variations of ξ we have $K^\mu = \text{const.}$ on small scales.

3. Toy supergravity cosmology

We now illustrate the result from Sec. 2 with a toy model. This model is motivated by pure $N = 4$ supergravity in four spacetime dimensions. Although unrealistic in detail, it is a limit of $N = 1$ supergravity in eleven dimensions, which is contained in M-theory. One thus expects that our model can illuminate generic aspects of a candidate fundamental theory.

Suppose that only one of the model's graviphotons, $F^{\mu\nu}$, is excited. Then, the bosonic part of pure $N = 4$ supergravity is given by [27,4]

$$\kappa\mathcal{L}_{\rm sg} = -\tfrac{1}{2}\sqrt{g}R + \sqrt{g}(\partial_\mu A\partial^\mu A + \partial_\mu B\partial^\mu B)/4B^2 \\ -\tfrac{1}{4}\kappa\sqrt{g}MF_{\mu\nu}F^{\mu\nu} - \tfrac{1}{4}\kappa\sqrt{g}NF_{\mu\nu}\tilde{F}^{\mu\nu}\,, \tag{1}$$

where M and N are known functions of the scalars A and B. The dual field-strength tensor is $\tilde{F}^{\mu\nu} = \varepsilon^{\mu\nu\rho\sigma}F_{\rho\sigma}/2$ and $g = -\det(g_{\mu\nu})$, as usual. Note that we can rescale $F^{\mu\nu} \to F^{\mu\nu}/\sqrt{\kappa}$ removing the explicit appearance of the gravitational coupling κ from the equations of motion.

For phenomenological reasons, we also include $\delta\mathcal{L} = -\frac{1}{2}\sqrt{g}(m_A^2A^2 + m_B^2B^2)$ into $\mathcal{L}_{\rm sg}$.[5] We further represent the model's fermions[27] by the energy–momentum tensor of dust $T_{\mu\nu} = \rho\, u_\mu u_\nu$ describing, e.g., galaxies. Here, u^μ is a unit timelike vector and ρ is the fermionic energy density.

The usual assumption of an isotropic homogeneous flat Friedmann–Robertson–Walker Universe implies that $F^{\mu\nu} = 0$ on large scales. Our cosmology is then governed by the Einstein equations and the equations of motion for the scalars A and B. At tree level, the fermionic matter is not coupled to the scalars, so we can take $T_{\mu\nu}$ as covariantly conserved separately. Searching for solutions with this input yields a nontrivial dependence of A and B on the comoving time t.

4. Lorentz violation

Consider small localized excitations of $F_{\mu\nu}$ in the scalar background $A_{\rm b}$ and $B_{\rm b}$ from Sec. 3. Since experiments are usually confined to a small spacetime region, it is appropriate to work in local inertial coordinates. The effective Lagrangian $\mathcal{L}_{\rm cosm}$ for such situations follows from Eq. (1) and is

$$\mathcal{L}_{\rm cosm} = -\tfrac{1}{4}M_{\rm b}F_{\mu\nu}F^{\mu\nu} - \tfrac{1}{4}N_{\rm b}F_{\mu\nu}\tilde{F}^{\mu\nu}\,, \tag{2}$$

where $A_{\rm b}$ and $B_{\rm b}$ imply the time dependence of $M_{\rm b}$ and $N_{\rm b}$. Comparison with the usual Maxwell Lagrangian $\mathcal{L}_{\rm em} = -\frac{1}{4e^2}F_{\mu\nu}F^{\mu\nu} - \frac{\theta}{16\pi^2}F_{\mu\nu}\tilde{F}^{\mu\nu}$ shows that $e^2 \equiv 1/M_{\rm b}$ and $\theta \equiv 4\pi^2 N_{\rm b}$. Thus, e and θ acquire time dependencies, as they are determined by the varying background $A_{\rm b}$ and $B_{\rm b}$.

The breakdown of Lorentz symmetry in our effective Lagrangian (2) is best exhibited by the resulting modified Maxwell equations:

$$\frac{1}{e^2}\partial^\mu F_{\mu\nu} - \frac{2}{e^3}(\partial^\mu e)F_{\mu\nu} + \frac{1}{4\pi^2}(\partial^\mu\theta)\tilde{F}_{\mu\nu} = 0\,. \tag{3}$$

In our toy cosmology, the gradients of e and θ are nonzero, approximately constant locally, and act as a nondynamical external background. Thus, the gradients select a preferred direction in local inertial frames violating Lorentz invariance. The term containing $\partial^\mu\theta$ can be identified with the k_{AF} operator in the minimal SME. This term has recently received a lot attention.[28] For instance, it can lead to vacuum Čerenkov radiation.[29]

Next, we consider small localized excitations δA and δB of the scalar background in local inertial coordinates at the point x_0. With the ansatz $A(x) = A_{\rm b}(x) + \delta A(x)$ and $B(x) = B_{\rm b}(x) + \delta B(x)$ in the equations of motion for A and B, we find the linearized equations

$$\begin{aligned} 0 &= \left[\Box - 2B^\mu\partial_\mu + 2m_A^2 B_{\rm b}^2\right]\delta A - \left[2A^\mu\partial_\mu - 2A^\mu B_\mu - 4m_A^2 A_{\rm b} B_{\rm b}\right]\delta B\ , \\ 0 &= \left[2A^\mu\partial_\mu\right]\delta A + \left[\Box - 2B^\mu\partial_\mu + 6m_B^2 B_{\rm b}^2 - A^\mu A_\mu + B^\mu B_\mu\right]\delta B\ , \end{aligned} \tag{4}$$

where $A_{\rm b}$, $B_{\rm b}$, $A^\mu \equiv B_{\rm b}^{-1}\partial^\mu A_{\rm b}$, and $B^\mu \equiv B_{\rm b}^{-1}\partial^\mu B_{\rm b}$ are evaluated at x_0.

Equation (4) governs the propagation of δA and δB in our cosmological background. Note that A^μ and B^μ are external nondynamical vectors violating Lorentz symmetry. This result also applies to quantum theory: the excitations δA and δB would be seen as the effective particles corresponding to A and B, so that such particles would break Lorentz invariance.

Acknowledgments

This work was funded by the Fundação Portuguesa para o Desenvolvimento.

References

1. R.R. Caldwell *et al.*, Phys. Rev. Lett. **80**, 1582 (1998); A. Masiero *et al.*, Phys. Rev. D **61**, 023504 (2000); Y. Fujii, Phys. Rev. D **62**, 044011 (2000).
2. C. Armendáriz-Picón *et al.*, Phys. Lett. B **458**, 209 (1999).
3. A.H. Guth, Phys. Rev. D **23**, 347 (1981).
4. V.A. Kostelecký *et al.*, Phys. Rev. D **68**, 123511 (2003).
5. O. Bertolami *et al.*, Phys. Rev. D **69**, 083513 (2004).
6. V.A. Kostelecký and R. Potting, Nucl. Phys. B **359**, 545 (1991); Phys. Lett. B **381**, 89 (1996); Phys. Rev. D **63**, 046007 (2001); V.A. Kostelecký *et al.*, Phys. Rev. Lett. **84**, 4541 (2000).
7. G. Amelino-Camelia *et al.*, Nature (London) **393**, 763 (1998).
8. D. Sudarsky *et al.*, Phys. Rev. D **68**, 024010 (2003).
9. F.R. Klinkhamer, Nucl. Phys. B **578**, 277 (2000).
10. J. Alfaro *et al.*, Phys. Rev. Lett. **84**, 2318 (2000).
11. S.M. Carroll *et al.*, Phys. Rev. Lett. **87**, 141601 (2001); A. Anisimov *et al.*, Phys. Rev. D **65**, 085032 (2002).

12. D. Colladay and V.A. Kostelecký, Phys. Rev. D **55**, 6760 (1997); **58**, 116002 (1998); V.A. Kostelecký and R. Lehnert, Phys. Rev. D **63**, 065008 (2001); V.A. Kostelecký, Phys. Rev. D **69**, 105009 (2004).
13. OPAL Collaboration, R. Ackerstaff *et al.*, Z. Phys. C **76**, 401 (1997); DELPHI Collaboration, M. Feindt *et al.*, preprint DELPHI 97-98 CONF 80 (1997); BELLE Collaboration, K. Abe *et al.*, Phys. Rev. Lett. **86**, 3228 (2001); BaBar Collaboration, B. Aubert *et al.*, hep-ex/0303043.
14. V.A. Kostelecký and R. Potting, Phys. Rev. D **51**, 3923 (1995).
15. D. Colladay and V.A. Kostelecký, Phys. Lett. B **344**, 259 (1995); Phys. Rev. D **52**, 6224 (1995); V.A. Kostelecký and R. Van Kooten, Phys. Rev. D **54**, 5585 (1996); O. Bertolami *et al.*, Phys. Lett. B **395**, 178 (1997); N. Isgur *et al.*, Phys. Lett. B **515**, 333 (2001).
16. V.A. Kostelecký, Phys. Rev. Lett. **80**, 1818 (1998); Phys. Rev. D **61**, 016002 (2000); Phys. Rev. D **64**, 076001 (2001).
17. D. Bear *et al.*, Phys. Rev. Lett. **85**, 5038 (2000); D.F. Phillips *et al.*, Phys. Rev. D **63**, 111101 (2001); M.A. Humphrey *et al.*, Phys. Rev. A **68**, 063807 (2003); Phys. Rev. A **62**, 063405 (2000); V.A. Kostelecký and C.D. Lane, Phys. Rev. D **60**, 116010 (1999); J. Math. Phys. **40**, 6245 (1999).
18. R. Bluhm *et al.*, Phys. Rev. Lett. **88**, 090801 (2002).
19. F. Canè *et al.*, Phys. Rev. Lett. **93**, 230801 (2004).
20. H. Dehmelt *et al.*, Phys. Rev. Lett. **83**, 4694 (1999); R. Mittleman *et al.*, Phys. Rev. Lett. **83**, 2116 (1999); G. Gabrielse *et al.*, Phys. Rev. Lett. **82**, 3198 (1999); R. Bluhm *et al.*, Phys. Rev. Lett. **82**, 2254 (1999); Phys. Rev. Lett. **79**, 1432 (1997); Phys. Rev. D **57**, 3932 (1998).
21. L.-S. Hou *et al.*, Phys. Rev. Lett. **90**, 201101 (2003); R. Bluhm and V.A. Kostelecký, Phys. Rev. Lett. **84**, 1381 (2000).
22. H. Müller *et al.*, Phys. Rev. D **68**, 116006 (2003); R. Lehnert, Phys. Rev. D **68**, 085003 (2003); J. Math. Phys. **45**, 3399 (2004).
23. S.M. Carroll *et al.*, Phys. Rev. D **41**, 1231 (1990); V.A. Kostelecký and M. Mewes, Phys. Rev. Lett. **87**, 251304 (2001); J. Lipa *et al.*, Phys. Rev. Lett. **90**, 060403 (2003); Q. Bailey and V.A. Kostelecký, Phys. Rev. D **70**, 076006 (2004).
24. V.W. Hughes *et al.*, Phys. Rev. Lett. **87**, 111804 (2001); R. Bluhm *et al.*, Phys. Rev. Lett. **84**, 1098 (2000); E.O. Iltan, JHEP **0306**, 016 (2003).
25. V. Barger *et al.*, Phys. Rev. Lett. **85**, 5055 (2000); V.A. Kostelecký and M. Mewes, Phys. Rev. D **70**, 031902 (2004).
26. D.L. Anderson *et al.*, Phys. Rev. D **70**, 016001 (2004).
27. E. Cremmer and B. Julia, Nucl. Phys. B **159**, 141 (1979).
28. R. Jackiw and V.A. Kostelecký, Phys. Rev. Lett. **82**, 3572 (1999); B. Altschul, Phys. Rev. D **69**, 125009 (2004).
29. R. Lehnert and R. Potting, Phys. Rev. Lett. **93**, 110402 (2004); Phys. Rev. D **70**, 125010 (2004).

SCALAR FIELD MODELS: FROM THE PIONEER ANOMALY TO ASTROPHYSICAL CONSTRAINTS

J. PÁRAMOS*

Instituto Superior Técnico, Departamento de Física,
Avenida Rovisco Pais 1, 1049-001 Lisboa, Portugal
E-mail: x_jorge@fisica.ist.utl.pt

In this work we study how scalar fields may affect solar observables, and use the constraint on the Sun's central temperature to extract bounds on the parameters of relevant models. Also, a scalar field driven by a suitable potential is shown to produce an anomalous acceleration similar to the one found in the Pioneer anomaly.

1. Introduction

Scalar fields play a fundamental role in contemporary physics, with applications ranging from particle physics to cosmology and condensed matter. In cosmology, these include inflation[1], vacuum energy evolving and quintessence models, the Chaplygin gas dark energy-dark matter unification model and dark matter candidates (for a complete list, see Ref. 2). A scalar field has recently been suggested as a possible explanation for the anomalous acceleration measured by the Pioneer spacecraft[3]. On an astrophysical context, a scalar field could be present as, for example, a mediating boson of an hypothetical fifth force. This can be modelled by a Yukawa potential, given by $V_Y(r) = Ae^{-mr}/r$, where A is the coupling strength and m is the mass of the field, setting the range of the interaction $\lambda_Y \equiv m^{-1}$. The bounds on parameters A and $\lambda_Y \equiv m^{-1}$ can be found in Ref. 4 and references therein. In this work, a scalar field with an appropriate potential is shown to account for the anomalous Pioneer acceleration. Also, the effects of this and other scalar field models on stellar equilibrium is reported.

*Work partially supported by the Fundação para a Ciência e Tecnologia, under the grant BD 6207/2001.

2. Scalar field induced Pioneer anomaly

The hypothesis that the Pioneer anomaly is due to new physics has been steadily gaining momentum. Fundamentally, this arises because is seems unfeasible that an engineering explanation, affecting spacecrafts with different trajectories and designs, could lead to similar effects. The lack of a simple explanation could hint that this anomalous acceleration is the manifestation of a new force.

Given the broad range of applicability of theories with scalar fields, it is possible that such an entity is responsible for the anomaly[3]. A natural candidate would be the radion, a scalar field present due to oscillations of inter-brane distance in braneworld models. However, this does not account for the phenomena[3]. Fortunately, one can put down a model with a scalar field that doesn't stray away much from the already accepted theories arising from cosmology. Specifically, we assume a scalar field with dynamics ruled by a potential of the form $V(\phi) = -A^2\phi^{-2}$, where A is a constant. The scalar field obeys the equation of motion

$$\phi''(r) + \frac{2}{r}\phi'(r) = 2A^2\phi^{-3} \quad , \tag{1}$$

and admits the solution

$$\phi(r) = \left(\frac{8}{3}\right)^{\frac{1}{4}} \sqrt{Ar} \equiv \beta^{-1} r^{\frac{1}{2}} \quad . \tag{2}$$

The $0-0$ linearized Einstein's equations reads

$$f''(r) + \frac{2}{r}f'(r) = 2\kappa\left(1 - \frac{2C}{r}\right) A^2\beta^2 r^{-1} \tag{3}$$

whose solution is

$$f(r) = \sqrt{\frac{3}{2}}\frac{A\kappa r}{2} - \sqrt{6}AC\kappa \, log\left(\frac{r}{1\ m}\right) \quad . \tag{4}$$

The resulting acceleration felt by a test body is given by

$$a_r = -\frac{C}{r^2} - \sqrt{\frac{3}{2}}\frac{A\kappa}{4} + \sqrt{\frac{3}{2}}\frac{AC\kappa}{r} \quad . \tag{5}$$

The first term is the Newtonian contribution, and identifying the second with the anomalous acceleration $a_A = 8.5 \times 10^{-10}\ ms^{-2}$, sets $A = 3.26\ a_A/\kappa = 4.7 \times 10^{42}\ m^{-3}$. The last term is much smaller than the anomalous acceleration for $4C/r \ll 1$, that is, for $r \gg 6\ km$; it is also much smaller than the Newtonian acceleration for $r \ll 2.9 \times 10^{22}\ km \approx 100\ Mpc$. It has been shown[3] that the effect of this model on the Doppler effect used to calculate the acceleration of the spacecrafts is negligible, indicating that the anomaly is real, an not due to a misinterpreted light propagation.

3. Variable Mass Particle models

The variable mass particle (VAMP) proposal[5] assumes the presence of yet unknown fermions coupled to a dark-energy scalar field with dynamics ruled by a monotonically decreasing potential. Although this has no minima, the coupling of the scalar field ϕ to this exotic fermionic matter yields an effective potential of the form $V_{eff}(\phi) = V(\phi) + \lambda n_\psi \phi$, where n_ϕ is the number density of fermionic VAMPs and λ is a Yukawa coupling. In the present work we approach a potential of the quintessence-type form, $V(\phi) = u_0 \phi^{-p}$. As a result, the effective potential acquires a minimum and the related vacuum expectation value (*vev*) is responsible for the mass term of the exotic VAMP. Since the number density n_ψ depends on the scale factor $a(t)$, this mass varies on a cosmological timescale. A noticeable shortcoming of VAMP models is the presence of weakly or unconstrained parameters: the relative density $\Omega_{\psi 0}$, the scalar field coupling constant and the potential strength.

One can attempt to overcome this drawback by assuming that all fermions couple to the quintessence scalar field. This coupling should not substitute the usual Higgs coupling, but merely add a small correction to the Higgs-mechanism induced mass. Since the effects of the quintessence scalar field coupling crucially depend on the particle number density n_ψ, one expects this variable mass term to play a more relevant role in a stellar environment than in the vacuum. This hints that VAMP model parameters could be constrained from stellar physics observables.

3.1. *The polytropic gas stellar model*

The polytropic gas model assumes an equation of state of the form $P = K\rho^{n+1/n}$, where n is the polytropic index, defining intermediate processes between the isothermic and adiabatic thermodynamical cases, and K is the polytropic constant, which depends on the star's mass M and radius R. This model leads to scaling laws for thermodynamical quantities, given by $\rho = \rho_c \theta^n(\xi)$, $T = T_c \theta(\xi)$ and $P = P_c \theta(\xi)^{n+1}$, where ρ_c, T_c and P_c is the density, temperature and pressure at the center of the star[6]. The dimensionless function $\theta(\xi)$ depends on the dimensionless variable ξ, related to the distance to the star's center by $r = \alpha\xi$, where α also depends on the star's mass M and radius R. The hydrostatic equilibrium condition enables a differential equation ruling the behavior of the scaling function $\theta(\xi)$, the Lane-Emden equation:

$$\frac{1}{\xi^2}\frac{\partial}{\partial\xi}\left(\xi^2\frac{\partial\theta}{\partial\xi}\right) = -\theta^n \quad . \tag{6}$$

Since the physical characteristics of a star appear only in the definitions of the constants K and α, its stability is independent of these quantities, and different polytropic indexes n label different types of stars. This scale-independence can be related to the homology symmetry enclosed in the Lane-Emden equation. The boundary conditions of this differential equation are, from the definition $\rho = \rho_c\theta(\xi)$, $\theta(0) = 1$; also, the hydrostatic equilibrium condition implies that $|d\theta/d\xi|_{\xi=0} = 0$. The first solar model ever considered corresponds to a polytropic star with $n = 3$ and was studied by Eddington in 1926. Although somewhat incomplete, this simplified model gives rise to relevant constraints on the physical quantities.

The following results are based on the luminosity constraint on the Sun's central temperature, $\Delta T_c/T_c \leq 0.4\%$. The central temperature can be computed from $T_c = Y_n\mu GM/R$, where k is the Boltzmann constant and Y_n depends on the star's mass M and radius R, as well as on the dimensionless quantity ξ_1, which is defined through $\theta(\xi_1) = 0$ and signals the surface of the star.

3.2. *Results*

Assuming isotropy, a variable mass term leads to both a radial, anomalous acceleration $a_A = \phi'/\phi < 0$ plus a time-dependent drag force $a_D = -\dot{\phi}/\phi < 0^2$. The time-dependent component should vary on cosmological timescales, and can thus can be absorbed in the usual Higgs mass term. Hence, one considers only the perturbation to the Lane-Emden equation given by the radial force. We start by defining the dimensionless scalar field $\Phi \equiv \phi/\phi_c^*$, with $\phi_c^* = \rho_{crit}/2n_V$, where Ω_V is the energy density due to the potential driving ϕ and $\rho_{crit} \simeq 9.48 \times 10^{-28}\ g\ cm^{-3}$ is the critical density. Assuming for simplicity a potential with $p = 1$, we obtain a perturbed Lane-Emden equation

$$\frac{1}{\xi^2}\frac{d}{d\xi}\left[\xi^2\frac{d\theta}{d\xi}\right] = -\theta^n(\xi) - \frac{C_n}{U}\frac{1}{\Phi(\xi)}\left[\Phi''(\xi) + \frac{2}{\xi}\Phi'(\xi) - \frac{\Phi'^2(\xi)}{\Phi(\xi)}\right] \quad . \tag{7}$$

where one has defined the dimensionless quantities

$$U \equiv \frac{GM}{Rc^2} = 2.12 \times 10^{-6} \quad , \quad C_n^{-1} \equiv (n+1)N_n^{n/(n+1)}W_n^{1/(n+1)} \quad , \tag{8}$$

with $W_n = [4\pi(n+1)\left(\frac{d\theta}{d\xi}\right)^2_{\xi_1}]^{-1}$.

Furthermore, one assumes that the scalar field is only weakly perturbed in relation to the cosmological *vev* of the effective potential. Denoting this small "astrophysical" contribution as $\Phi_a(\xi)$, one has $\Phi(\xi) = \Omega_V/\lambda + \Phi_a(\xi)$. Hence, the Klein-Gordon equation becomes

$$\Phi''_a(\xi) + \frac{2}{\xi}\Phi'_a(\xi) \approx \frac{2\alpha^2\lambda n_V}{\rho_{crit}}\frac{\rho_c}{\mu}\theta^3(\xi) - \frac{2\alpha^2\lambda^2 n_V^2}{\rho_{crit}}\left[1 - \frac{2\lambda\Phi_a(\xi)}{\Omega_V}\right] , \tag{9}$$

inside the star, and

$$\Phi''_a(\xi) + \frac{2}{\xi}\Phi'_a(\xi) \approx \frac{2\alpha^2\lambda n_V^2}{\rho_{crit}} - \frac{2\alpha^2\lambda^2 n_V^2}{\rho_{crit}}\left[1 - \frac{2\lambda\Phi_a(\xi)}{\Omega_V}\right] , \tag{10}$$

in the outer region. In the above, $n_\psi \equiv n_V = 3\ m^{-3}$ is the number density of fermions in the vacuum. μ is the mean molecular weight of Hydrogen. The boundary conditions for the perturbed Lane-Emden are the same as in the unperturbed case. For the scalar field, we assume both $\Phi(\xi)$ and its derivative vanish beyond the Solar System (about $10^5\ AU$). Following Ref. 2, one gets that the maximum deviation $\Delta T_c/T_c = 2.82 \times 10^{-8}$ occurs for $\Omega_V = 0.7$, $\lambda = 2.82 \times 10^{-14}$. Hence, the luminosity constraint is always respected as long as one assumes the bound arising from the assumption that the "astrophysical" component of the scalar field is much smaller than its cosmological *vev*, $\lambda < 10^{-14}$.

4. Yukawa potential induced perturbation

One now looks at the hydrostatic equilibrium equation with a Yukawa potential which, after a small algebraic manipulation, implies the perturbed Lane-Emden equation

$$\frac{1}{\xi^2}\frac{d}{d\xi}\frac{d\theta}{d\xi} = -\theta^n\left[1 + Ae^{-\gamma\xi/\xi_1}\right] - \gamma A C_n \frac{d\theta}{d\xi}e^{-\gamma\xi/\xi_1} , \tag{11}$$

where we have defined the dimensionless quantities

$$C_n \equiv \left(\frac{n+1}{4\pi}\right)^{1/2} N_n^{n/(n+1)} W_n^{(1-n)/2(n+1)} , \quad \gamma \equiv mR \tag{12}$$

and ξ_1 signals the surface of the star (more accurately, a surface of zero temperature, but the difference is negligible. It can be shown that the boundary conditions are unaffected by the perturbation[2].

One can study the variation of the central temperature as a function of $\lambda \equiv m^{-1}$ and A, and constraint these so that $\Delta T_c/T_c \leq 0.4\%$. The

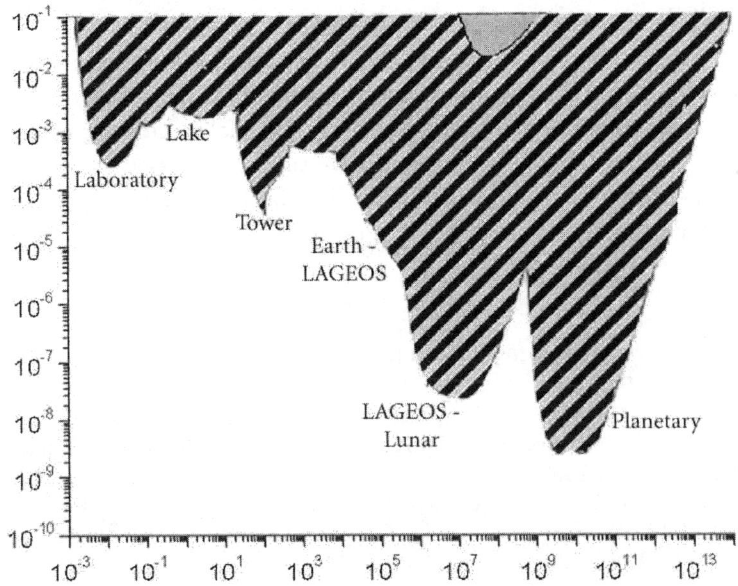

Figure 1. Exclusion plot for the relative deviation from unperturbed central temperature T_c, for A ranging from 10^{-3} to 10^{-1}, and γ from 10^{-1} to 10 (tip at the top), superimposed on the available bounds.

parameters were chosen for Yukawa interactions in the range $0.1R < \lambda_Y < 10R$; the Yukawa coupling A was chosen so that the variation of T_c is of the same order $O(10^{-4})$) as the luminosity constraint. Numerical integration of Eq. 11 is then used to derive the exclusion plot of Figure 1, superimposed on the accepted bound[4]. Notice that the central temperature is not precisely known and it is clear that constraining its uncertainty below 10^{-4} would yield a larger exclusion region in the parameter space.

4.1. *The Pioneer anomaly*

Following a similar procedure to the one depicted in the above two sections, one can prove that a constant, anomalous acceleration a_A inside the Sun yields a relative deviation of the central temperature which scales linearly with a_A as $\delta T_c \sim a_A/a_\odot$, where $a_\odot = 274\ m.s^{-2}$[2]. Thus, the bound $\delta T_c < 4 \times 10^{-3}$ is satisfied for values of this constant anomalous acceleration up to $a_{Max} \sim 10^{-4} a_\odot$. The reported value is then well within the allowed region, and has a negligible impact on the astrophysics of the Sun.

5. Conclusion

In this work we have developed a study of the impact of some exotic physics models on stellar equilibrium, enabling the extraction of constraints on the relevant parameters of the theories. Also, a model exhibiting a scalar field with a attractive quintessence-like potential is discussed, which can account for the anomalous acceleration felt by the Pioneer 10/11, Ulysses and Galileo spacecrafts.

References

1. See e.g. A. Linde, hep-th/0402051 and K.A. Olive, *Phys. Rev.* **190** (1990) 307.
2. O. Bertolami, J. Páramos *Phys. Rev.* **71**, (2005) 23521.
3. O. Bertolami, J. Páramos, *Class. Quantum Gravity* **21** (2004) 3309.
4. "The search for non-Newtonian gravity", E. Fischbach, C.L. Talmadge (Springer, New York 1999).
5. G.W. Anderson, S.M. Carroll, astro-ph/9711288.
6. "Textbook of Astronomy and Astrophysics with Elements of Cosmology", V.B. Bhatia (Narosa Publishing House, Delhi 2001).
7. "Theoretical Astrophysics: Stars and Stellar Systems", T. Padmanabhan (Cambridge University Press, Cambridge 2001).
8. J.N. Bahcall, *Phys. Rev.* **D33** (2000) 47.
9. J.D. Anderson, P.A. Laing, E.L. Lau, A.S. Liu, M.M. Nieto and S.G. Turyshev, *Phys. Rev.* **D65** (2002) 082004, *Phys. Rev. Lett.* **81** (1998) 2858.

BRANEWORLDS, CONFORMAL FIELDS AND DARK ENERGY

RUI NEVES
Departamento de Física, Faculdade de Ciências e Tecnologia
Universidade do Algarve
&
Centro Multidisciplinar de Astrofísica-CENTRA
Campus de Gambelas, 8005-139 Faro, Portugal
E-mail: rneves@ualg.pt

In the Randall-Sundrum scenario we analize the dynamics of a spherically symmetric 3-brane when matter fields propagate in the bulk. For a well defined class of conformal fields of weight -4 we determine a new set of exact 5-dimensional solutions which localize gravity in the vicinity of the brane and are stable under radion field perturbations. Geometries which describe the dynamics of inhomogeneous dust, generalized dark radiation and homogeneous polytropic dark energy on the brane are shown to belong to this set.

1. Introduction

In the Randall-Sundrum (RS) scenario [1,2] the visible Universe is a 3-brane world of a Z_2 symmetric 5-dimensional (5D) anti-de Sitter (AdS) space. In the RS1 model [1] there is a compact fifth dimension and two brane boundaries. The gravitational field is localized near the hidden positive tension brane and decays towards the visible negative tension brane. In this model the hierarchy problem is reformulated as an exponential hierarchy between the weak and Planck scales [1]. In the RS2 model [2] there is a single positive tension brane in an infinite fifth dimension. Then gravity is bound to the positive tension brane now interpreted as the visible brane.

At low energies the theory of gravity on the observable brane is 4-dimensional (4D) general relativity and the cosmology may be Friedmann-Robertson-Walker [1-10]. In the RS1 model this is only possible if the radion mode is stabilized using for example a bulk scalar field [3,6,9,10]. Gravitational collapse was also analyzed in the RS scenario [11-16]. However, an exact 5D solution representing a stable black hole localized on a 3-brane has not yet been discovered. So far, the only known static black holes localized on a brane remain to be those found for a 2-brane in a 4D AdS space [12]. A solution to this problem requires the non-singular localization of both gravity and matter [11,13–16] and could be connected to quantum black holes on the brane [15]. This is an extra motivation to look for 5D collapse solutions localized on a brane. In addition, the effective covariant Gauss-Codazzi approach [17,18] has permitted the discovery of many braneworld solutions which have not yet been associated with exact 5D spacetimes [19–22].

In this paper we continue the research on the dynamics of a spherically symmetric RS 3-brane when 5D conformal matter fields propagate in the bulk [16,23] (see also [24]). In our previous work [16,23] we have found a new class of exact 5D dynamical solutions for which gravity is localized near the brane by the exponential RS warp. These solutions were shown to describe on the brane the dynamics of inhomogeneous dust, generalized dark radiation and homogeneous polytropic matter. However, the density and pressures of the conformal bulk fluid increase with the coordinate of the fifth dimension. In the RS2 model this generates a divergence at the AdS horizon as in the Schwarzschild black string solution [11]. In the RS1 model this is not a problem. However, the solutions turn out to be unstable under radion field perturbations [25]. In this work we report on a new set of exact 5D braneworld solutions which have a stable radion mode and still describe on the observable brane the dynamics of inhomogeneous dust, generalized dark radiation and homogeneous polytropic matter.

2. 5D Einstein Equations and Conformal Fields

On the 5D RS orbifold the non-factorizable metric consistent with the Z_2 symmetry in z and with 4D spherical symmetry on the brane corresponds to the 5D line element $d\tilde{s}_5^2 = \Omega^2\left(dz^2 - e^{2A}dt^2 + e^{2B}dr^2 + R^2 d\Omega_2^2\right)$, where $\Omega = \Omega(t,r,z)$, $A = A(t,r,z)$, $B = B(t,r,z)$ and $R = R(t,r,z)$ are Z_2 symmetric functions. $R(t,r,z)$ represents the physical radius of the 2-spheres and Ω is the warp factor which defines a global conformal transformation on the metric.

The classical dynamics is defined by the 5D Einstein equations,

$$\tilde{G}_\mu^\nu = -\kappa_5^2 \left\{ \Lambda_B \delta_\mu^\nu + \frac{1}{\sqrt{\tilde{g}_{55}}} \left[\lambda \delta\left(z - z_0\right) + \lambda' \delta\left(z - z'_0\right)\right] \left(\delta_\mu^\nu - \delta_5^\nu \delta_\mu^5\right) - \tilde{T}_\mu^\nu \right\}, \quad (1)$$

where Λ_B is the negative bulk cosmological constant, λ and λ' are the brane tensions, $\kappa_5^2 = 8\pi/M_5^3$ with M_5 the fundamental 5D Planck mass and $\tilde{T}_\mu^\nu$

is the stress-energy tensor associated with the matter fields. In 5D $\tilde{T}_\mu^\nu$ is conserved, $\tilde{\nabla}_\mu \tilde{T}_\nu^\mu = 0$.

Let us consider the special class of conformal bulk matter defined by $\tilde{T}_\mu^\nu = \Omega^{-2} T_\mu^\nu$ and assume that T_μ^ν depends only on t and r. The conformal stress-energy tensor $\tilde{\mathcal{T}}^{\mu\nu}$ may be separated in two sectors $\tilde{T}^{\mu\nu}$ and $\tilde{U}^{\mu\nu}$ with the same weight s, $\tilde{\mathcal{T}}^{\mu\nu} = \tilde{T}^{\mu\nu} + \tilde{U}^{\mu\nu}$ where $\tilde{T}^{\mu\nu} = \Omega^s T^{\mu\nu}$ and $\tilde{U}^{\mu\nu} = \Omega^s U^{\mu\nu}$. Assuming that T_μ^ν and U_μ^ν are conserved tensor fields, $A = A(t,r)$, $B = B(t,r)$, $R = R(t,r)$ and $\Omega = \Omega(z)$ we obtain

$$G_a^b = \kappa_5^2 T_a^b, \quad G_5^5 = \kappa_5^2 T_5^5 \quad \nabla_a T_b^a = 0, \quad \nabla_a U_b^a = 0, \tag{2}$$

$$6\Omega^{-2}(\partial_z \Omega)^2 + \kappa_5^2 \Omega^2 \Lambda_B = k_5^2 U_5^5 \tag{3}$$

$$\left\{3\Omega^{-1}\partial_z^2\Omega + \kappa_5^2\Omega^2\left\{\Lambda_B + \Omega^{-1}\left[\lambda\delta(z-z_0) + \lambda'\delta(z-z'_0)\right]\right\}\right\}\delta_a^b = k_5^2 U_a^b, \tag{4}$$

where the latin indices represents the 4D coordinates t, r, θ and ϕ. On the other hand we also find the following equations of state $2T_5^5 = T_c^c, 2U_5^5 = U_c^c$. Note that U_μ^ν must be a diagonal tensor field, $U_\mu^\nu = diag\left(-\bar{\rho}, \bar{p}_r, \bar{p}_T, \bar{p}_T, \bar{p}_5\right)$ with constant density and pressures satisfying $\bar{\rho} = -\bar{p}_r = -\bar{p}_T$ and $\bar{p}_5 = -2\bar{\rho}$. On the other hand if $T_\mu^\nu = diag\left(-\rho, p_r, p_T, p_T, p_5\right)$ where ρ, p_r, p_T and p_5 denote bulk matter density and pressures then its equation of state is re-written as

$$\rho - p_r - 2p_T + 2p_5 = 0, \tag{5}$$

where ρ, p_r, p_T and p_5 must be independent of z but may be functions of t and r. The bulk matter is, however, inhomogeneously distributed along the fifth dimension because the physical energy density, $\tilde{\rho}(t,r,z)$, and pressures, $\tilde{p}(t,r,z)$, are related to $\rho(t,r)$ and $p(t,r)$ by the scale factor $\Omega^{-2}(z)$. Note also that the warp depends on the conformal bulk fields only through U_μ^ν. So the role of U_μ^ν is to influence how the gravitational field is warped around the branes. On the other hand T_μ^ν determines the dynamics on the branes. In the RS1 model the two branes have identical cosmological evolutions and gravity will be localized on the Planck brane and not on the visible one.

3. Exact 5D Warped Solutions

The dynamics of the AdS_5 braneworlds is defined by the solutions of equations (2) to (5). Let us first solve the warp equations (3) and (4). If $\bar{p}_5 = 0$ then $U_\mu^\nu = 0$ and we obtain the usual RS warp equations. A solution is the exponential RS warp [1,2]. Using the coordinate y related to z by $z = le^{y/l}$ for $y > 0$ we find

$$\Omega(y) = \Omega_{\text{RS}}(y) = e^{-|y|/l}, \tag{6}$$

where l is the AdS radius given by $l = 1/\sqrt{-\Lambda_B \kappa_5^2/6}$. If $\bar{p}_5$ is non-zero then there is a new set of warp solutions to be considered. Integrating Eq. (3) and taking into account the Z_2 symmetry we find

$$\Omega(y) = e^{-|y|/l}\left(1 + \frac{\bar{p}_5}{4\Lambda_B} e^{2|y|/l}\right). \tag{7}$$

This set of solutions which depends on the 5D pressure $\bar{p}_5$ must also satisfy Eq. (4) which contains the Israel conditions. This may only happen if the brane tensions λ and λ' are given by

$$\lambda = \frac{6}{l\kappa_5^2} \frac{1 - \frac{\bar{p}_5}{4\Lambda_B}}{1 + \frac{\bar{p}_5}{4\Lambda_B}}, \quad \lambda' = -\frac{6}{l\kappa_5^2} \frac{1 - \frac{\bar{p}_5}{4\Lambda_B} e^{2\pi r_c/l}}{1 + \frac{\bar{p}_5}{4\Lambda_B} e^{2\pi r_c/l}}, \tag{8}$$

where r_c is the RS compactification scale.

The conformal factor $\Omega(y)$ defines how the gravitational field is warped around the brane. To find the dynamics on the brane we need to consider solutions of Eq. (2) when the diagonal bulk matter T_μ^ν satisfies Eq. (5). For inhomogeneous dust, generalized dark radiation and homogeneous polytropic matter such solutions were determined in Refs. [16] and [23]. The latter describes the dynamics on the brane of dark energy in the form of a polytropic fluid. The diagonal conformal matter may be defined by $\rho = \rho_{\text{P}}, p_r + \eta \rho_{\text{P}}{}^\alpha = 0, p_T = p_r, p_5 = -\left(\rho_{\text{P}} + 3\eta\rho_{\text{P}}{}^\alpha\right)/2$, where ρ_{P} defines the polytropic energy density and the parameters $(\alpha,\ \eta)$ characterize different polytropic phases. For $-1 \leq \alpha < 0$ the fluid is in its generalized Chaplygin phase (see also [26]). The 5D polytropic solutions are [23]

$$d\tilde{s}_5^2 = \Omega^2\left[-dt^2 + S^2\left(\frac{dr^2}{1 - kr^2} + r^2 d\Omega_2^2\right)\right] + dy^2, \tag{9}$$

where the Robertson-Walker brane scale factor S satisfies

$$\dot{S}^2 = -k + \frac{\kappa_5^2}{3} S^2 \left(\eta + \frac{a}{S^{3-3\alpha}}\right)^{\frac{1}{1-\alpha}}. \tag{10}$$

4. Radion Stability

To analyze how these solutions behave under radion field perturbations we apply a saddle point expansion procedure based on the action [27,28]. Let us write the most general metric consistent with the Z_2 symmetry in y and with 4D spherical symmetry on the brane in the form $d\tilde{s}^2 = a^2 ds_4^2 + b^2 dy^2$ with $ds_4^2 = -dt^2 + e^{2B} dr^2 + R^2 d\Omega_2^2$. The metric functions $a = a(t, r, y)$, $B = B(t, r, y)$, $R = R(t, r, y)$ and $b = b(t, r, y)$ are Z_2 symmetric. Now a is the warp factor and b is related to the radion field. The 5D dynamical RS action is given by

$$\tilde{S} = \int d^4x dy \sqrt{-\tilde{g}} \left\{ \frac{\tilde{R}}{2\kappa_5^2} - \Lambda_B - \frac{1}{\sqrt{\tilde{g}_{55}}} \left[\lambda \delta\left(y\right) + \lambda' \delta\left(y - \pi r_c\right) \right] + \tilde{L}_B \right\}. \quad (11)$$

Our braneworld backgrounds correspond to the metric functions $b = 1$, $B = B(t, r)$, $R = R(t, r)$ and $a = \Omega(y)$.

To calculate the radion potential we consider the dimensional reduction of (11). Using the metric with $a(t, r, y) = \Omega(y)e^{-\beta(t,r)}$ and $b(t, r) = e^{\beta(t,r)}$ we obtain in the Einstein frame [25]

$$\tilde{S} = \int d^4x \sqrt{-g_4} \left(\frac{R_4}{2\kappa_4^2} - \frac{1}{2} \nabla_c \gamma \nabla_d \gamma g_4^{cd} - \tilde{V} \right), \quad (12)$$

where $\gamma = \beta/(\kappa_4 \sqrt{2/3})$ is the canonically normalized radion field. The function $\tilde{V} = \tilde{V}(\gamma)$ is the radion potential and it is given by

$$\tilde{V} = \frac{2}{\kappa_5^2} \chi^3 \left[3 \int dy \Omega^2 (\partial_y \Omega)^2 + 2 \int dy \Omega^3 \partial_y^2 \Omega \right] + \chi \int dy \Omega^4 \left(\Lambda_B - \tilde{L}_B \right)$$

$$+ \chi^2 \int dy \Omega^4 \left[\lambda \delta\left(y\right) + \lambda' \delta\left(y - \pi r_c\right) \right], \quad (13)$$

where the field χ is defined as $\chi = e^{-\sqrt{(2/3)}\, \kappa_4 \gamma}$. The integration in the fifth dimension is performed in the interval $[-\pi r_c, \pi r_c]$ and that we have chosen $\int dy \Omega^2 = \kappa_5^2/\kappa_4^2$.

To analyze the stability of the AdS_5 braneworld solutions we consider a saddle point expansion of the radion field potential V. If $\bar{p}_5 = 0$ then $\Omega = \Omega_{\rm RS}$. The critical extremum corresponding to our braneworlds is $\chi = 1$ [25]. Stable solutions must be associated with a positive second variation of the radion potential. If the equation of state of the conformal bulk fields is independent of the radion perturbation then for $\chi = 1$ the second variation is negative and so the corresponding braneworlds are unstable [25]. If the equation of state is kept invariant under the radion perturbations it is possible to find stable solutions at $\chi = 1$ if the warp is changed. Indeed, the new relevant warp functions are given in Eq. (7) and stability exists for a range of the parameters if $\bar{p}_5 < 0$. As an example consider the interval $4\Lambda_B e^{-2\pi r_c/l} < \bar{p}_5 < 0$ which corresponds to a brane configuration with $\lambda > 0$ and $\lambda' < 0$. Then the stability conditions are $l > 3r_c, 4\Lambda_B e^{-2\pi r_c/l} < \bar{p}_5 < \bar{p}_5^*(l)$ and $\bar{p}_5^*(l) < \Lambda_B/2$.

5. Conclusions

In this paper we have analized exact 5D dynamical solutions with gravity localized near the brane which are associated with conformal bulk fields of weight -4 and describe the dynamics of inhomogeneous dust, generalized dark radiation and homogeneous polytropic matter on the brane. We have discussed their behaviour under radion field perturbations and shown that they are extrema of the radion potential. We have also shown that if the equation of state characterizing the conformal fluid is independent of the perturbation then the radion may be stabilized by a sector of the conformal fields while another sector generates the dynamics on the brane. Stabilization requires a bulk fluid with a constant negative pressure and involves new warp functions. On the brane these solutions also describe the dynamics of inhomogeneous dust, generalized dark radiation and homogeneous polytropic matter. Whether gravity is suficiently localized on the brane is an open problem for future research.

Acknowledgments

We would like to thank the financial support of *Fundação para a Ciência e a Tecnologia* (FCT) and *Fundo Social Europeu* (FSE) under the contract SFRH/BPD/7182/2001 (*III Quadro Comunitário de Apoio*) as well as of *Centro Multidisciplinar de Astrofísica* (CENTRA).

References

1. L. Randall and R. Sundrum, Phys. Rev. Lett. **83**, 3370 (1999).
2. L. Randall and R. Sundrum, Phys. Rev. Lett. **83**, 4690 (1999).
3. W. D. Goldberger and M. B. Wise, Phys. Rev. D. **60**, 107505 (1999); Phys. Rev. Lett. **83**, 4922 (1999); Phys. Lett. B **475**, 275 (2000).
4. N. Kaloper, Phys. Rev. D **60**, 123506 (1999); T. Nihei, Phys. Lett. B **465**, 81 (1999); C. Csáki, M. Graesser, C. Kolda and J. Terning, Phys. Lett. B **462**, 34 (1999); J. M. Cline, C. Grojean and G. Servant, Phys. Rev. Lett. **83**, 4245 (1999).
5. P. Kanti, I. I. Kogan, K. A. Olive and M. Pospelov, Phys. Lett. B **468**, 31 (1999); Phys. Rev. D **61**, 106004 (2000).
6. O. DeWolf, D. Z. Freedman, S. S. Gubser and A. Karch, Phys. Rev. D **62**, 046008 (2000).
7. J. Garriga and T. Tanaka, Phys. Rev. Lett. **84**, 2778 (2000).
8. S. Giddings, E. Katz and L. Randall, J. High Energy Phys. **03**, 023 (2000).
9. C. Csáki, M. Graesser, L. Randall and J. Terning, Phys. Rev. D **62**, 045015 (2000).
10. T. Tanaka and X. Montes, Nucl. Phys. **B582**, 259 (2000).

11. A. Chamblin, S. W. Hawking and H. S. Reall, Phys. Rev. D **61**, 065007 (2000).
12. R. Emparan, G. T. Horowitz and R. C. Myers, J. High Energy Phys. **0001**, 007 (2000); J. High Energy Phys. **0001**, 021 (2000).
13. P. Kanti, K. A. Olive and M. Pospelov, Phys. Lett. B **481**, 386 (2000); T. Shiromizu and M. Shibata, Phys. Rev. D **62**, 127502 (2000).
14. P. Kanti and K. Tamvakis, Phys. Rev. D **65**, 084010 (2002); P. Kanti, I. Olasagasti and K. Tamvakis, Phys. Rev. D **68**, 124001 (2003).
15. R. Emparan, A. Fabbri and N. Kaloper, J. High Energy Phys. **0208**, 043 (2002).
16. R. Neves and C. Vaz, Phys. Rev. D **68**, 024007 (2003).
17. B. Carter, Phys. Rev. D **48**, 4835 (1993); R. Capovilla and J. Guven, Phys. Rev. D **51**, 6736 (1995); Phys. Rev. D **52**, 1072 (1995).
18. T. Shiromizu, K. I. Maeda and M. Sasaki, Phys. Rev. D **62**, 024012 (2000); M. Sasaki, T. Shiromizu and K. I. Maeda, Phys. Rev. D **62**, 024008 (2000).
19. J. Garriga and M. Sasaki, Phys. Rev. D **62**, 043523 (2000); R. Maartens, D. Wands, B. A. Bassett and I. P. C. Heard, Phys. Rev. D **62**, 041301 (2000); H. Kodama, A. Ishibashi and O. Seto, Phys. Rev. D 62, 064022 (2000); D. Langlois, Phys. Rev. D **62**, 126012 (2000); C. van de Bruck, M. Dorca, R. H. Brandenberger and A. Lukas, Phys. Rev. D **62**, 123515 (2000); K. Koyama and J. Soda, Phys. Rev. D **62**, 123502 (2000).
20. N. Dadhich, R. Maartens, P. Papadopoulos and V. Rezania, Phys. Lett. B **487**, 1 (2000); N. Dadhich and S. G. Ghosh, Phys. Lett. B **518**, 1 (2001); C. Germani and R. Maartens, Phys. Rev. D **64**, 124010 (2001); M. Bruni, C. Germani and R. Maartens, Phys. Rev. Lett. **87**, 231302 (2001).
21. R. Maartens, Phys. Rev. D **62**, 084023 (2000).
22. R. Neves and C. Vaz, Phys. Rev. D **66**, 124002 (2002).
23. R. Neves and C. Vaz, Phys. Lett. B **568**, 153 (2003).
24. E. Elizalde, S. D. Odintsov, S. Nojiri and S. Ogushi, Phys. Rev. D **67**, 063515 (2003); S. Nojiri and S. D. Odintsov, JCAP **06**, 004 (2003).
25. R. Neves, in TSPU Vestnik, edited by S. D. Odintsov, vol. 7 (2004) p. 94.
26. M. C. Bento, O. Bertolami and S. S. Sen, Phys. Rev. D **66**, 043507 (2002); Phys. Rev. D **67**, 063003 (2003); Phys. Rev. D **70**, 083519 (2004); M. C. Bento, O. Bertolami, N. M. C. Santos and S. S. Sen, Phys. Rev. D **71**, 063501, 2005.
27. R. Hofmann, P. Kanti and M. Pospelov, Phys. Rev. D **63**, 124020 (2001); P. Kanti, K. A. Olive and M. Pospelov, Phys. Lett. B **538**, 146 (2002).
28. S. M. Carroll, J. Geddes, M. B. Hoffman and R. M. Wald, Phys. Rev. D **66**, 024036 (2002).

SUN AND STARS AS COSMOLOGICAL TOOLS: PROBING SUPERSYMMETRIC DARK MATTER

ILÍDIO LOPES*

Centro de Geofisica de Évora, Departamento de Física, Universidade de Évora, Colégio Luís António Verney, 7002-554 Évora - Portugal

CENTRA, Departamento de Física, Instituto Superior Técnico Av. Rovisco Pais 1, 1049-001 Lisboa - Portugal

Department of Physics, University of Oxford, Denys Wilkinson Building, Keble Road, OX1 3RH Oxford - United Kingdom

Observational cosmology has shown that a quarter of the matter of the universe is composed by nonbaryonic dark matter. This dark matter may well constitute halos of massive and annihilating particles, weakly interacting and if so, these particles would be trapped in the interior of stars like our Sun. Whence solar evolution, in combination with helioseismic data, can provide clues as to the existence of nonbaryonic dark matter and some of their intrinsic properties. The current status of the solar standard model opens the way for using the current solar observations, namely, the neutrino flux detections and the solar acoustic data to test and probe the new physics behind the standard model of particle physics and cosmology. In this article we make a brief presentation of the physics behind the presence of dark matter inside stars.

1. The present Sun: The solar standard model

The Sun is a unique star for research because its proximity allows a superb quality of solar data, enabling precision measures of its luminosity, mass, radius and chemical composition. The Sun therefore naturally becomes a privileged target for testing stellar evolution theory. In recent years, different groups around the world have produced solar models in the framework of classical stellar evolution, taking into account the best known physics as well as all the available observational seismic data. This has led to the determination of a well-established model for the Sun, the so-called standard solar model [17, 4, 12, 14, 3], for which the acoustic modes are in

*Work partially supported by grant pocti/fis/57568/2004 of the Fundação para a Ciência e Tecnologia.

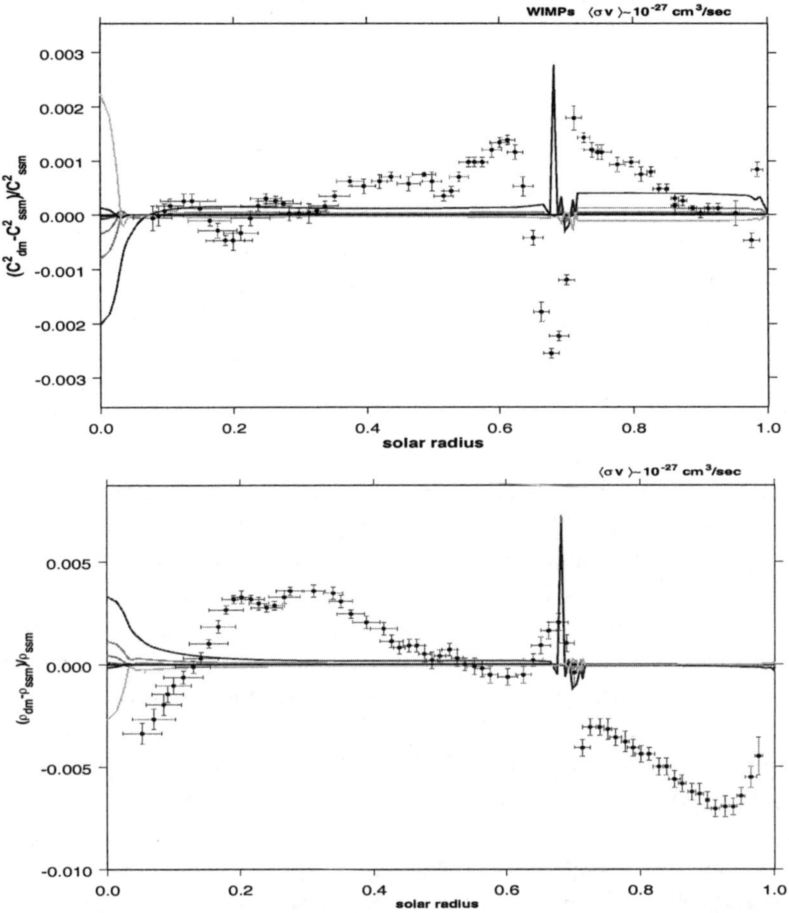

Figure 1. *The relative differences between the square of the sound speed* (**a**) *and the density* (**b**) *of the solar model embedded in dark matter halos and the standard solar model. The continuous curves correspond to models of WIMPs with masses m_x of 15 GeV, 30 GeV and 60 GeV with annihilation cross-section of $10^{-27} cm^3/s$, and scalar scattering cross-section σ_s of $10^{-38} cm^2$ or $10^{-40} cm^2$. The curves are as follows: $m_x \sim 60$ GeV, $\sigma_s \sim 10^{-38} cm^2$ (green curve) and $\sigma_s \sim 10^{-40} cm^2$ (magenta curve); $m_x \sim 30$ GeV, $\sigma_s \sim 10^{-38} cm^2$ (cyan curve) and $\sigma_s \sim 10^{-40} cm^2$ (dark blue curve); $m_x \sim 15$ GeV, $\sigma_s \sim 10^{-38} cm^2$ (dark green curve) and $\sigma_s \sim 10^{-40} cm^2$ (pink curve).*

very good agreement with observation. The potential diagnostic of observed frequencies of solar acoustic modes has allowed the determination of some of the basic properties of the microscopic physics of the solar interior [e.g. 15]. This solar model has established considerable consensus among the different research groups, concerning the predictions of the solar neutrino fluxes, and has unambiguously helped define the difference between the theoretical predictions and the experimental results. In the previous three decades, solar neutrino experiments have measured fewer neutrinos than were predicted by the solar models. One explanation for the deficit is the transformation of the Sun's electron-type neutrinos into other active flavors. Recently, the Sudbury Neutrino Observatory (SNO) has measured the ^8B solar neutrinos. The results obtained establish direct evidence for the non-electron flavor component in the solar neutrino flux and yield the first unequivocal determination of the total flux of ^8B neutrinos produced by the Sun [2, 3].

The combined effort between helioseismology and stellar astrophysics has contributed to strongly constrain the internal structure of the present Sun. Indeed, among the different seismic diagnostics that the solar model is tested against, the radial profile of the square of the sound speed is known with a precision as high as 0.3%, and in particular in the nuclear region this precision goes up to 0.2% (see Fig 1). This progress has been achieved through a detailed description of the microscopic physics, such as the nuclear reaction rates, the equation of state, the coefficient of opacities and the gravitational settling of chemical elements. More recently, the macroscopic mixing induced by differential rotation, seems to present a determining role in the evolution of the star. These physical processes take place in the thin shear layer, located between the radiative interior and the convection zone, the so-called tachocline [5].

The standard stellar evolution of the Sun assumes that the star is in hydrostatic equilibrium, is spherically symmetric and that the effects of rotation and magnetic fields are negligible. The present solar structure and evolution are computed starting from an initially homogeneous star of a given composition. The main physical inputs are as follows. The microphysics data are obtained from different sources: the solar composition [7], the OPAL96 tables [8], the equation of state [13], the nuclear reaction rates [1], the prescription for the treatment of the nuclear screening rates [10], and the microscopic diffusion coefficients suggested by [9]. The present structure of the Sun is obtained by evolving a initial star from the pre-main sequence, around 0.05 Gyr from the ZAMS, until its present age, 4.6 Gyr. The present solar model is obtained by choosing the initial Helium content, as well as the mixing length parameter of convection that best predicts the present solar luminosity and radius. The helioseismology requires that the present structure of the Sun, including the global quantities, should be determined with a very high precision. In particular, the calibration of the solar radius is done with a precision of 10^{-5}.

If our understanding of the basic equilibrium structure of the present Sun seems quite-well established, the same cannot be said about the dynamical processes acting in its interior, such as the turbulent convection, the transport of angular momentum, and the mechanism excitation of acoustic-gravity waves. These physical processes are particularly critical in regions of the star that somewhat regulate their evolution but also in regions where important dynamical processes are taking place in a much more short scale, from some minutes to several years, which is the case of the base of the convection zone, where the magnetic field is believed to be generated [6], and in the super-adiabatic region where the modes are believed to be excited [11].

If the stellar evolution theory still presents some unsolved problems, which is particularly true in the case of the Sun, it is fair to say that we have obtained a good understanding of the basic processes behind the evolution of the Sun and stars. This will allow us to use the Sun and stars as an instrument to test new physical ideas, namely to research the particle candidates for dark matter.

2. Helioseismology and Dark Matter

The knowledge of the Sun has improved with the advent of helioseismology. Since different modes of oscillations penetrate different depths in its interior, the measured frequencies impose severe constraints on the models of the present Sun, ruling out most of those which have been put forward to reconcile the difference between the theoretical and the observed spectra [17]. A similar strategy is used in the research for dark matter. To obtain an understanding of the physical processes engaged in stellar interiors by SUSY particles, we need different solar models for a wide range of particles with different properties, and to consider different abundances in its interior. The measured oscillation frequencies by the three helioseismology instruments on board the Solar and Heliospheric Observatory (SOHO) have put severe constraints on the variation of sound speed with radius and also with density, especially in the central regions which have the greatest importance for determining the impact of dark matter in stellar evolution: they also enhance our knowledge of the nature of the dark

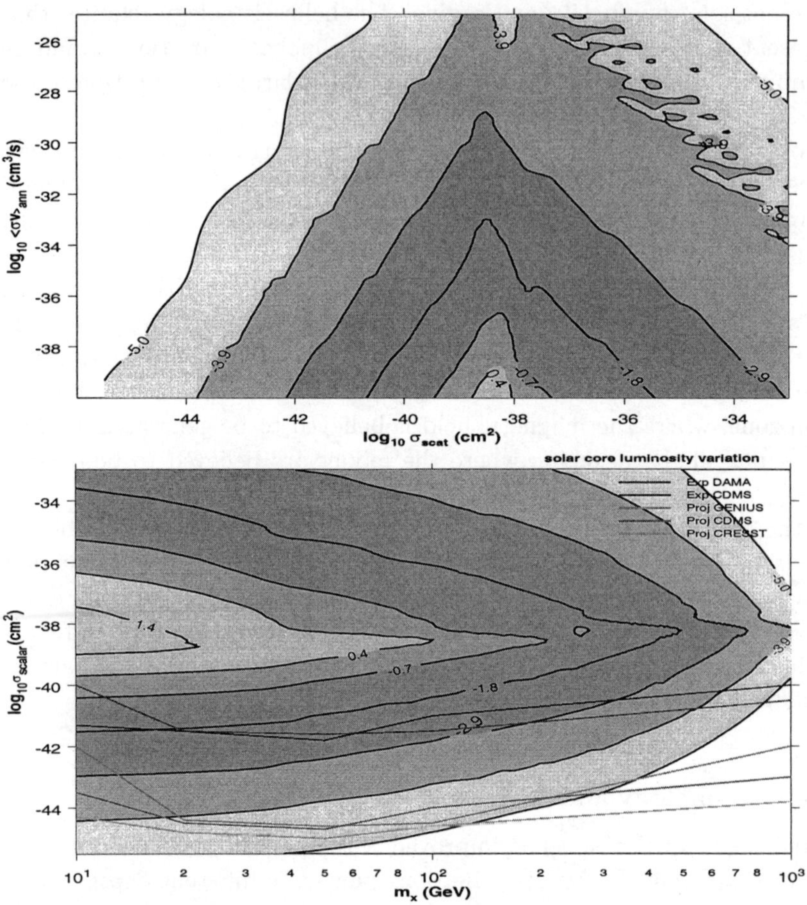

Figure 2. *Relative variation of the luminosity in the core of the Sun, due to the concentration of WIMPs in the solar core, as a function of scalar scattering cross section and annihilation cross section (**a**), and as a function of WIMP mass and scattering cross-section (**b**). The iso-curves represent the decimal logarithm of the ratio of the luminosity of WIMPs against the Sun's luminosity in the inner core, within 5% of the solar radius. The computation was made using the structure of the present Sun. The solid curves represent the current and future experimental bounds placed by different experiments.*

matter particles. It follows from Fig. 1 that the sound speed and density in the interior of the Sun is presently known with a precision of much less than a few per cent. This level of accuracy in constraining the solar interior has been achieved by a systematic study of the differences between the acoustic spectrum obtained from helioseismology experiments and the theoretical spectrum. Presently, this difference is less than several micro Hertz for almost all of the three thousand acoustic modes that probe the interior of the Sun. Nevertheless, a higher accuracy in probing the structure of the nuclear region is expected in the coming years, provided by more accurate measurements of acoustic modes, but also, for the first time, by the detection and measurement of gravity modes [16]. Even if the answers to some questions about the stellar physics are still unclear, it seems very likely that there is a class of annihilating weak interacting massive particles (10^{-27} cm^3/sec) and relatively small masses (60 GeV) that is already excluded by the present seismic results.

This type of analysis will enable comparisons to be made with a range of theoretical models built on the basis of different assumptions in the nature of interacting massive particles, leading to improvements in the origin and nature of dark matter. Even in the cases of distant stars, from which the data will be of relatively low quality compared with that from the Sun, measurements of oscillation frequencies of a number of stars in our Galaxy can be used to calibrate, constrain, test and improve our knowledge about the distribution of dark matter in the Milky Way.

The thermodynamic structure of the interior of the Sun, namely in the nuclear region, is presently known with a precision of much less than a few per cent. In particular, the luminosity in the solar core is known with a precision of one part in thousand. In the coming years, it is very likely that the new seismic data available from the SOHO experiments, will allow us to obtain a solar model with seismic constraints by as much as one part in a hundred thousand. Alternatively, as it follows from Fig. 2, the contribution of the WIMPs luminosity for the total luminosity of the Sun's core is above the accuracy of seismic data. In such conditions, the Sun can and should be used as a probe for dark matter in the solar neighborhood, spanning the parameter space of SUSY particle theory (cf. Fig. 2). This order of magnitude in the luminosity produced by the solar core can be tested through seismological data, leading to the prediction of the annihilating neutrino flux with an accuracy not available with the present SUSY particle models.

References

1. E. G. *et al.* Adelberger. Solar fusion cross sections. *Reviews of Modern Physics*, 70:1265–1291, October 1998.
2. Q. R. *et al.* Ahmad. Measurement of the Rate of ν_e + d : p + p + e^- Interactions Produced by ^{8}B Solar Neutrinos at the Sudbury Neutrino Observatory. *Physical Review Letters*, 87:071301–+, August 2001.
3. J. N. Bahcall. High-energy physics: Neutrinos reveal split personalities. *Nature*, 412:29–31, July 2001.
4. J. Christensen-Dalsgaard, W. Dappen, S. V. Ajukov, E. R. Anderson, H. M. Antia, S. Basu, V. A. Baturin, G. Berthomieu, B. Chaboyer, S. M. Chitre, A. N. Cox, P. Demarque, J. Donatowicz, W. A. Dziembowski, M. Gabriel, D. O. Gough, D. B. Guenther, J. A. Guzik, J. W. Harvey, F. Hill, G. Houdek, C. A. Iglesias, A. G. Kosovichev, J. W. Leibacher, P. Morel, C. R. Proffitt, J. Provost, J. Reiter, E. J. Rhodes, F. J. Rogers, I. W. Roxburgh, M. J. Thompson, and R. K. Ulrich. The Current State of Solar Modeling. *Science*, 272:1286–+, 1996.
5. J. R. Elliott and D. O. Gough. Calibration of the Thickness of the Solar Tachocline. *ApJ*, 516:475–481, May 1999.
6. D. Gough. Some theoretical remarks on solar oscillations. In *Nonradial and Nonlinear Stellar Pulsation*, pages 273–299, 1980.
7. N. Grevesse, A. Noels, and A. J. Sauval. Standard Abundances. In *ASP Conf. Ser. 99: Cosmic Abundances*, pages 117–+, 1996.
8. C. A. Iglesias and F. J. Rogers. Updated Opal Opacities. *ApJ*, 464:943–+, June 1996.
9. G. Michaud and C. R. Proffitt. Particle transport processes. In *ASP Conf. Ser. 40: IAU Colloq. 137: Inside the Stars*, pages 246–259, 1993.
10. H. E. Mitler. Thermonuclear ion-electron screening at all densities. I - Static solution. *ApJ*, 212:513–532, March 1977.
11. R. Nigam and A. G. Kosovichev. Source of Solar Acoustic Modes. *ApJ*, 514:L53–L56, March 1999.
12. J. Provost, G. Berthomieu, and P. Morel. Low-frequency p- and g-mode solar oscillations. 353:775–785, January 2000.
13. F. J. Rogers, F. J. Swenson, and C. A. Iglesias. OPAL Equation-of-State Tables for Astrophysical Applications. *ApJ*, 456:902–+, January 1996.
14. S. Turck-Chièze, S. Couvidat, A. G. Kosovichev, A. H. Gabriel, G. Berthomieu, A. S. Brun, J. Christensen-Dalsgaard, R. A. García, D. O. Gough, J. Provost, T. Roca-Cortes, I. W. Roxburgh, and R. K.

Ulrich. Solar Neutrino Emission Deduced from a Seismic Model. *ApJ*, 555:L69–L73, July 2001.

15. S. Turck-Chièze, W. Däppen, E. Fossat, J. Provost, E. Schatzman, and D. Vignaud. The solar interior. *Phys. Rep.*, 230:57–235, 1993.
16. S. Turck-Chièze, R. A. García, S. Couvidat, R. K. Ulrich, L. Bertello, F. Varadi, A. G. Kosovichev, A. H. Gabriel, G. Berthomieu, A. S. Brun, I. Lopes, P. Pallé, J. Provost, J. M. Robillot, and T. Roca Cortés. Looking for Gravity-Mode Multiplets with the GOLF Experiment aboard SOHO. *ApJ*, 604:455–468, March 2004.
17. S. Turck-Chieze and I. Lopes. Toward a unified classical model of the sun - On the sensitivity of neutrinos and helioseismology to the microscopic physics. *ApJ*, 408:347–367, May 1993.

ZEPLIN III: A XENON DETECTOR FOR WIMP SEARCHES

HENRIQUE ARAÚJO, on behalf of the ZEPLIN Collaboration[1]

Blackett Laboratory, Imperial College London, SW7 2BW, UK.

ZEPLIN III is a two-phase xenon detector soon to be deployed underground at the Boulby mine to search for dark matter in the form of Weakly Interacting Massive Particles (WIMPs). It exploits both scintillation and ionisation signals produced in liquid xenon in order to discriminate between gamma-like interactions and the nuclear recoils expected from the scattering of WIMPs off Xe atoms. This discrimination of particle species, together with the low radioactivity of its components, should extend the search for WIMP-nucleon interactions down to a spin-independent cross-section of 10^{-8} pb.

1. Introduction

ZEPLIN III is a two-phase (liquid/gas) xenon experiment which aims to detect galactic dark matter in the form of WIMPs [1]. It operates on the principle that electron and nuclear recoils, produced in liquid xenon by different particle species, generate different amounts of scintillation light and ionisation charge for a given deposited energy. WIMPs are expected to scatter elastically off Xe atoms in the same way as neutrons. The resulting nuclear recoils produce a different signature to electron recoils caused by gamma-ray interactions, which are the dominant source of background in the target.

Two-phase emission detectors using condensed noble gases were proposed some three decades ago [2]. ZEPLIN II, a two-phase system operating at low fields [3], is presently being installed underground at the Boulby mine (North Yorkshire, UK). ZEPLIN III follows extensive R&D into two-phase operation at high electric fields [4,5]. In these so-called hybrid detectors, the vacuum ultraviolet (VUV) photons produced by the scintillation and, indirectly, the ionisation channels are measured by the same array of photomultiplier tubes (PMTs) immersed in the liquid. The ability to identify rare nuclear recoils also relies on a low-background construction using radio-pure materials, a thin-slab geometry which improves light collection and ensures uniform electric fields,

---

[1] Edinburgh University, Imperial College London, ITEP-Moscow, LIP-Coimbra, Rochester University, Rutherford Appleton Laboratory, Sheffield University, Texas A&M, UCLA.

and capability for 3-dimensional event reconstruction.

2. The ZEPLIN III detector

The Xe target and PMT array are located in a cold vessel (–100°C) which sits above a liquid N_2 reservoir. Both are mounted inside a vacuum vessel ≈0.7 m in diameter and 1.1 m in height, as shown in Fig. 1 (left). High-purity copper is the main construction material, which reduces potential radioactive backgrounds.

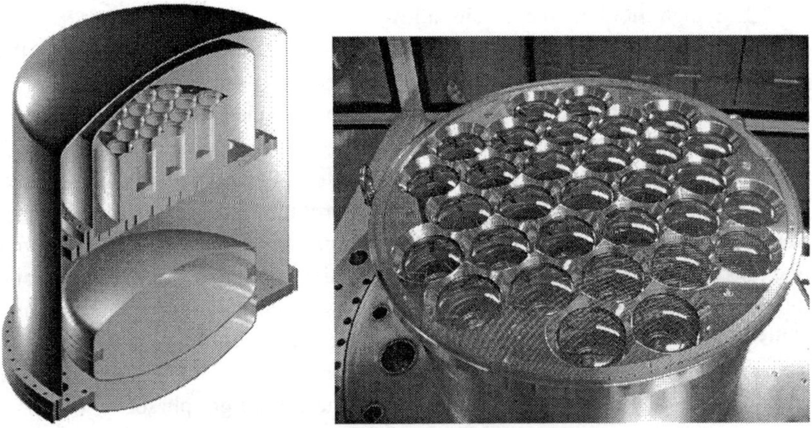

Fig. 1: Left: Cut-out view of ZEPLIN III copperwork. Right: Fully-assembled PMT array.

Inside the target chamber, a 40-mm thick disk of liquid xenon is viewed from below by 31 PMTs immersed in the liquid [6]. The assembled array is shown in Fig. 1 (right). The PMTs are biased by a connector-network of copper plates located in the liquid, thus reducing the number of required feedthroughs considerably. A strong electric field (~8 kV/cm) is applied to the target between a top electrode mirror in the gas phase and a wire grid above the PMTs. This field extracts the ionisation from the interaction site into the 5-mm thick gas layer which exists above the liquid. A reverse-field region is created by a second grid located just above the phototubes in order to suppress ionisation signals from the bottom 5 mm of liquid, and so reduce some of the gamma-ray background from the PMTs.

Submerging the photomultipliers in the liquid ensures good light collection for the primary scintillation, which is critical given the small number of photoelectrons (phe) expected from WIMP interactions. The simulated light collection efficiency, shown in Fig. 2 (left), is ≈3.5 phe/keV in the active region above the cathode grid, except near the field-shaping rings, and is quite constant over the z (vertical) coordinate [7]. This yield would be observed for gamma-rays in the absence of field; high-field operation reduces the number of VUV photons to ≈1/3 of the zero-field value due to partial suppression of the recombination of the ionisation charge generated at the interaction site [8,9,12]. Lower suppression is expected for nuclear recoils, although these generate only a fraction (~20%) of the prompt scintillation light relative to an electron of the same energy [10-12].

The high field in the liquid allows ionisation to be extracted from the primary tracks. More charge can be collected from electron interactions than from nuclear recoil ones, owing to the higher ionisation density around the primary track in the latter case [13]. Any charge which escapes recombination will drift upward in the liquid and, at such high fields, can be extracted into the gas phase with nearly unity efficiency [14]. Here, the emitted electrons acquire enough energy to produce excitation and electroluminescence of gaseous xenon, resulting in ample generation of 175 nm photons. At nominal field (17 kV/cm in the gas) a single electron extracted from the interaction site can generate 500 VUV photons in the 5 mm gas layer in this proportional scintillation regime. This gives an average ≈26 phe detected in the PMTs [8,15] — highlighting the sensitivity which can be achieved for the ionisation channel. Figure 2 (right) shows the light collection efficiency expected across the gas phase.

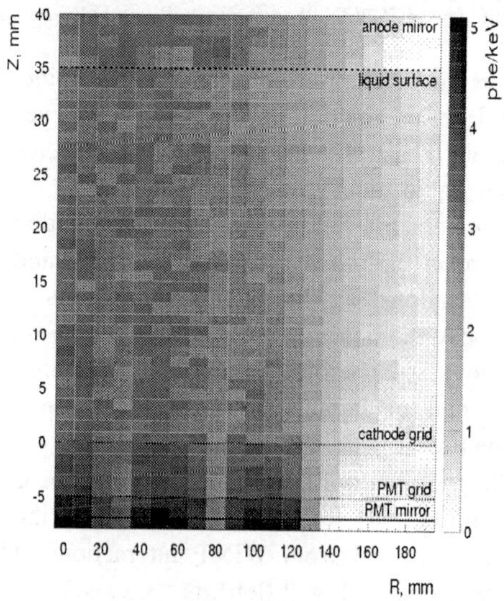

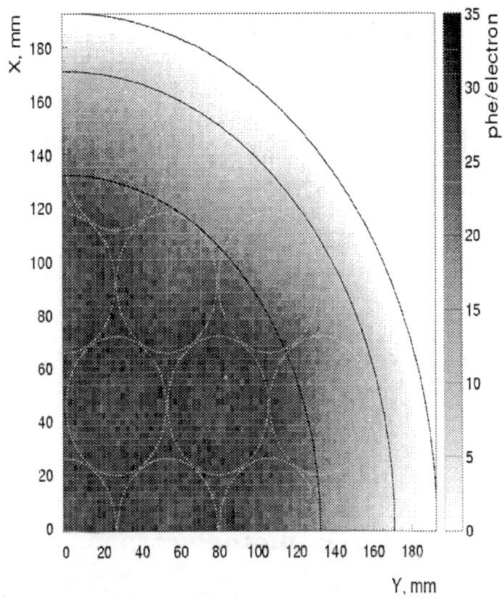

Fig. 2: Light yield from the primary scintillation (*S1*) at zero electric field (left) and yield from electroluminescence (*S2*) at nominal field (right).

The two signals generated by every interaction in the active region provide both discrimination of the type of interacting particle and 3-dimensional information of the interaction point. When combined, these features allow excellent discrimination of gamma-ray and other reducible backgrounds. The ratio between the electroluminescence (*S2*) and the prompt scintillation (*S1*) signals will be much higher for gamma-ray interactions than for nuclear recoils. An analysis of the discrimination ability of ZEPLIN III [16] has shown that a gamma-ray rejection fraction of 10^{-5} can be achieved whilst maintaining good detection efficiency for nuclear recoils down to a few S1 photoelectrons, as shown in Fig. 3. This suggests that a (zero-field) threshold of ~2 keV in visible energy (10 keV recoils) is feasible. Also, we note that ZEPLIN III is sensitive down to 1 electron extracted from the liquid. Since the system is triggered by the larger *S2* signals, the effective threshold for *S1* is therefore quite low.

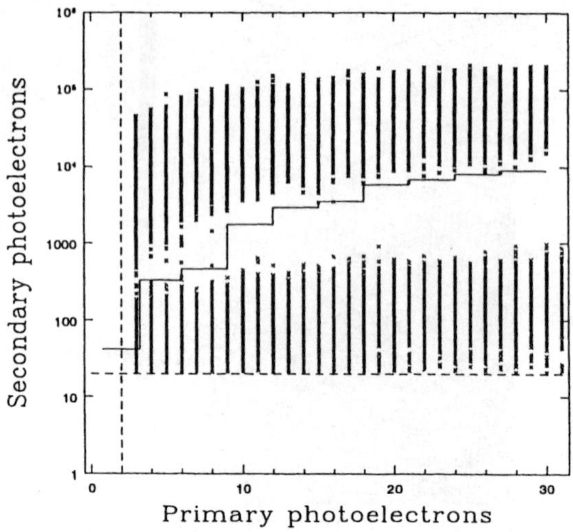

Fig. 3: Simulation of particle discrimination in ZEPLIN III [16]: scatter plot of the primary and secondary harvest of photoelectrons from electrons (upper distribution) and nuclear recoils (lower). The latter interactions are assumed to be α-like with a quenching factor of 0.2. The horizontal segments follow a 10^{-5} rejection boundary.

Three-dimensional information allows the identification of peripheral events and the fiducialisation of the active volume. Particles interacting in outlying regions, where the electric field may not be uniform, can therefore be identified. Small energy deposits from α, β and nuclear recoils arising from radioactive contamination of detector surfaces can also be excluded. The (x,y) coordinates of the interaction can be reconstructed from *S2* alone. Sub-centimetre accuracy (FWHM) can be obtained for small signals (a few ionisation electrons extracted from the liquid). The z coordinate is calculated from the time difference between *S1* and *S2*; the saturation drift velocity at high field is approximately 3 mm/μs in liquid xenon [17]. A Xe fiducial mass in excess of 6 kg should be achieved, corresponding to a disk 3.5 cm thick and ≈28 cm in diameter.

3. WIMP-nucleon sensitivity

PMT radioactivity is the dominant internal source of background in the detector. Gamma-rays from the ^{238}U and ^{232}Th decay chains and from ^{40}K were measured at just over 10^5 /day/PMT, setting a detector trigger rate of a few events per second. An interaction rate of around 10 dru (1 dru=1 evt/kg/day/keV) is expected in the low-energy region in the active volume. A 10^{-5} rejection efficiency should remove this major background. Neutrons from (α,n) reactions in the PMT glass will cause an irreducible background of nuclear recoils. This is estimated at ~10^{-3} dru at 2 keV threshold (10 keV recoil energy), totalling ≈15 single recoil events per year above threshold in a 6 kg fiducial mass.

A judicious choice of the materials and techniques used in the detector construction ensures that other potential neutron and electron/gamma backgrounds are kept to a minimum. Xenon contamination with ^{85}Kr, a β emitter with $T_{1/2}$=10.76 yr, should be minimised with the sourcing of xenon some four decades old (present equivalent Kr concentration ~10 ppb Kr). External backgrounds, mainly gamma-rays and neutrons emanating from the Boulby rock, are mitigated by extensive passive shielding. For this purpose, the detector is enclosed in a 15 cm-thick lead castle, containing a further 30 g/cm^2 of hydrocarbon material for neutron moderation.

The neutron background due to the PMTs will set a sensitivity limit for the spin-independent WIMP-nucleon cross-section below 10^{-7} pb, which should be achieved within a few months of deployment underground. An active neutron veto has been planed for retro-fitting around the detector. This hydrocarbon scintillator veto will tag internal neutrons in coincidence with the target. With this system in place, we expect to achieve the design sensitivity of 10^{-8} pb within a year (an exposure of 2000 kg days). Replacement of the phototubes by new models with even lower radioactivity is being studied, which could improve our WIMP-nucleon sensitivity even further.

4. Conclusion

ZEPLIN III is in the final stages of commissioning, and will soon undergo the first surface tests. It will then be installed at the Boulby Underground Laboratory. A sensitivity to the spin-independent WIMP-nucleon cross-section as low as 10^{-8} pb should be achieved, which is well inside the SUSY parameter space where WIMP interactions are more likely to be found.

References

1. T. J. Sumner, *Proc. 3rd Int. Workshop on the Identification of Dark Matter*, ed. N. J. C. Spooner & V. Kudryavtsev, Singapore: World Scientific, p.452 (2001). T. J. Sumner *et al.*, New Astronomy Reviews (2005) (in press).
2. B. A. Dolgoshein, V. N. Lebedenko & B. U. Rodionov, JETP Lett. 11 (1970) 513.
3. D. Cline, Nuc. Phys. B Proc. Sup.) 87 (2000) 114.
4. A. S. Howard *et al.*, *Proc. 3rd Int. Workshop on the Identification of Dark Matter*, ed. N. J. C. Spooner & V. Kudryavtsev, Singapore: World Scientific, p.457 (2001).
5. D. Yu. Akimov, *Proc. 4rd Int. Workshop on the Identification of Dark Matter*, ed. N. J. C. Spooner & V. Kudryavtsev, Singapore: World Scientific, p.371 (2003).

6. H. M. Araújo *et al.*, Nucl. Instrum. & Meth. A521 (2004) 407.
7. H. M. Araújo, *ZepIII: A (GEANT4) Simulation Tool for ZEPLIN III*, Internal Report, Imperial College London (2001).
8. S. Kubota, Phys. Rev. B17 (1978) 2762.
9. J. Dawson *et al.*, Nucl. Instrum. & Meth. A, in press; physics/0502026.
10. D. Akimov *et al.*, Phys. Lett. B 524 (2002) 245.
11. M. I. Lopes *et al.*, presented at XeSAT2005, Tokyo, Japan, 8-10 March, 2005.
12. E.Aprile *et al.* (2005), astro-ph/0503621.
13. E. Aprile *et al.* (2005), astro-ph/0502279.
14. E. M. Gushchin *et al.*, Sov. Phys. JETP 49(5) (1979) 856.
15. D. Davidge, PhD Thesis, Imperial College London (Un. London), 2004.
16. J. Dawson, PhD Thesis, Imperial College London (Un. London), 2003.
17. L. S. Miller *et al.*, Phys. Rev. 166(3) (1968) 871.

DARK MATTER DETECTABILITY WITH ČERENKOV TELESCOPES

F. PRADA
Instituto de Astrofisica de Andalucia (CSIC)
Camino Bajo de Huetor, 50
18008 Granada, Spain
E-mail: fprada@iaa.es

If the dark matter (DM) is made of supersymmetric particles, the central region of the Milky Way should emit gamma rays produced by their annihilation. We are using detailed models of our Galaxy to make accurate predictions of continuum gamma-ray fluxes. We discuss that the most important effect is the compression of the dark matter due to the infall of baryons to the galactic center. Our models predict that the signal could be detected at high confidence levels by imaging atmospheric Čerenkov telescopes assuming that neutralinos make up most of the DM in the Universe.

1. Introduction

There is an increasing hope that the new generation of Imaging Atmospheric Čerenkov Telescopes (IACTs) would detect in the very near future the γ-rays coming from the annihilation products of the SUSY DM in galaxy halos. The success of such a detection will solve one of the most fundamental questions in Astrophysics and Particle Physics: the nature of the dark matter. The lightest supersymmetric particle (LSP) has been proposed to be a suitable candidate for the non-baryonic cold DM. The LSP is stable in SUSY models where R-parity is conserved and its annihilation cross section and mass leads appropriate relic densities in the range allowed by WMAP, i.e. $0.095 < \Omega_{CDM}h^2 < 0.129$. Most of SUSY breaking scenarios yields the lightest neutralino ($\tilde{\chi}_1^0$) as the leading candidate for the cold DM.

The number of $\tilde{\chi}_1^0$ annihilations in galaxy halos and therefore the expected γ signal arriving at the Earth depends not only on the adopted SUSY model but also strongly depends on the DM density $\rho_{dm}(r)$. This is why the central region $r < 200\,\text{pc}$ of the MW, where the density is the largest, is the favorite site to search for this signal. The expected total number of continuum γ-ray photons received per unit time and per unit area, from a circular aperture on the sky of width σ_t (the resolution of the telescope) observing at a given direction Ψ_0 relative to the center of the MW can be written as:

$$F(E > E_{\rm th}) = \frac{1}{4\pi} f_{\rm SUSY} \cdot U(\Psi_0), \tag{1}$$

$$f_{\rm SUSY} = \frac{N_\gamma \langle \sigma v \rangle}{2m_\chi^2}, \; U(\Psi_0) = \int J(\Psi) B(\Omega) d\Omega.$$

The factor $f_{\rm SUSY}/4\pi$ represents the isotropic probability of γ-ray production per unit of DM density and depends only on the physics of annihilating $\tilde{\chi}_1^0$ particles. It can be determined for any SUSY scenario given the neutralino mass m_χ, the number of continuum γ-ray photons N_γ emitted per annihilation, with energy above the IACT energy threshold (E_{th}), and the thermally average cross section $\langle \sigma v \rangle$.

All the astrophysical properties (such as the DM distribution and geometry considerations) appear only in the factor $U(\Psi_0)$. This factor also accounts for the beam smearing, where $J(\Psi) = \int_{l.o.s} \rho_{dm}^2(r)\, dl$, is the integral of the line-of-sight of the square of the DM density along the direction Ψ, and $B(\Omega) d\Omega$ is the Gaussian beam of the telescope.

A cuspy DM halo $\rho_{dm}(r) \propto {\rm r}^{-\alpha}$ predicted by the simulations of the Cold Dark Matter with the cosmological constant (ΛCDM) is often assumed for the calculations of $U(\Psi_0)$. Cosmological $N-$body simulations indicate that the distribution of DM in relaxed halos varies between two shapes: the Navarro, Frenk & White (NFW) density profile with asymptotic slope $\alpha = 1$ and the steeper Moore *et al.* profile with slope 1.5.

2. Milky Way mass models with adiabatic compression

The predictions for the DM halos are valid *only* for halos without baryons. When normal gas ("baryons") loses its energy through radiative processes, it falls to the central region of forming galaxy. As the result of this redistribution of mass, the gravitational potential in the center changes substantially. The DM must react to this deeper potential by moving closer to the center and increasing its density. This increase in the DM density is often treated using adiabatic invariants. This is justified because there is a limit to the time-scale of changes in the mass distribution: changes of the potential at a given radius cannot happen faster than the dynamical time-scale defined by the mass inside the radius. Adiabatic contraction of dark matter in a collapsing protogalaxy was used already by Zeldovich *et al.* (1980) to set constraints of properties of elementary particles (annihilating massive neutrinos). The present form of analytical approximation (circular orbits) was introduced in Blumenthal *et al.* (1986). If $M_{\rm in}(r_{\rm in})$ is the initial distribution of mass (the one predicted by cosmological simulations), then the final (after compression and formation of galaxy) mass distribution is given by $M_{\rm fin}(r)r = M_{\rm in}(r_{\rm in})r_{\rm in}$, where $M_{\rm fin} = M_{\rm DM} + M_{\rm bar}$. The approximation assumes that orbits are circular and, thus $M(r)$ is the mass inside the orbit. This is not true for elongated orbits: mass $M(r)$ is smaller than the real mass, which a particle "feels" when it travels along elongated trajectory. This difference in masses requires a relatively small correction: mass M should be replaced by the mass inside time-averaged radius of trajectories passing through given radius r: $M_{\rm fin}(\langle r\rangle)r = M_{\rm in}(\langle r_{\rm in}\rangle)r_{\rm in}$. We find the correction using Monte Carlo realizations of trajectories in the NFW equilibrium halo and finding the time-averaged radii $\langle x\rangle \approx 1.72x^{0.82}/(1+5x)^{0.085}$, $x \equiv r/r_s$.

In order to make realistic predictions for annihilation rates, we construct two detailed models of the MW Galaxy by redoing the full analysis of numerous observational data collected in Klypin *et al.* 2001. The models are compatible with the available observational data for the MW and their main parameters are given in Table 1 of Prada *et al.* (2004). Models assume that without cooling the density of baryons is proportional to that of the DM and the final baryon distribution is constrained by the observational data.

3. Computation of $f_{\rm SUSY}$ in the mSUGRA scenario

To illustrate the expected variations of $f_{\rm SUSY}$ we focus on the mSUGRA scenario. We are using the *Darksusy* package (Gondolo *et al.* 2002) to compute $\tilde{\chi}_1^0$ relic densities and annihilation properties for 10^6 mSUGRA random models. We also check that results are not ruled out by present accelerator bounds. We compute a scan of models, which covers the relevant regimes in mSUGRA parameter space. All models within WMAP allowed relic densities leads to $\tilde{\chi}_1^0$ masses from 70 GeV up to 1400 GeV, approximately. The $\langle\sigma v\rangle$ lies in the range $1{\cdot}10^{-29}$ to $3{\cdot}10^{-26}$ cm^{-3}s^{-1}. We also compute the $f_{\rm SUSY}/10^{-32}$ dependence with the IACT $E_{\rm th}$. For a given $E_{\rm th}$, we compute all the m_χ, $\langle\sigma v\rangle$ and N_γ intervals. This gives the allowed $f_{\rm SUSY}$ region for $\tilde{\chi}_1^0$ which can be detected with the IACT, i.e, those with $m_\chi \geq E_{th}$.

4. Gamma-ray annihilation observability in the Milky Way

The expected $\tilde{\chi}_1^0$ annihilation γ flux can be computed from Eq. 1 for the compressed DM density profile provided by our MW models as a function of the angular distance Ψ_0 from the Galactic Center. In Fig. 1 we show the predicted fluxes in units of $f_{\rm SUSY}/10^{-32}$. We also show as a comparison the expected flux for the uncompressed NFW density profile. The flux profiles were determined for a typical IACT of resolution $\sigma_t = 0.1°$ ($\Delta\Omega = 10^{-5} sr$). We have multiplied the flux profiles by a factor of 1.7 quoted by Stoehr *et al.* (2003) to account for the presence of substructure inside the MW halo.

The success of a detection requires that the minimum detectable γ flux $F_{\rm min}$ for an exposure of t seconds, given an IACT of effective area $A_{\rm eff}$, angular resolution $\sigma_{\rm t}$ and threshold energy $E_{\rm th}$ exceeds a significant number M_s of standard deviations ($M_s\sigma$) the background noise $\sqrt{N_b}$, i.e. $F_{\rm min} A_{\rm eff} t/\sqrt{N_b} \geq M_s$. The background counts (N_b) due to electronic and hadronic (cosmic protons and helium nucleids) cosmic ray showers have been estimated using the expressions from Bergstrom *et al.* (1998). As an additional background, one has to consider also the contamination due to isolated muons which depending on the f.o.v. and altitude location of the telescope may be even the dominant background at some energy range. The diffuse galactic and extragalactic gamma radiation are negligible compare to this background. The $E_{\rm th}$ of an IACT depends on the zenith angle of observation. The Galactic Center is visible at different zenith angles by all present IACTs (e.g. CANGAROO-III, H.E.S.S.,MAGIC, VERITAS), but in the best case an $E_{\rm th}$ of about 100 GeV can be achieved. The $A_{\rm eff}$ is also sensible to the zenith angle of observation, here we choose a value of 1×10^9 cm^2. This detectability condition will allow us to compute the 5σ minimum detectable flux $F_{\rm min}$ in 250 hours of integration with a typical IACT of $A_{\rm eff} = 1 \times 10^9$ cm^2 and $E_{\rm th} = 100$ GeV (dashed line in Fig. 1). At a given distance from the Galactic Center only the flux values, for a particular model of the Milky Way, greater than $F_{\rm min}$ will be detected. The detection will be much harder and may result only in a central spot in the case of an IACT with higher $E_{\rm th}$, as the $f_{\rm SUSY}$ parameter decreases with $E_{\rm th}$. For an $E_{\rm th} = 100$ GeV the $f_{\rm SUSY}/10^{-32}$ factor varies from $6{\cdot}10^{-4}$ to 5, approximately.

The compressed MW models presented here likely would result in a detection of the annihilation signal for the discussed SUSY scenario. For the discussed IACTs and exposure times, this detection will be successful only for the very central regions, less than $\sim 0.4°$. The uncompressed

NFW DM profile of the Model A will not be detected even in the direction of the Galactic Center. The effect of the adiabatic compression, which was previously ignored, is a crucial factor: in the central ~ 3 kpc of the Milky Way, where the baryons dominate, it does not make sense to use the dark matter profiles provided by cosmological N-body simulations: the DM must fall into the deep potential well created by the collapsed baryons. Thus, the models presented here are not extreme: they are the starting point for realistic predictions of the annihilation fluxes.

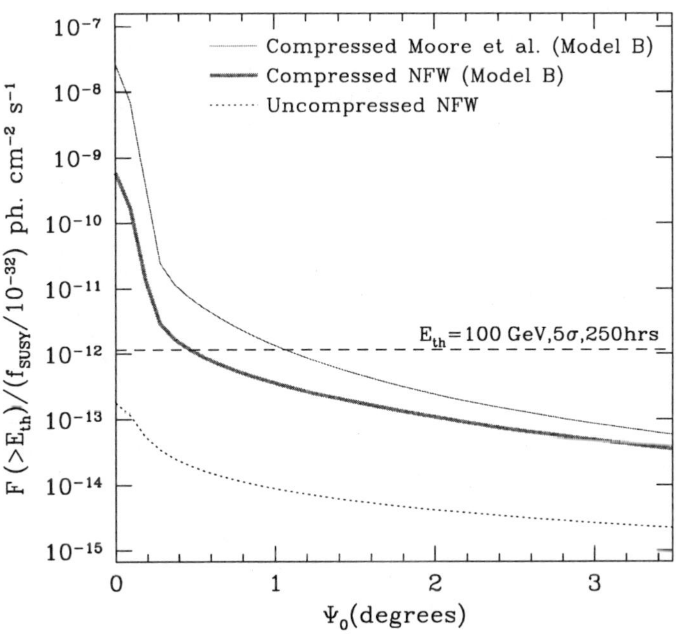

Figure 1. Predicted continuum γ fluxes as a function of distance Ψ_0 from the Galactic Center. The dashed line give the minimum detectable gamma flux F_{min} (see text).
Predicted continuum γ fluxes as a function of distance Ψ_0 from the Galactic Center. The dashed line give the minimum detectable gamma flux F_{min} (see text).

References

1. G.R. Blumenthal, S.M. Faber, R. Flores and J.R. Primack, Astrophys. J., 301, 27 (1986).
2. P. Gondolo *et al.*, astro-ph/0211238 (2002).
3. A. Klypin, H. Zhao, and R.S. Somerville, Astrophys. J., 573, 597 (2002).
4. F. Prada, A. Klypin *et al.* Phys. Rev. Letters, 93, 241301 (2004).
5. F. Stoehr, S.D.M. White, V. Springel, G. Tormen, and N. Yoshida, MNRAS, 345, 1313 (2003).
6. Ya.B. Zeldovich, A.A. Klypin, M.Yu. Khlopov, and V.M. Chechetkin, Soviet J. Nucl. Phys., 31, 664 (1980).

LIST OF PARTICIPANTS

Pedro Abreu
LIP
Av. Elias Garcia 14, 1°
1000-149 Lisboa
Portugal
abreu@lip.pt

Sofia Andringa
IFAE/UAB
Edifici de Ciences
08193 Bellaterra (Barcelona)
Spain
andringa@ifae.es

Pierre Antilogus
LPNHE/IN2P3
4 place Jussieu
Tour 33 - Rez de Chaussée
75252 Paris Cedex 05
France
P.Antilogus@in2p3.fr

Henrique Araújo
Imperial College London
Blackett Laboratory
Prince Consort RD
London SW7 2BW
United Kingdom
H.Araujo@imperial.ac.uk

Luísa Arruda
LIP
Av. Elias Garcia 14, 1°
1000-149 Lisboa
Portugal
luisa@lip.pt

Vladan Arsenijevic
Instituto Superior Técnico
CENTRA
Av. Rovisco Pais
1049-001 Lisboa
Portugal
arsenije@ist.utl.pt

Pedro Assis
LIP
Av. Elias Garcia 14, 1°
1000-149 Lisboa
Portugal
pedro.assis@lip.pt

Domingos Barbosa
Instituto Superior Técnico
CENTRA
Av. Rovisco Pais
1049-001 Lisboa
Portugal
barbosa@astro.up.pt

Gaspar Barreira
LIP
Av. Elias Garcia 14, 1°
1000-149 Lisboa
Portugal
gaspar@lip.pt

Orfeu Bertolami
Instituto Superior Técnico
Departamento de Física
Av. Rovisco Pais
1049-001 Lisboa
Portugal
orfeu@cosmos.ist.utl.pt

Gonçalo Borges
LIP
Av. Elias Garcia 14, 1°
1000-149 Lisboa
Portugal
goncalo@lip.pt

Pedro Brogueira
Instituto Superior Técnico
Departamento de Física
Av. Rovisco Pais
1049-001 Lisboa
Portugal
pedro@fisica.ist.utl.pt

Vitor Cardoso
McDonnell Center for the Space Sciences
Department of Physics
Washington University,
St. Louis
MO 63130
USA
vcardoso@wugrav.wustl.edu

Nuno Castro
LIP
Av. Elias Garcia 14, 1°
1000-146 Lisboa
Portugal
nuno.castro@cern.ch

Osvaldo Catalano
IASF – Palermo/CNR
Via Ugo la Malfa 153
90146 Palermo
Italy
osvaldo.catalano@pa.iasf.cnr.it

Ruben Conceição
LIP
Av. Elias Garcia 14, 1°
1000-149 Lisboa
Portugal
rconceicao@nfist.ist.utl.pt

João Costa
LIP
Av. Elias Garcia 14, 1°
1000-149 Lisboa
Portugal
jcosta@lip.pt

Eamonn Daly
ESA/ESTEC
Keplerlaan 1
Postbus 299
2200 AG Noordwijk
The Netherlands
eamonn.daly@esa.int

Alessandro de Angelis
University of Udine
Via delle Scienze, 208
I-33100 UDINE
Italy
deangelis@fisica.uniud.it

Jorge Dias de Deus
Instituto Superior Técnico
CENTRA
Av. Rovisco Pais
1049-001 Lisboa
Portugal
jdd@fisica.ist.utl.pt

Gonçalo Dias
Instituto Superior Técnico
CENTRA
Av. Rovisco Pais
1049-001 Lisboa
Portugal
gdias@fisica.ist.utl.pt

Óscar Dias
Perimeter Institute for
Theoretical Physics
31 Caroline St. N.,
Waterloo, Ontario N2L 2Y5
Canada
odias@perimeterinstitute.ca

Teresa Dova
Departamento de Física
Universidad Nacional de la
Plata
CC67 La Plata (1900)
Argentina
dova@fisica.unlp.edu.ar

Dominik Elsaesser
University of Wuerzburg
Institut Theoretische Physik
Astrophysik
Am Hubland, D-97074
Wuerzburg
Germany
elsaesser@astro.uni-wuerzburg.de

Catarina Espírito Santo
LIP
Av. Elias Garcia 14, 1º
1000-149 Lisboa
Portugal
catarina@lip.pt

Sèbastien Fabbro
Instituto Superior Técnico
CENTRA
Av. Rovisco Pais
1049-001 Lisboa
Portugal
seb@ist.utl.pt

Enrique Fernandez
IFAE, Edifici Cn.
Facultat Ciències UAB
E-08193 Bellaterra (Barcelona)
Spain
Enrique.fernandez@ifae.es

Manuel Fiolhais
Universidade de Coimbra
Departamento de Física
3004-516 Coimbra
Portugal
tmanuel@teor.fis.uc.pt

Margarida Fraga
LIP – Coimbra
Departamento de Física – Univ.
Coimbra
Rua Larga, 3004-516 Coimbra
Portugal
margarida@lipc.fis.uc.pt

Sijie Gao
Instituto Superior Técnico
CENTRA
Av. Rovisco Pais
1049-001 Lisboa
Portugal
sijie@fisica.ist.utl.pt

Nicola Giglietto
INFN – Bari
Via E. Orabona, 4
I-70126 Bari
Italy
giglietto@ba.infn.it

Patrícia Gonçalves
LIP
Av. Elias Garcia 14, 1°
1000-149 Lisboa
Portugal
patricia@lip.pt

Alfredo Barbosa Henriques
Instituto Superior Técnico
CENTRA
Av. Rovisco Pais
1049-001 Lisboa
Portugal
alfredo@fisica.ist.utl.pt

Antonio Insolia
University of Catania
Dipartimento di Física e Astronomia
Via. S. Sofia, 64
95123 Catania
Italy
Antonio.Insolia@ct.infn.it

Charles Jui
University of Utah
115 S. 1400 E. #201 JFB
Salt Lake City, Utah 84112
USA
jui@cosmic.utah.edu

Kostas Kokkotas
Aristotle University
Department of Physics
Section of Astrophysics, Astronomy and Mechanics
541 24 Thessaloniki - Macedonia
Greece
kokkotas@auth.gr

Ralf Lehnert
University of Vanderbilt
Department of Physics and Astronomy
6301 Stevenson Center
Nashville, Tennessee 37235
USA
ralf.lehnert@vanderbilt.edu

José Sande Lemos
Instituto Superior Técnico
CENTRA
Av. Rovisco Pais
1049-001 Lisboa
Portugal
lemos@fisica.ist.utl.pt

Cipriano Lomba
EFACEC
Sistemas de Electrónica, S.A.
Rua Eng. Frederico Ulrich, Apartado 3081
4471-907 Moreira – Maia
Portugal
cipriano.lomba@ent.efacec.pt

Ilídio Lopes
Instituto Superior Técnico –
CENTRA
& Universidade de Évora
Av. Rovisco Pais
1049-001 Lisboa
Portugal
lopes@fisica.ist.utl.pt

João Magueijo
Theoretical Physics Group
Department of Physics
Imperial College
Prince Consort RD.,
London SW7 2BZ
United Kingdom
j.magueijo@imperial.ac.uk

Dalmiro Maia
Universidade do Porto
Faculdade de Ciências
Observatório Astronómico
Alameda do Monte da Virgem
4130-146 Vila Nova de Gaia
Portugal
dmaia@fc.up.pt

Amélia Maio
LIP
Av. Elias Garcia 14, 1°
1000-149 Lisboa
Portugal
amelia@lip.pt

José Maneira
LIP
Av. Elias Garcia 14, 1°
1000-149 Lisboa
Portugal
maneira@lip.pt

Ana Maria Mourão
Instituto Superior Técnico
CENTRA
Av. Rovisco Pais
1049-001 Lisboa
Portugal
amourao@ist.utl.pt

Sergio Navas Concha
University of Granada
Departamento de Fisica Teórica
y del Cosmos
Av. Severo Ochoa s/n
18071 Granada
Spain
navas@ugr.es

Rui Neves
Universidade do Algarve
Fac. de Ciências e Tecnologia
Departamento de Física e
CENTRA
Campus de Gambelas
8005-139 Faro
Portugal
rneves@ualg.pt

António Onofre
LIP – Coimbra
Departamento de Física – Univ.
Coimbra
Rua Larga, 3004-516 Coimbra
Portugal
antonio.onofre@cern.ch

Catarina Ortigão
LIP
Av. Elias Garcia 14, 1°
1000-149 Lisboa
Portugal
ortigão@lip.pt

Jorge Páramos
Instituto Superior Técnico
Departamento de Física
Av. Rovisco Pais
1049-001 Lisboa
Portugal
x_jorge@fisica.ist.utl.pt

Miguel Paulos
LIP
Av. Elias Garcia 14, 1º
1000-149 Lisboa
Portugal
miguel_paulos@netcabo.pt

Rui Pereira
LIP
Av. Elias Garcia 14, 1º
1000-149 Lisboa
Portugal
pereira@lip.pt

Tiago Pereira
Instituto Superior Técnico
CENTRA
Av. Rovisco Pais
1049-001 Lisboa
Portugal
tiago.pereira@ist.utl.pt

Mário Pimenta
LIP
Av. Elias Garcia 14, 1º
1000-149 Lisboa
Portugal
pimenta@lip.pt

Robertus Potting
Universidade do Algarve
Fac. de Ciências e Tecnologia
Departamento de Física e
CENTRA
Campus de Gambelas
8005-139 Faro
Portugal
rpotting@ualg.pt

Francisco Prada
Instituto de Astrofísica de
Andalucia
Camino Bajo de Huetor, 50
18008 Granada
Spain
fprada@iaa.es

Pedro Rato
Universidade do Algarve
Fac. de Ciências e Tecnologia
LIP
Campus de Gambelas
8005-139 Faro
Portugal
rato@lip.pt

Luciano Rezzolla
SISSA/ISAS
Via Beirut 2-4
34014 Trieste
Italy
rezzolla@sissa.it

Pedro Rodrigues
LIP
Av. Elias Garcia 14, 1º
1000-149 Lisboa
Portugal
psilva@lip.pt

Jorge Romão
Instituto Superior Técnico
CFTP
Av. Rovisco Pais
1049-001 Lisboa
Portugal
jorge.romao@ist.utl.pt

Paulo M. Sá
Universidade do Algarve
Fac. de Ciências e Tecnologia
Departamento de Física e
CENTRA
Campus de Gambelas
8005-139 Faro
Portugal
pmsa@ualg.pt

Nuno Santos
Instituto Superior Técnico
CENTRA
Av. Rovisco Pais
1049-001 Lisboa
Portugal
nlsantos@fisica.ist.utl.pt

Rui Santos
LIP
Av. Elias Garcia 14, 1°
1000-149 Lisboa
Portugal
el.bimbo@sapo.pt

Fabio Scardigli
Instituto Superior Técnico
CENTRA
Av. Rovisco Pais
1049-001 Lisboa
Portugal
scardigli@fisica.ist.utl.pt

Peter Sonderegger
LIP/CERN
Av. Elias Garcia 14, 1°
1000-149 Lisboa
Portugal
peter.sonderegger@cern.ch

Andreia Trindade
LIP
Av. Elias Garcia 14, 1°
1000-149 Lisboa
Portugal
andreia@lip.pt

Bernardo Tomé
LIP
Av. Elias Garcia 14, 1°
1000-149 Lisboa
Portugal
bernardo@lip.pt

Brigitte Tomé
Universidade do Algarve
Fac. de Ciências e Tecnologia
CENTRA
Campus de Gambelas
8005-139 Faro
Portugal
btome@ualg.pt

Filipe Veloso
LIP
Av. Elias Garcia 14, 1°
1000-149 Lisboa
Portugal
filipe.veloso@cern.ch

Pedro Viana
CAUP
Rua das Estrelas, s/n
4150-762 Porto
Portugal
viana@astro.up.pt

Alan Watson
School of Physics and
Astronomy
University of Leeds
Leeds LS2 9JT
United Kingdom
a.a.watson@leeds.ac.uk